POLYMER SYNTHESES
Volume II

This is Volume 29 of
ORGANIC CHEMISTRY
A series of monographs
Editors: ALFRED T. BLOMQUIST and HARRY H. WASSERMAN

A complete list of the books in this series appears at the end of the volume.

POLYMER SYNTHESES
Volume II

Stanley R. Sandler

PENNWALT CORPORATION
KING OF PRUSSIA, PENNSYLVANIA

Wolf Karo

HUNTINGDON VALLEY, PENNSYLVANIA

ACADEMIC PRESS   New York   San Francisco   London   1977

A Subsidiary of Harcourt Brace Jovanovich, Publishers

ACADEMIC PRESS, INC.
111 Fifth Avenue, New York, New York 10003

United Kingdom Edition published by
ACADEMIC PRESS, INC. (LONDON) LTD.
24/28 Oval Road, London NW1

Library of Congress Cataloging in Publication Data

Sandler, Stanley R   Date
    Polymer syntheses.

    (Organic chemistry; a series of monographs, v. 29)
    Includes bibliographic references and indexes.
    1.   Polymers and polymerization.   I.   Karo, Wolf,
Date   joint author.   II.   Title.   III.   Series.
QD281.P6S27   547'.84   73-2073
ISBN 0−12−618502−6

CONTENTS

Chapter 9. **Polymerization of Acrylic Acids and Related Compounds**

Chapter 10. **Poly(vinyl chloride)**

PREFACE

This volume continues in the tradition of Volume I in presenting detailed laboratory instructions for the preparation of various types of polymers such as urea, melamine, benzoguanamine/aldehyde resins (amino resins–aminoplasts), phenol/aldehyde condensates, epoxy resins, silicone resins, alkyd resins, polyacetals/polyvinyl acetals, polyvinyl ethers, polyvinyl pyrrolidones, polyacrylic acids, and polyvinyl chloride. Polyvinyl acetate and related vinyl esters, allyl polymers, acetylene polymers, maleate and fumarate polymers, and several other addition-condensation polymer types will be covered at a later date in Volume III.

In all chapters the latest journal articles and patents have been reviewed. Each chapter has tables presenting compilations from several papers, and they are all appropriately referenced. Each contains an introductory section in which older preparative references are given in order to maintain the continuity of the subject. Procedures are chosen on the basis of safety considerations and ease of being carried out with standard laboratory equipment. Each chapter should be considered a good preparative introduction to the subject and not a final, definitive work. Space limitations have guided us in presenting only the core of the subject. This book should be especially useful to industrial chemists and students of polymer chemistry because it provides a ready source of preparative procedures for various polymer syntheses.

Safety hazards and precautions (see for example Chapter 10 on PVC) are stressed in all chapters, and the reader is urged not only to observe these but to constantly seek further up-to-date information on the appropriate monomer and from the chemical manufacturer.

This book is designed only to provide useful polymer synthesis information and not to override the question of legal patentability or to suggest allowable industrial use. The toxicological properties of the reagents in most cases have not been completely evaluated, and the reader is urged to exercise care in their use. We assume no liability for injuries, damages, or penalties resulting from the use of the chemical procedures described.

We express our appreciation to our wives and children for their under-standing and encouragement during the preparation of this manuscript. Special thanks are due to Miss Emma Moesta for typing our manuscript in a most professional fashion. Finally, we thank the staff of Academic Press for guiding the publication of the manuscript to its final book form.

Stanley R. Sandler
Wolf Karo

CONTENTS OF VOLUME I

Chapter 1

UREA, MELAMINE, BENZOGUANAMINE–ALDEHYDE RESINS (AMINO RESINS OR AMINOPLASTS)

I. INTRODUCTION

The condensation of compounds containing the amino groups with alde-hydes, and in particular formaldehyde, gives materials commonly known as amino resins or aminoplasts. This chapter will concentrate on resins derived from urea, melamine, or benzoguanamine and will describe other types in Section 5 (Miscellaneous Preparations). Formaldehyde reacts with the amino groups to give hydroxymethylol groups, which condense either with each other or with free amino groups to give resinous products. These products can be prepared to be either thermoplastic and soluble or cross-linked, insoluble infusible products known also as aminoplasts. Some other members of this class include formaldehyde condensation products of thiourea, ethyleneurea, guanamines, aniline, *p*-toluenesulfonamide, and acrylamide. Other aldehydes such as acetaldehyde, glyoxal, furfural, and acrolein have also been used in condensations with amino compounds to give resinous products.

In contrast to phenolic resins, the urea, melamine, and benzoguanamine–formaldehyde resins are colorless, odorless, and lightfast. Amino resins find use in decorative applications because of their clarity and colorless bonds. The arc resistance of amino resins makes them important components in electrical circuits. The amino resins, in contrast to phenolic resins, do not carbonize readily by the application of higher voltages between current-conducting parts and require 100 sec or more for development of an arc.

The basic raw material, urea, paved the way to the development of most of the other amino resin starting materials as described below, and is synthe-sized by the reaction of carbon dioxide with ammonia or the hydrolysis of cyanamide.

$$CO_2 + 2NH_3 \longrightarrow NH_2CONH_2 + H_2O \rightleftharpoons H_2NCOONH_4 \qquad (1)$$

Thiourea is prepared by heating cyanamide with hydrogen sulfide:

$$H_2NCN + H_2O \longrightarrow H_2NCONH_2 \qquad (2)$$
$$\text{(m.p. } 132.7°C)$$

$$H_2NCN + H_2S \longrightarrow H_2N\overset{\overset{\text{S}}{\|}}{C}NH_2 \qquad (3)$$
$$\text{(m.p. } 173°C)$$

Urea is also used to produce other important amino compounds that are used as raw materials for amino resin production [1].

$$
\underset{\substack{\text{H}_2\text{C}\text{——}\text{CH}_2 \\ \text{HN}\diagdown_{\text{C}}\diagup\text{NH} \\ \|\ \\ \text{O}}}{} \xleftarrow{\ \text{NH}_2\text{—CH}_2\text{CH}_2\text{—NH}_2\ } \text{H}_2\text{NCONH}_2 \longrightarrow \text{H}_2\text{N—C}\equiv\text{N} \qquad (4)
$$

Benzoguanamine is obtained by the reaction (heat and pressure) of ammonia dicyandiamide in the presence of aniline or by the sodium metal-catalyzed reaction of benzonitrile with dicyandiamide in ethylene glycol monomethyl ether.

Cyanuric chloride can be reacted with various amines to give substituted melamines used in speciality amino resin production.

Other important amino compounds used for resin production are *p*-toluenesulfonamide, aniline, and acrylamide. Some of these preparations are found in Section 5.

Several important reviews and key references are worthwhile sources for additional details and historical background material on amino resins [2,2a].

The major urea–melamine producers and their current and projected capacity are shown in Table I.

The production of amino resins basically hinges on ammonia and urea crystal production as seen in Tables II–IV.

$$
3\ \text{lb ammonia} \longrightarrow 1\ \text{lb urea resin} \qquad (5)
$$
$$
1.45\ \text{lb urea crystal} \longrightarrow 1\ \text{lb melamine resin} \qquad (6)
$$

Amino resins are used in the manufacture of adhesives, molding compounds, paper and protective coatings, textile-treating resins, electrical devices, and melamine, the latter especially as buttons, dinnerware, and sanitary ware. The pattern of consumption of these resins is summarized in Tables II–IV. Note that bonding resins used for adhesives, laminating, and plywood account for 72% of the consumption pattern.

CAUTION: As described later in Chapter 2, the reaction of formaldehyde with hydrogen chloride has been shown to lead to the spontaneous production of the now known carcinogen bis(chloromethyl) ether. In addition, other

TABLE I
MAJOR CURRENT PRODUCERS OF UREA–MELAMINE RESINS[a,b]

Producer	Current resin capacity, 1000 tons	Expected capacity, Dec. 31, 1975, 1000 tons	Future expansion, 1000 tons	Expected starting date
Gulf	68	68	23	1976
Allied Chemical	55	55		
Monsanto	45	45	c	
Reichold	45	45		
American Cyanamid	36	54		
Pacific Resin	32	32		
Borden Chemical	30	30		
Jersey State Chemical	30	30		
National Casein	23	23		
Sun Chemical	14	14		
Koppers	11	11		
Ashland Chemical	9	9		
Celanese	9	9		
Georgia-Pacific	9	9		
Perstorp AB	9	18		
Others[d]	300	300		
Total	725[e]	752	23	

[a] Source: Individual urea and melamine producers.

[b] Reprinted from M. Slovick, *Mod. Plast.* **51**, 44 (1974). Copyright 1974 by Modern Plastics. Reprinted by permission of the copyright owner.

[c] Amount and date not announced; some additions could be made in 1975.

[d] Twenty-one other merchant suppliers produce an estimated 100,000 tons. Between 300 and 400 captive producers account for remaining 200,000 tons, Modern Plastics estimates.

[e] Modern Plastics estimates 80 to 85% is urea.

aldehydes (crotonaldehyde, acetaldehyde, acrolein, etc.) are also toxic [3]. Toxicity limits for some representative compounds in air are tabulated below [3]. Although urea, melamine, and benzoguanamine are not considered highly

Compound	Toxicity limit (ppm)
Formaldehyde	2
Acetaldehyde	100
Crotonaldehyde	2
Acrolein	0.1
Aniline	5

TABLE II
UREA AND MELAMINE:
PATTERN OF CONSUMPTION[a]

Market	1000 metric tons	
	1972[b]	1973[b]
Bonding and adhesive resins for:		
Fibrous and granulated wood	232	262
Laminating	24	24
Plywood	40	40
Molding compounds	40	44
Paper treating and coating resins	16	22
Protective coatings	28	33
Textile treating and coating resins	23	26
Exports	5	10
Other	3	3
Total	411	464

[a] Reprinted from M. Slovick, *Mod. Plast.* **51**, 44 (1974). Copyright 1974 by Modern Plastics. Reprinted by permission of the copyright owner.

[b] Due to incomplete reporting (figures not available), data in this table are not fully representative of consumption.

toxic, little information is available on the toxicity of all their condensation products. Amino compounds such as acrylamide and aniline are considered toxic and should be handled with care. All other amino compounds should be considered suspect unless toxicity data are available. The use of efficient hoods and good personal hygiene (gloves, lab coats, etc.) are essential when amino resins are prepared.

2. UREA–ALDEHYDE CONDENSATIONS

In 1920, Jahn [4] provided the groundwork that lead to the modern developments in urea–formaldehyde adhesives, casting compositions, and the textile-treating materials [5]. Urea–formaldehyde (U–F) resins have the disadvantage that they are not resistant to moisture, so that clear castings

TABLE III

UREA:

MOLDING POWDER MARKETS[a]

| Market | 1000 metric tons | |
	1972[b]	1973[b]
Closures	6.8	7.9
Electrical devices	10.9	12.6
Other	2.3	2.7
Total	20.0	23.2

[a] Reprinted from M. Slovick, *Mod. Plast.* **51**, 44 (1974). Copyright 1974 by Modern Plastics. Reprinted by permission of the copyright owner.

[b] Due to incomplete reporting (figures not available), data in this table are not fully representative of consumption.

TABLE IV

MELAMINE:

MOLDING POWDER MARKETS[a]

| Market | 1000 metric tons | |
	1972[b]	1973[b]
Buttons	0.8	0.9
Dinnerware	18.2	19.1
Sanitary ware	0.5	0.6
Other	0.5	0.6
Total	20.0	21.2

[a] Reprinted from M. Slovick, *Mod. Plast.* **51**, 44 (1974). Copyright 1974 by Modern Plastics. Reprinted by permission of the copyright owner.

[b] Due to incomplete reporting (figures not available), data in this table are not fully representative of consumption.

tend to craze and surfaces are too soft and scratch easily. They have the advantage that they form clear, ultraviolet absorption resistant coatings, in contrast to cast phenolic resins which are not lightfast and yellow on exposure to light. The addition of thiourea confers better gloss and water resistance to the molded articles or coatings of U–F resins. U–F resins find wide use in the adhesive industry, where they are used to prepare interior grade plywood. Urea can also react with other aldehydes and is used in cocondensations with phenol–formaldehyde and melamine–formaldehyde resins.

A. Formaldehyde Condensations

Holzer [6] (1884) and Ludy [7] (1889) isolated a white precipitate by the reaction of urea with formaldehyde under acid conditions.

Goldschmidt in 1896 [8] studied the reaction of urea with formaldehyde in various strength acid solutions and obtained a granular white deposit of the empirical formula $C_5H_{10}N_4O_3$. In 1908, Einhorn and Hamburger [9] studied the same reaction in the presence of hydroxyl ions, and depending on the mole ratio of formaldehyde to urea isolated mono- or dimethylolurea.

In 1927, Schiebler *et al.* [10] found that in acid solution the methylolureas are converted to insoluble substances similar to Goldschmidt's compound.

Today the polymerization mechanisms involved are similar to those discussed for other methylol compounds such as phenolic resins (Chapter 2) or melamine resins (see Section 3 of this chapter). It is interesting to note that because urea has four active hydrogens and three sites for polymerization, linear, branched, and cyclic structures are possible. In fact, Kadowaki [11] has isolated several low molecular weight condensation products of urea–formaldehyde and has described their properties. The cyclic structures commonly called urones, such as dimethylolurone (*N,N'*-dimethyloltetrahydro-4*H*-1,3,5-oxidiazin-4-one), have also been prepared by Kadowski.

DeJong and DeJong [12] studied the kinetics of the U–F reaction and conclude as does Smythe [13] that the reaction is bimolecular, and the rate is directly proportional to the concentration of hydroxyl or hydrogen ions. The reaction of aminoethylol groups with amino groups is said to also give methylene linkages between urea fragments. Glutz and Zollinger [14] found that in the rate-limiting transition state, urea, formaldehyde, and the catalyst were present. In addition, at pH 11, the Cannizzaro reaction was competitive with the formation of methylol.

The possible reaction of urea with formaldehyde is summarized in Scheme 1.

These reactions give evidence for the methylol, methylene, and methylene ether linkages formed in U–F reactions. Although tri- and tetramethylolureas have not been isolated, there is evidence that they exist in solution [15]. Even at urea:formaldehyde ratios of 1:20, only 2.81 molar equivalents of formaldehyde react per mole of urea [16]. The isolation of dimethylolurone is evidence for the existence of tetramethylolurea. Methylolureas are ordinarily formed under slightly alkaline conditions at room temperature. When the crystalline monomers are heated at their fusion points they lose some formaldehyde and set up to a resinous mass. Polymerization may be initiated by the catalytic influence of hydrogen ions. Thiourea has been found to give the same type of methylene-linked polymers as the urea system [17]. Various methods exist for the reaction of urea–formaldehyde resins and for the production of alcohol-soluble resins. The important variables are (1) purity of reagents, (2) molar ratio of reagents, and (3) pH control. The use of formaldehyde with 0.05–0.1% formic acid content is satisfactory, but higher percentages are not acceptable since they may cause pregellation reactions. Close control of the pH is also maintained by the use of buffers. In the absence of buffers the pH tends to decrease below 7 as a result of the Cannizzaro reaction or by oxidation of formaldehyde. Some typical urea molding resins are made by the reaction of 1.6:1 molar ratio of aqueous formaldehyde to urea at pH 7–8 at 50°C for 3–24 hr. The turbid solution is used to impregnate the filler to prepare the molding resin. The resin can also be prepared by using excess formaldehyde:urea, such as 2.2:1, and reacting at 90°C. After 10–30 min the exothermic reaction ceases, and the reaction is completed under reflux with the

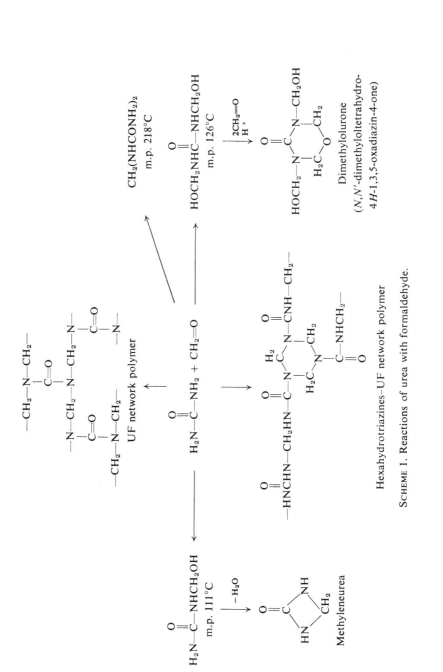

SCHEME 1. Reactions of urea with formaldehyde.

aid of an acid catalyst to give the solution increased viscosity. At the right viscosity, the solution is cooled to 30°C and urea is added in an amount that gives a final ratio of formaldehyde to urea of 1.6–1.7:1. The pH is adjusted to 7.5–8.0 to stabilize the resin while filling material is added to prepare the molding resin. Some typical preparations which illustrate these and other applications of urea–formaldehyde resins are given in Table V and in the various preparations described below.

2-1. Preparation of Urea–Formaldehyde Textile Resins (F : U Ratio 2:1) [18]

$$H_2NCONH_2 + 2CH_2O \longrightarrow HOCH_2NHCNHCH_2OH \overset{\overset{\displaystyle O}{\displaystyle \|}}{} \qquad (7)$$

To a 3-liter, 3-necked resin flask equipped with a mechanical stirrer, condenser, and thermometer are added 810 gm of 37% (10.0 moles) of formaldehyde, 4.0 gm of sodium acetate, and 8.0 gm of concentrated ammonia (28% NH_3). The mixture is stirred until all reactants have gone into solution and then 300 gm (5.0 moles) of urea is added. Heat is then applied to raise the temperature to 90°C over a $\frac{1}{2}$-hr period; this temperature is maintained for 2 hr. Approximately 350 gm of water is removed under reduced pressure to yield a white turbid resin dispersion of 75% solids content with a pH of 5.5 and an acid number of less than 1.0. The pH is adjusted to 7.4 (with sodium hydroxide, sodium carbonate, or ammonium hydroxide) to stabilize the resin.

This resin is not only useful as a textile resin but also is of value in the manufacture of adhesives (with starch), wood glues for hot- and cold-pressed plywood, and in the paper wet strength resin area.

2-2. Preparation of a Urea–Formaldehyde Adhesive [19]

To a resin kettle are added 286 gm of a 70% solids urea–formaldehyde resin (see note below) of the type described in Section 2-1, 80 gm (0.80 mole) furfuryl alcohol, 44 gm wood flour, 1.1 gm triethanolamine, and 2 gm tricalcium phosphate. The mixture is stirred (pH approximately 8) and slowly heated to about 90°C over a $\frac{1}{2}$-hr period. The temperature is kept at 90°C for about 15 min and then cooled to 30°C over an 80-min period. The product is now ready to be cured with either one of the following catalysts. (Before addition of the catalyst, the adhesive resin is stable for over 3 months at 80°F.)

6-hr working life catalyst: 3 gm ammonium chloride and 10 gm water
2-hr working life catalyst: 2 gm ammonium chloride, 1 gm ferric chloride, and 7 gm water

TABLE V

PREPARATION OF UREA–FORMALDEHYDE RESIN

Urea (moles)	Formaldehyde (37% aqueous, moles)	Other	Solvent (gm or ml)	Reaction conditions				Ref.
				pH	Temp. (°C)	Time (hr)		
18.9	35.0 (less than 2% CH_3OH content)	—	H_2O, triethanolamine (2 gm)	7.2	98–102	$1\frac{1}{2}$–$2\frac{1}{2}$, pH adjusted with triethanolamine to 7.5–7.7		a
5.0	10	—	Conc. NH_4OH (8 gm)	7.5	90	$2\frac{1}{2}$		b
4.0	7.9	Hexamethylenetetramine (24 gm)	—	—	90–100	—		c
0.4	0.8	Melamine (0.02 mole)	—	—	90–100	—		c
1.0	3.0	n-C_4H_9OH (1.0 mole)	—	7.5	90–100	—		d
0.57	1.0	—	—	7.0 (Na citrate)	90–100	—		e

[a] L. Boldizar, *Macromol. Synth.* **3**, 45 (1969).
[b] P. S. Hewett, U.S. Patent 2,456,191 (1948).
[c] K. Ripper, U.S. Patent 2,056,142 (1936).
[d] T. S. Hodgins and A. G. Hovey, *Ind. Eng. Chem.* **30**, 1021 (1938).
[e] N. V. Philips-Gloeilampenfabrieken, British Patent 773,349 (1957).

NOTE: The urea–formaldehyde resin is one prepared from a formaldehyde:urea molar ratio of 1:1 to 2:1 and is about 70% solids. However, a solid resin may be used to give similar results.

The adhesive gives the following result according to tests as provided by the "Army-Navy Aeronautical Specification Glue: Cold Setting Resins," An-G-8, dated April 25, 1942:

Plywood shear strength
    Dry: 510 psi, 63% wood failure
    Wet: 387 psi, 38% wood failure
Block shear strength
    3637 psi, 57% wood failure
    Glue line pH: 2.9–3.2

2-3. *Preparation of a Butanol-Modified Urea–Formaldehyde Resin* [20]

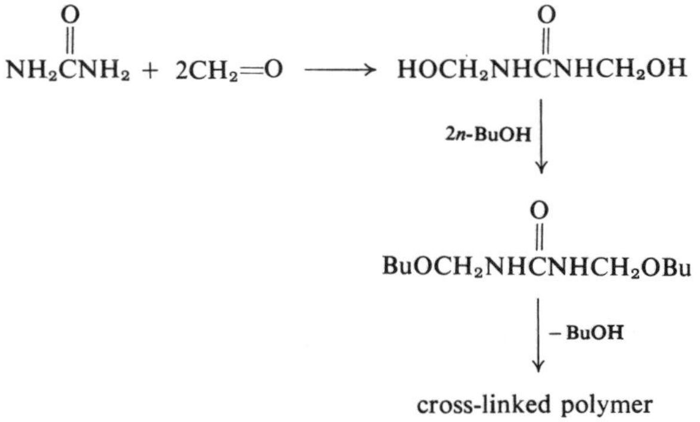

$$NH_2\overset{\overset{O}{\|}}{C}NH_2 + 2CH_2{=}O \longrightarrow HOCH_2NH\overset{\overset{O}{\|}}{C}NHCH_2OH$$

$$\Big\downarrow 2n\text{-BuOH}$$

$$BuOCH_2NH\overset{\overset{O}{\|}}{C}NHCH_2OBu \qquad (8)$$

$$\Big\downarrow -BuOH$$

cross-linked polymer

To a resin flask equipped with a reflux condenser, mechanical stirrer, and thermometer is added 243 gm of 37% (3.0 moles) formaldehyde. The pH is adjusted to 7.5–8.5 with 4–6 oz of concentrated ammonium hydroxide and then 60 gm (1.0 mole) of urea is added with stirring. The mixture is heated to 100°C over a 1 hr period and maintained at this point for $\frac{1}{2}$ hr. Then 148 gm (2.0 moles) of *n*-butanol is added along with enough phosphoric acid to adjust the pH to 5.5. The reaction mixture is heated and stirred for $\frac{1}{2}$ hr at 100°C and the temperature is then lowered to 60°–70°C. The resin is concentrated under reduced pressure (100–200 mm Hg) to give a viscous material that is tacky at room temperature. The resin can be cured by heating to 150°C for $\frac{1}{2}$ hr to give a hard, clear film.

B. Other Urea–Aldehyde Condensations

Urea reacts with a variety of mono- [21] and dialdehydes [22] to give either linear or cyclic products as shown in Eqs. (9) and (10) and Table VI.

$$(9)$$

$$(10)$$

2-4. Preparation of Isobutylene Diurea [23]

$$(CH_3)_2CH-CH=O + H_2N-\underset{\underset{O}{\|}}{C}-NH_2 \xrightarrow[-H_2O]{HCl} (CH_3)_2CH-CH(NHCONH_2)_2 \quad (11)$$

A mixture of 40.0 gm (0.66 mole) of powdered urea and 25 gm (0.35 mole) isobutyraldehyde is treated during grinding with 120 ml gaseous hydrochloric acid over a 1–2-min period to give 30.2 gm (53%) product, m.p. 198°–200°C (30.5% N).

2-5. Preparation of an Acetaldehyde–Urea Resin [24]

$$2CH_3CH=O + H_2N\underset{\underset{O}{\|}}{C}NH_2 \xrightarrow{OH^-} CH_3\underset{\underset{OH}{|}}{CH}-NH\underset{\underset{O}{\|}}{C}-NH\underset{\underset{OH}{|}}{CH}CH_3 \xrightarrow{H^+} \text{resin} \quad (12)$$

To 120 gm of 50% (1.0 mole) aqueous urea, adjusted to pH 8.0 (1 N NaOH) at 3°–5°C, is added dropwise with stirring 27 gm of 80% aqueous (2.3 moles) acetaldehyde. The reaction mixture is stirred at this temperature for 6 hr to give a viscous solution. Evaporation of the solution under reduced pressure (5–10 mm Hg) and purification with methanol gives hygroscopic crystals of molecular weight of 147–148, corresponding to diethylolurea. Heating the viscous solution before evaporation with a small portion of concentrated HCl gives a clear, tough resin insoluble in ethanol. Similar reactions at pH 4 at 20°–25°C, or at 3°–5°C, leads to ethylidenediurea which on hardening affords a brittle resin of molecular weight 700.

2-6. Preparation of Trichloroethylidene Urea [25]

$$CCl_3CH{=}O + 2NH_2{-}\overset{\overset{\displaystyle O}{\|}}{C}{-}NH_2 \xrightarrow[-H_2O]{H^+} CCl_3CH(NH{-}\overset{\overset{\displaystyle O}{\|}}{C}{-}NH_2)_2 \quad (13)$$

To 100 gm (1.7 moles) urea dissolved in 3 ml of 38% hydrochloric acid, 2.0 gm of ammonium chloride in 100 ml water, and 80 ml of 96% ethanol at 60°C is added portionwise with stirring 470 gm of 96.4% (3.1 moles) chloral. The reaction mixture is stirred for 2 hr at 60°C and then kept for 24 hr at room temperature to yield 242.1 gm (90.5%) of product, m.p. 191°–192°C.

2-7. Preparation of 4,5-Dihydroxy-2-imidazolidinone [26]

$$H_2N{-}\overset{\overset{\displaystyle O}{\|}}{C}{-}NH_2 + \overset{\overset{\displaystyle O}{\|}}{H}C{-}\overset{\overset{\displaystyle O}{\|}}{C}H \xrightarrow{OH^-} \begin{array}{c} \overset{O}{\overset{\|}{\underset{\displaystyle}{}}} \\ NH{-}C{-}NH \\ | \qquad | \\ HC{-}\!\!-\!\!CH \\ / \qquad \backslash \\ OH \qquad OH \end{array} \quad (14)$$

To 100 gm of 32% (0.55 mole) glyoxal solution at pH 7.0 (sodium carbonate) is added 50 gm (0.83 mole) of urea. The temperature rises to 35°–40°C in the course of about ½ hr, and the solution is then stirred for 12 hr while being cooled to about 0°C. Approximately 25 gm of the product precipitates, and evaporation to half volume at 35°C under reduced pressure affords another 15 gm of product. The total yield is 61%, m.p. 140°–142°C.

$$\begin{array}{c} \overset{O}{\overset{\|}{}} \\ HN{-}C{-}NH \\ | \qquad | \\ HC{-}\!\!-\!\!CH \\ | \qquad | \\ OH \quad OH \end{array} + H_2N{-}\overset{\overset{\displaystyle O}{\|}}{C}{-}NH_2 \xrightarrow{H^+} \begin{array}{c} \overset{O}{\overset{\|}{}} \\ HN{-}C{-}NH \\ | \qquad | \\ HC{-}\!\!-\!\!CH \\ | \qquad | \\ HN{-}C{-}NH \\ \overset{\|}{\underset{\displaystyle O}{}} \end{array} + H_2O \quad (15)$$

TABLE VI
Reaction of Urea with Aldehydes Other than Formaldehyde

Urea (moles)	Aldehyde (RCH=O) R=	Moles	Catalyst or solvent	Temp. (°C)	Time (hr)	Yield (%)	M.p. (°C)	Ref.
5.0	CH_3	5.2	200 gm 0.5 N HCl	20–30	48–168	—	240–245	a
			510 gm H_2O	20–25	24–48	—	182–183	a
16.6	CH_3	18.4	30 gm 50% H_2SO_4	70	$1\frac{3}{4}$	95	—	a
2.0	CH_3	2.3	1 N NaOH to pH 8.0	3–5	6	—	(mol. wt. 147–148)	b
2.0	C_2H_5	2.3	NaOH to pH 8	3–5	6	—		b
2.0	C_3H_7	2.3	NaOH to pH 8	3–5	6	—		b
2.0	CCl_3	2.3	NaOH to pH 8	60	2	—		b
1.67	CCl_3	3.1	3 ml 38% HCl	25	24	90.5	191–192	c
			2 gm NH_4Cl in 100 ml H_2O	—				c
			8 gm 96% ethanol					c
6.0	$(CH_3)_2CH$—	4.9	HOAc, NH_4OH to pH 6.5	10–70	0.5	52	—	d
1.0	$CH_2=CH$—	1.0	Azobis(isobutyronitrile), pH 8 (1 N NaOH)	3–5	5	—	(mol. wt. 2300)	e
1.0	$CH_2=CH$— (with O, O)	1.0	10 gm 10% acetic acid + 20 gm H_2O	40	1	—	—	f
1.5	HC—CH (40%) (with O, O)	1.0	50 ml C_2H_5OH, pH 8–9 (2% NaOH)	−12 to −20	72	—	133–135	g

Thiourea (moles)	CH$_2$=CH (moles)	Catalyst or solvent	Reaction conditions		Product		Ref.
			Temp. (°C)	Time (hr)	Yield (%)	M.p. (°C)	
2.0	O=HC—CH=O (40%)	1.0	Conc. HCl	50	24		250	g,h
1.67	(furan structure)	3.0	26% aq. NH$_4$OH 4	105	2	—	—	i
0.1	0.2	3 gm H$_2$O	40-50	—	—	—	f

[a] G. Barsky and H. P. Wohnsiedler, U.S. Patent 1,896,276 (1933); K. Yoshida, O. Senda, K. Sawano, and H. Ito, Japanese Patent 70/23,154 (1970); Y. Ogata, A. Kawasaki, and N. Okumura, *J. Org. Chem.* **30**, 1636 (1965).

[b] K. Mihara and K. Takaoka, Japanese Patent 15,635 (1967).

[c] F. Bihari, A. Madarasz, R. Veroczei, and O. Szatala, Hungarian Patent 156,805 (1969).

[d] S. Kano, N. Yanai, and N. Osako, Japanese Patent 70/07,313 (1970); Badische Anilin- und Soda Fabrik A. G., French Patent 1,555,361 (1969).

[e] K. Mihara and K. Takaoka, Japanese Patent 8944 (1967).

[f] A. Gams and G. Widmer, U.S. Patent 1,654,215 (1927).

[g] S. L. Vail, R. H. Barker, and P. G. Mennitt, *J. Org. Chem.* **30**, 2179 (1965); H. Pauly and H. Sauter, *Ber. Dtsch. Chem. Ges. B* **63**, 2063 (1930); Kalle & Co., A. G., British Patent 488,686 (1938); B. Reibnitz, U.S. Patent 2,731,472 (1956).

[h] H. Biltz, *Ber. Dtsch. Chem. Ges.* **40**, 4806 (1907); J. Nemallahi and R. Ketcham, *J. Org. Chem.* **28**, 2378 (1963).

[i] E. E. Novotny and W. W. Johnson, U.S. Patent 1,827,824 (1931).

The filtrate is diluted with an equal volume of water, the pH adjusted to 2–3, and 25 gm of urea is added. Heating to 95°–97°C affords approximately 25 gm of glycouril [27], m.p. > 300°C (decomposition).

The condensation of glyoxal with substituted ureas gives N-substituted 4,5-dihydroxy-2-imidazolidinones [28]. The dimethylol derivative of 4,5-dihydroxy-2-imidazolidinone is readily prepared by the addition of formaldehyde at pH 4–8.0 [29]. The latter compound is widely used as a durable press agent for textiles [30].

3. MELAMINE–ALDEHYDE CONDENSATIONS

Melamine was discovered by Liebig [31] in 1834 and commercially produced [32] in the 1930's from dicyandiamide, but only with the publication of a patent by Soc. pour l'ind. chim. a' Bale [33] and one by Widmer and Fisch [34] were the reactions of melamine with formaldehyde described under a variety of conditions. Soon after, many publications and symposia appeared describing the latter reactions [2a,35]. Methylolmelamines undergo resinification reaction via esterification or self-condensation as in the case of the phenolic alcohols.

Melamine resin formation is therefore similar to phenolic resin production in that the A-stage resin is water soluble. The resin is partially dehydrated and the water solution that it forms is used to impregnate materials. Molding resins are filled with cellulose and impregnated in a vacuum mixer. The subsequent drying step converts the resin to the B stage and it is then ground to a given particle size. Melamine–formaldehyde resins find use in many thermosetting resin applications such as molding resins; adhesives (mainly for plywood and furniture); laminating resins for counter, cabinet, and table tops; textile resins to impart crease resistance, stiffness, shrinkage control, water repellency, and fire retardance; wet-strength resins for paper; and in alkyd resin preparations to give baking enamels such as for automotive finishes.

A. Formaldehyde Condensations

Melamine reacts with a calculated amount of formaldehyde at pH 8.0 by heating rapidly to 70°–80°C for a few minutes, cooling to crystallize the products, washing the products with alcohol, and then air drying. Depending on the amount of formaldehyde used, the products are usually either di-, tri-, tetra-, penta-, or hexamethylolmelamines. The monomethylol compound has not been reported, probably because it is too unstable. All the methylol

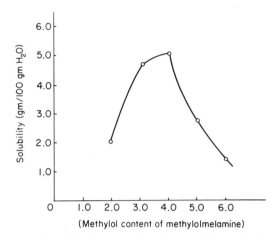

FIG. 1. Solubility in water of various methylolmelamines.

compounds are difficult to purify because of their solubility and instability toward heat. Hexamethylolmelamine is the most stable and easiest to purify. It can be prepared by heating for 10 min at 95°C a combination of melamine with 8 equivalents of formaldehyde. The product is isolated by cooling and filtering the white crystalline cake. Heating hexamethylolmelamine to 150°C converts it to an insoluble glass. Similarly, trimethylolmelamine can be prepared by stirring melamine with 3 moles of formaldehyde in the cold for 15–20 min. The solubility of methylolmelamines reaches a maximum for tetramethylolmelamine as shown in Fig. 1.

The basic dissociation constants of melamine and some representative compounds are shown in Table VII. The dissociation constant data were obtained using electrometric titration and UV absorption data.

Melamine–formaldehyde polymers belong to the general class of condensation polymers, a subject reviewed by Flory [36] and Billmeyer [37].

The following four mechanisms have been suggested by which polymer formation occurs [38]:

$$RNHCH_2OH + H_2NR \longrightarrow RNHCH_2NHR + H_2O \qquad (16)$$

$$2RNHCH_2OH \longrightarrow \underset{\underset{R}{|}}{RNHCH_2NCH_2OH} + H_2O \qquad (17)$$

$$2RNHCH_2OH \longrightarrow RNHCH_2OCH_2NHR + H_2O \qquad (18)$$

$$3RNHCH_2OH \longrightarrow 3RN{=}CH_2 \longrightarrow \underset{\underset{R}{|}}{\overset{CH_2}{\underset{H_2C}{RN}}\diagdown\underset{N}{\diagup}\overset{NR}{\underset{CH_2}{}}} + 3H_2O \qquad (19)$$

TABLE VII

Summary of Data on Dissociation Constants of Melamine Compounds at 25° ± 3°C[a]

Compound	Identification	Moles HCHO: moles melamine	% Methylation	Mol. wt.	% N obs.	pK_b[b] Titration	pK_b[b] UV	K_s[c] sp. extn. coeff. Acidic	K_s[c] sp. extn. coeff. Basic
Melamine	1	0	0	126	66.67	9.0	8.9	73	10
Dimethylolmelamine	180G	1.99	0	185	43.1	9.2	9.5	70	17
Dimethylolmelamine	410B	2.29	0	195	—	9.6	—	—	—
Trimethylolmelamine	410A	3.00	0	216	—	10.1	—	—	—
Trimethylolmelamine	OPR1001	2.97	0	215	—	9.9	9.9	62	14
Methylated trimethylmelamine	A-2	3.17	70	252	26.77	10.4	10.4	74	21
Methylated pentamethylolmelamine	A-5	4.52	79	312	—	10.9	11.5	66	18
Methylated pentamethylolmelamine	S-2	5.00	74	328	—	11.0	—	—	—
Methylated pentamethylolmelamine	L-1	5.35	79	346	23.90	11.8	12.0	62	17
Methylated hexamethylolmelamine	401C	6.00	95	386	—	12.3	11.8	44	14

[a] Reprinted from J. K. Dixon, N. T. Woodberry, and G. W. Costa, *J. Am. Chem. Soc.* **69**, 599 (1947). Copyright 1947 by the American Chemical Society. Reprinted by permission of the copyright owner.

[b] $pK_a = 14.0 - pK_b$ at 25°C.

[c] Concentration in gm per liter; length in cm, K_s is for curve maximum a 4150 mm^{-1}.

Okano and Ogata [39] have studied the rates of condensation of melamine with formaldehyde at 35°, 40°, and 70°C and pH 3–10.6. Their results showed that methylol formation occurs at 35°–40°C in the absence of acid and that the reaction is reversible throughout the pH range. At 70°C, condensation in neutral or acid media is rapid and irreversible [40]. In the case of the preparation of hexamethylolmelamine, where 1.0 mole of melamine is reacted with 8.0 moles of formaldehyde, essentially only one product is formed. However, the others consist of mixtures of products. For example, in the preparation of monoethylolmelamine, paper chromatography revealed that the mono product as well as the methylol products are formed within 5 min. Recently the kinetics of the hydroxymethylation of melamine with formaldehyde in dimethyl sulfoxide has been reported [41].

As is seen from the above equations (16–19) the moles of water produced per mole melamine serve both as a measure of the degree of condensation as well as a measure of the cross-linking index. The mole ratio Lm is defined below as

$$Lm = \frac{\text{moles condensation water}}{\text{moles melamine}} \tag{20}$$

The degree of reaction p is defined as

$$p = \frac{\text{moles condensation water}}{\text{moles formaldehyde}} \tag{21}$$

Lm is related to p as follows:

$$Lm = p\left(\frac{\text{moles formaldehyde}}{\text{moles melamine}}\right) \tag{22}$$

The degree of polymerization DP is related to p as follows:

$$p = \frac{DP - 1}{DP} \tag{23}$$

and therefore

$$Lm = \frac{DP - 1}{DP}\left(\frac{\text{moles formaldehyde}}{\text{moles melamine}}\right) \tag{24}$$

Various model systems for linking units together for a formaldehyde:melamine mole ratio of 2 are shown in Fig. 2.

Curve A is for a linear to branched-chain polymer with methylene or dimethylene ether linkages. In curves B, C, and D the linking unit is methylene, and the cross-linking is greatest in D. Curve E is for a linear polymer series with hexahydrotriazine linkages. The condensation of hexamethylolmelamine has been interpreted in a similar fashion [38].

This graphic information has been used to interpret the structures found in molding compositions as described in Fig. 3.

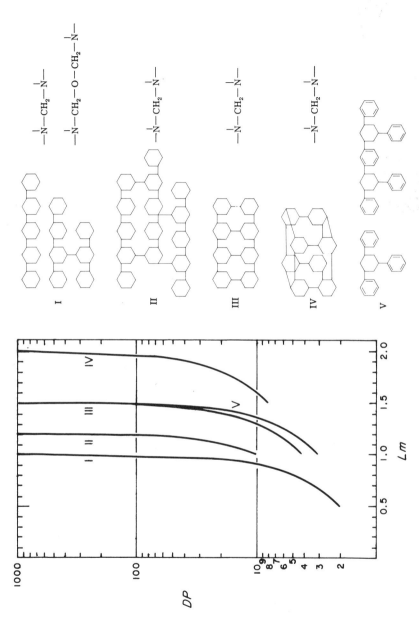

FIG. 2. Degree of polymerization vs. Lm values for theoretical polymers. [Reprinted from H. P. Wohnsiedler, *Ind. Eng. Chem.* **45**, 2307 (1953). Copyright 1953 by the American Chemical Society. Reprinted by permission of the copyright owner.]

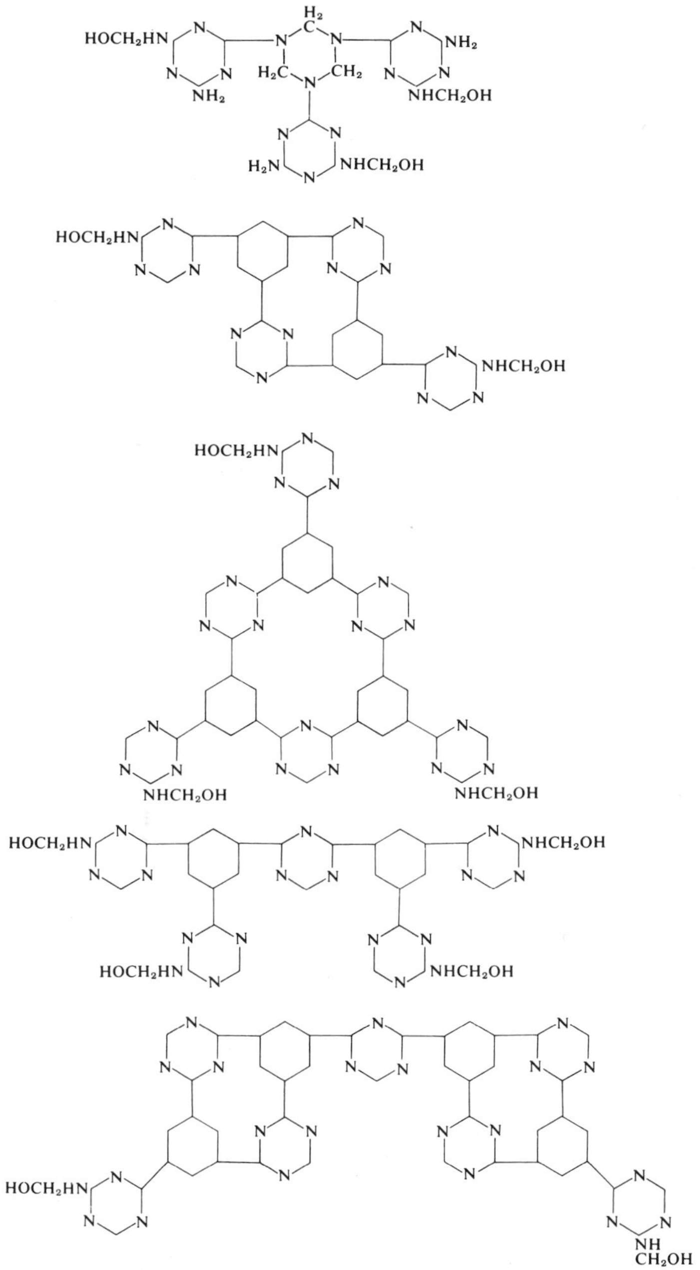

FIG. 3. Polymers with hexahydrotriazine linkages. [Reprinted from H. P. Wohnsiedler, *Ind. Eng. Chem.* **45**, 2307 (1953). Copyright 1953 by the American Chemical Society. Reprinted by permission of the copyright owner.]

3-1. Preparation of Dimethylolmelamine Molding Composition [38]

$$
\begin{array}{c}
\text{H}_2\text{N} \quad \text{N} \quad \text{NH}_2 \\
\\
\text{N} \quad \text{N} \\
\\
\text{NH}_2
\end{array}
+ 2\text{CH}_2{=}\text{O} \longrightarrow
\begin{array}{c}
\text{HOCH}_2\text{HN} \quad \text{N} \quad \text{NHCH}_2\text{OH} \\
\\
\text{N} \quad \text{N} \\
\\
\text{NH}_2
\end{array}
\quad (25)
$$

To a resin kettle equipped with a mechanical stirrer and condenser are added 126.0 gm (1.0 mole) of melamine and 161.0 gm (37.2%), of formaldehyde containing 7.5% methanol by wt. (2.0 moles). The mixture is heated to dissolve the melamine and the pH is adjusted to 8.5. The solution is refluxed and the solution is then cooled to 75°C to continue the reaction until a slightly hydrophobic condition is reached. This is detected by adding several drops of the reaction solution to 25 ml of cold water and noting a permanent cloudiness. The pH of the reaction mixture is raised to 10.0. At this point the solution is concentrated to 80% solids, and 0.93 gm of an alkanolamine buffer is added. The solution is poured into trays and oven-dried at 70°C to a thin resin layer. Before the resin is completely dried, it is reduced to granules of < 20 mesh in size. This resin is suitable as a molding resin. Molding resins are cured in molds at 4000 lb/in.² for 3 min at 155°C. Some typical analytical results for these resins are shown in Table VIII.

Methylolmelamines can be prepared which are either water soluble, hydrocarbon soluble, or insoluble depending on their preparative reaction conditions.

Water-soluble resins are prepared under carefully controlled melamine–formaldehyde condensations [42,42a]. Trimethylolmelamine may be dissolved in the hydrochloride using one equivalent of aqueous hydrochloric acid. On further reaction this solution gradually assumes a Tyndall blue haziness due to polymer formation. This solution is suitable for improving the wet strength of paper [43] or for treating textiles to impart durable press-shrinkage control [44]. The preparation of the entire range of methylolmelamines is described in Table IX and detailed examples of a few are presented.

The effect of hydrogen ion concentration in the condensation of melamine (M) and formaldehyde (F) (to give trimethylolmelamine) is shown in Table IX and Preparations 3-3.

3-2. Preparation of Hexamethylolmelamine [45,46]

$$
\begin{array}{c}
\text{H}_2\text{N} \quad \text{N} \quad \text{NH}_2 \\
\\
\text{N} \quad \text{N} \\
\\
\text{NH}_2
\end{array}
+ 6\text{CH}_2{=}\text{O} \longrightarrow 2
\begin{array}{c}
(\text{HOCH}_2)\text{N} \quad \text{N} \quad \text{N}(\text{CH}_2\text{OH})_2 \\
\\
\text{N} \quad \text{N} \\
\\
\text{N}(\text{CH}_2\text{OH})_2
\end{array}
\quad (26)
$$

To a resin flask equipped with a mechanical stirrer and condenser are added 37.8 gm (0.3 mole) of melamine and 195 gm of 37% (2.4 moles) formaldehyde

TABLE VIII
ANALYTICAL RESULTS FOR MELAMINE RESINS AND CALCULATED POLYMER Lm
VALUES[a]

	Commercial melamine		Recrystallized melamine	
	Molding composition[b]	Molded composition	Molding composition[c]	Molded composition
Analysis on wet basis				
Nitrogen, %	45.94	46.13	46.42	46.52
	45.92	46.15	46.60	46.64
Formaldehyde, %	32.0	32.0	32.2	32.2
	32.0	31.8	32.4	32.1
Moisture, %	5.14	9.26	3.12	7.61
	5.12	9.20	3.05	7.68
Methoxyl, %	0.48	—	0.42	—
	0.42	0.45 (calc.)	0.51	0.47 (calc.)
Ash, %	0.05	0.09	0.04	0.57
	0.05	0.09	0.07	0.57
Buffer, % (calculated)	0.51	0.51	0.52	0.52
Formaldehyde–melamine (molar)	1.95	1.94	1.95	1.93
Analysis on dry and true melamine-formaldehyde polymer basis[d]				
Nitrogen, %	48.80	51.28	48.39	51.16
Melamine equivalent of nitrogen, %	73.20	76.93	72.58	76.74
Formaldehyde, %	34.00	35.46	33.60	35.37
Condensation water/100 gm	7.20	12.39	6.18	12.11
Lm	0.69	1.13	0.60	1.10

[a] Reprinted from H. P. Wohnsiedler, *Ind. Eng. Chem.* **45**, 2307 (1953). Copyright 1953 by the American Chemical Society. Reprinted by permission of the copyright owner.

[b] Molding composition #148C marked by "stiff" plasticity.

[c] Molding composition #154C "free" in plasticity.

[d] Corrections made for free moisture, buffer, ash, and CH_3 vs. H. Nitrogen introduced with buffer insignificant.

solution. The pH of the mixture is adjusted with caustic to pH 8.3. The reaction mixture is heated on a water bath with stirring at 50°C for about 70–80 min, at which time all the melamine has gone into solution. On cooling the product separates as a thick white precipitate and is filtered to afford the monohydrate (after drying at 50°C). Yield: 62.5–69.0 gm (68–75%), m.p. 147°–153°C.

TABLE IX
PREPARATION OF METHYLOLMELAMINES

Melamine (moles)	37% $CH_2{=}O$[a] (moles)	Time (min)	Temp. (°C)	pH	Yield (%)	M.p.[b] (°C)
1	8	70	50	8.3	85	147–153
1	8	2 weeks	25	8.4	100	150–154
1	5	10	50	7.9	100	115–117
1	4	20	50	8.4	49	142–143
1	3	70	50	7.8	90–100	151–153[c]
1	2	Reaction does not clear	50	8.2	100	147–155[d]
1	1	Reaction does not clear	50	8.3	72	> 200

[a] Paraformaldehyde can also be used, as was demonstrated in the preparation of trimethylolmelamine. See H. P. Wohnsiedler, *Ind. Eng. Chem.* **44**, 2679 (1952).

[b] The products from the pentamethylol, dimethylol, and monomethylol derivatives are not single molecular species and probably consist of mixtures of products.

[c] M.p. reported as 130°–160°C by H. Schoenenberger and A. Adam, *Arzneim. Forsch.* **15**, 30 (1965); see also J. K. Dixon, N. T. Woodberry, and G. W. Costa, *J. Am. Chem. Soc.* **69**, 599 (1947).

[d] A melamine resin of M:F ratio 1:2 had been reported by Boldizar to be prepared at 98°–102°C (10 min) and 80°–85°C (1–2 hr). The product from this high-temperature procedure affords a mixture consisting of 74% monomer and 26% dimer (No. Av. mol. wt. = 230). For details, see L. Boldizar, *Macromol. Synth.* **3**, 49 (1969).

NOTE: Approximately one-third of the product melts at 120°–140°C.

This reaction can also be carried out at room temperature over a 2-week period to give a quantitative yield, m.p. 150°–154°C.

Infrared absorption data (KBr) are (microns): 2.93 (s), 3.37 (w), 6.40 (s), 6.65 (m), 6.90 (w), 7.20 (m), 7.50 (w), 7.82 (s), 8.35 (w), 10.10 (s), 10.90 (w), 11.45 (w), 12.30 (w) (where s = strong, m = medium and w = weak)

3-3. *Preparation of Trimethylolmelamine under Various Conditions* [42]

$$\text{(melamine)} + 3CH_2{=}O \longrightarrow \text{(trimethylolmelamine)} \qquad (27)$$

*Procedure for Preparation 1 (Table X).** Paraformaldehyde (91.8 gm,

* Procedures for preparations and Table X from H. P. Wohnsiedler, *Ind. Eng. Chem.* **44**, 2679 (1952). Copyright 1952 by The American Chemical Society. Reprinted by permission of the copyright owner.

equivalent to 3 moles of formaldehyde) and 173 ml of water were heated to 60°C in a flask equipped with a stirrer and reflux condenser, and 2.0 ml of 0.5 N sodium hydroxide was added. The temperature fell and when it was restored to 60°C within 4 min the paraformaldehyde dissolved almost completely. One hundred and twenty-six gm (1 mole) of melamine was added, followed by 3.0 ml of 0.5 N sodium hydroxide, and heating was continued for a 20-min interval at 60°C. After the first 10 min the melamine dissolved. The pH value measured after the full interval with a glass electrode-type pH meter (Leeds and Northrop No. 7662) was 10.0 at 25°C. At the completion of this stage, considered as the trimethylolmelamine stage, the buffer was added in the form of a solution of the glycine and sodium chloride (Table X) in 42.8 ml of 0.5 N sodium hydroxide. Heating was continued for 10 min at 60°C and the solution cooled promptly. On standing one day it changed in form to a thick, granular paste. The concentration of nonaqueous material was 50% by weight. After standing several days 150 gm of the product was mulled with 64 ml of water to lower the concentration to 35% and the slurry was stirred 30 min to equilibrate it. The precipitate was collected on a Büchner funnel and washed with 50 ml of water. Drying was carried out in an evacuated desiccator over phosphorus pentoxide.

The same procedure was used for Preparations 2 to 5 (Table X), inclusive, except that the concentration was closer to 50% than 55%, as in the foregoing case, during the trimethylolmelamine stage. Reaction in the presence of the buffer was at 50 $\pm$ 1% in all five cases.

*Procedure for Preparations 6 and 7 (Table X).** In these two cases it was impractical to operate at 50% concentration throughout the reaction because the high reaction rate leads to gelation. The concentration therefore had to be lowered and the reaction time curtailed. For Preparation 6 the trimethylolmelamine solution was developed in the same way as in Preparation 1, that is, by heating 20 min at 60°C. Two hundred and eighteen grams of the 50% solution (equivalent to 0.5 mole of melamine) was diluted at once with 56.5 ml of water followed by 35.4 gm of 36% hydrochloric acid, held 2 min at 50°C, and then cooled. During reaction with the acid in Preparation 7 as well as 6 the concentration of melamine plus formaldehyde was 35%.

Methylolmelamine readily reacts with alcohols in the presence of acid catalysts to give ether derivatives [47]. For example,

$$+ 3H_2O \quad (28)$$

TABLE X

SUMMARY OF DATA FOR MELAMINE-FORMALDEHYDE REACTION PRODUCTS [a,b]

Identification	Buffer [c]	pH Sought	pH At 50% concentration	Wet product at 35% concentration pH	Wet product at 35% concentration Appearance	Precipitates Microscopic appearance	Precipitates Solubility	Yield (%)	X-Ray diffraction	Chemical analysis (%) Nitrogen	Chemical analysis (%) Formaldehyde	Chemical analysis (%) Moisture (over P_2O_5)	Fusion point (°C) [d]	Assigned composition
1	0.86 NaOH 0.13 Glycine 0.10 NaCl	13	11.6	11.8	Thick slurry	Anisotropic spherulites, particle size 100–50 μm, no obvious differences	Soluble in boiling water	91	Crystalline	39.07	40.0	0.60	152–157	Trimethylol-melamine
2	0.31 NaOH 0.78 $Na_2B_4O_7$	11	10.0	10.5	Thick slurry	—	Soluble in boiling water	97	Crystalline	38.75	39.7	0.51	160–165	—
3	1.02 $Na_2B_4O_7$ 0.07 HCl	9	8.8	8.9	Thick slurry	—	Soluble in boiling water	88	Crystalline	38.50	40.5	0.20	159–163	—
4	0.67 Na_2HPO_4 0.41 KH_2PO_4	7	7.6	7.4	Thick slurry	—	Soluble in boiling water	71	Crystalline	39.21	40.7	0.39	155–159	—
5	0.66 Na_2HPO_4 0.42 Citric acid 0.46 HCl	6	6.3	6.4	Thick slurry [e]	Faintly anisotropic particles <1 to 25 μm	Coalesces in boiling water, soluble in 0.1 N HCl	75	Amorphous	42.3	42.6	0.62	No fusion at or <225	Polymer [f] av. DP 2–4
6	0.7 mole HCl/mole Melamine	3	—	2.4	Sl. hazy syrup 1 hr, then gel	—	—	—	—	—	—	—	—	—
7	1.0 mole HCl/mole melamine	1	—	0.6	Hazy syrup 1 day, then gel	—	—	—	—	—	—	—	—	—

[a] Melamine:Formaldehyde mole ratio = 1:3.

[b] Reprinted from H. P. Wohnsiedler, *Ind. Eng. Chem.* **44**, 2679 (1952). Copyright 1952 by the American Chemical Society. Reprinted by permission of the copyright owner.

[c] In grams per mole melamine, except where indicated otherwise.

[d] Dennis bar.

[e] 50% product after 1 day was composed of two phases, a small liquid phase and an opaque, granular plastic phase. 50% products at higher pH values after aging 1 day had appearance of thick, granular pastes.

[f] Chemical analysis does not distinguish between various possible polymer structures. Degree of polymerization is therefore assigned on basis of calculated degree of condensation and structures derivable through methylene, ether, and azomethine mechanisms.

A typical preparation is given for the case of hexamethoxymethylmelamine and others are described in Table XI.

3-4. Preparation of Hexamethoxymethylmelamine

$$(HOCH_2)_2N \underset{N(CH_2OH)_2}{\overset{N(CH_2OH)_2}{\diagup}} + 6CH_3OH \xrightarrow{H^+}$$

$$(CH_3OCH_2)_2N \underset{N(CH_2OCH_3)_2}{\overset{N(CH_2OCH_3)_2}{\diagup}} + 6H_2O \quad (29)$$

Method A [48,48a]: To a flask are added 3.1 gm (0.01 mole) of hexa-methylolmelamine, 20.0 gm (0.065 mole) methanol, and 1.0 ml concentrated hydrochloric acid. The reaction mixture is stirred for 8 min at 25°C, then neutralized and concentrated under reduced pressure to afford 2.2 gm (56.4%) of product, m.p. 30°–35°C.

Method B [49]: To a resin flask equipped with a mechanical stirrer, con-denser, thermometer, and gas sparge are added 70 gm of 20–50 mesh de-hydrated (16 hr at 80°C in a vacuum oven) sulfonated polystyrene and cation exchange resin (Dowex 50W-2). Methanol is added to wash the resin and then the methanol is removed through the sparger with stirring for 10 min at reflux. A slurry of 153 gm (0.5 mole) of hexamethylolmelamine in 500 gm (15.6 moles) of methanol is then added and the mixture is heated with stirring to about 50°C under a nitrogen atmosphere. The temperature is maintained at 45°–50°C for 2 hr and 15 min, at which time it appears that all the hexa-methylolmelamine has dissolved. The solution is then drawn off and evapo-rated under reduced pressure at 45°C over a 3 hr period to afford 116 gm (82%) of product, m.p. 79°C.

Using Method A or B the preparation of the ethyl, propyl, butyl, and other alkoxymethyl derivatives of melamine can be prepared. The ethyl derivative is a low-melting solid (approximately 30°C), whereas the others are liquids. Some examples are shown in Table XI.

The methylolmelamine ethers are soluble in all common organic solvents. The hexamethoxy derivative is also soluble in water up to 15% at 40°C.

Hexamethoxymethylmelamine [50] is commercially available from Ameri-can Cyanamid, as are several butylated melamine–formaldehyde resin solu-tions (xylene) [51]. The latter are used in enamel formulations to obtain faster

TABLE XI

PREPARATION OF ALKOXYMETHYLMELAMINES

Methylolmel-amine (moles)	Alcohol (moles)	Catalyst (gm)	Temp. (°C)	Time (min)	Yield (%)	M.p. (°C)	Ref.
Hexa (1.0)	C_2H_5OH (37.5)	200 Dowex 50W-2 (20–50 mesh)	78–80	120–150	91	liq.	a
Hexa (0.1)	C_2H_5OH (6.1)	HCl gas	0–5	5–10	44	approx. 30	b
Hexa (0.04)	n-C_3H_2OH (2.5)	4 ml conc. HCl	25	180–250	68	liq.	b
Hexa (0.04)	n-C_4H_9OH (2.5)	4 ml conc. HCl	25	300–400	80	liq.	b
Hexa (1.0)	i-C_3H_2OH (26.2)	200 Dowex 50W-2 (20–50 mesh)	80	120–150	77	liq.	a
Tri (2.2)	CH_3OH (24.8)	140 Dowex 50W-2 (20–50 mesh)	64	120–130	98.5	liq.	a
Tri (1.0)	CH_3OH (86.4)	H_3PO_4(pH 6–7)	64	—	—		c
Tri (0.25)	C_2H_5OH (250)	Anhydrous HCl, 80°C	—	60		134–135	d
Tetra (0.5)	i-C_3H_7OH (13.1)	100 Dowex 50W-2 (20–50 mesh)	82	150	93	liq.	a
Tetra (1.0)	i-C_4H_9OH (25.3)	170 Dowex 50W-2 (20–50 mesh)	80	120–150	88	liq.	a
Tetra (0.66)	$CH_2{=}CH{-}CH_2OH$ (6.9)	38.5 gm H_3PO_4	49	15	56	liq.	e

[a] M. M. Donaldson, U.S. Patent 3,488,350 (1970).

[b] Procedure essentially that described for benzoguanamine derivatives in S. R. Sandler, *J. Appl. Polym. Sci.* **13**, 555 (1969).

[c] W. N. Oldbam, U.S. Patent 2,378,724 (1945); *Chem. Abstr.* **39**, 5088 (1945).

[d] E. I. du Pont de Nemours & Co., British Patent 580,359 (1946); *Chem. Abstr.* **41**, 2077 (1947); F. C. McGrew, U.S. Patent 2,454,078 (1948); *Chem. Abstr.* **43**, 3473 (1949).

[e] E. R. Atkinson and A. H. Bump, *Ind. Eng. Chem.* **44**, 333 (1952).

cure and chemically resistant coatings. The use of these resins in styrenated alkyd resins in baking finishes also gives improved solvent, chemical, and mar resistance.

Other Aldehyde Condensations

Melamine reacts with several other aldehydes as well as with alcohols to give etherified products. Most other aldehydes give resinous products which have not been fully characterized. Rudelburgher reacted melamine (2 moles) with glucose (1.0 mole) to obtain a derivative melting at 218°C [52]. The synthesis of a curable melamine–acetaldehyde resin was reported by Henkel & Co. [53]. The reaction of melamine with aldehydes such as benzaldehyde, cinnamaldehyde, hexahydrobenzaldehyde, terephthalaldehyde, and acrolein have been reported [54].

Some typical examples are described in Table XII.

3-5. *Preparation of N^2,N^4-Bis(2,2,2-trichloro-1-hydroxyethyl)melamine* [55]

$$\begin{array}{c} \text{H}_2\text{N} \diagup \text{N} \diagdown \text{NH}_2 \\ \text{N} \diagdown \text{N} \\ | \\ \text{NH}_2 \end{array} + 2\text{CCl}_3\text{CHO}\cdot\text{H}_2\text{O} \xrightarrow[\Delta]{\text{H}_2\text{O}} \begin{array}{c} \text{CCl}_3\text{—CHHN} \diagup \text{N} \diagdown \text{NHCH—CCl}_3 \\ | \qquad\qquad\qquad | \\ \text{OH} \quad \text{N} \diagdown \text{N} \quad \text{OH} \\ | \\ \text{NH}_2 \end{array}$$

$$(30)$$

To a flask equipped with a stirrer and condenser is added a solution of 132 gm (0.8 mole) of chloral hydrate in 125 ml of water. The solution is stirred at 70°–80°C while 12.6 gm (0.1 mole) of pure melamine is added. Within about 2 min the melamine is all dissolved and a granular solid begins to precipitate. After heating for another 5 min the suspension is cooled, and the product is filtered, washed with water, and dried to afford 35.0 gm (83%). Heating the product on a spatula causes it to decompose to chloral and melamine. However, a determination of the melting point on a Fisher–John melting point apparatus indicates it melts at 327°–329°C (see Table XII). The literature [55] reports that it melts with decomposition.

Other aldehydes reported to condense with melamine are furfural [56], terephthalic aldehyde [56], cinnamaldehyde [56], cyclohexane, carboxaldehyde [56], buten-2-al [57], and butyraldehyde [57].

B. Other Melamine–Formaldehyde Condensations

Melamine is also employed in condensation reactions with urea, thiourea, phenol, or other amino resin starting materials to give resins with particular

TABLE XII

REACTION OF MELAMINE WITH ALDEHYDES OTHER THAN FORMALDEHYDE

Melamine (moles)	Aldehyde (moles)	Alcohol (or water) (gm)	Temp. (°C)	Time (min)	Yield (%)	M.p. (°C)	Ref.
0.2	C_6H_5CHO 0.6	—	100	60	—	Brittle resin	a
0.1	$CCl_3CHO \cdot H_2O$ 0.8	H_2O 12.5	70–80	5	83–90	Decomp.	b
0.1	0.8	12.5	70–80	5	93	327–329	c
0.3	C_3H_7CHO 2.4	C_2H_5OH 200	70–80	60	39	111–115	d
0.05	C_2H_5CHO 0.4	CH_3OH 50	60	60	76	135–138	e
0.3	CH_3CHO 1.2	C_2H_5OH (95%) 200	78–80	150	—	Brittle resin	a
0.3	1.2	200	78–80	150	79	100–105	f
0.3	2.4	200	78–80	150	28	147–150	g

[a] G. Widmer and W. Fisch, U.S. Patent 2,328,592 (1943); Soc. pour l'ind. Chim. a' Bale, British Patent 468,746 (1937); Chem. Abstr. 32, 687 (1938).

[b] E. R. Atkinson and A. H. Bump, *J. Am. Chem. Soc.* **72**, 629 (1950).

[c] Analysis calculated for structure

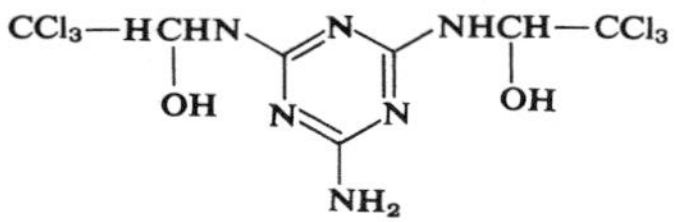

CCl₃—HCHN—NHCH—CCl₃
OH   OH
N   N
NH₂

Chemical Abstracts names this compound as follows: 1,1′-[[(6-amino-s-triazine-2,4-diyl)diimo]bis-2,2,2-trichloroethanol; see *Chem. Abstr.* **61**, 8311f. (1964), and formula index under $C_7H_8Cl_6N_6O_2$.

Calculated: C, 19.95; H, 1.90; N, 19.95
Found: C, 19.98; H, 2.06; N, 19.03

[d] Analyses in agreement with the following structure:

C₃H₇—CH—HN—N—NHCHC₃H₇
OC₂H₅   OC₂H₅
N   N
NH—CH—C₃H₇
OC₂H₅

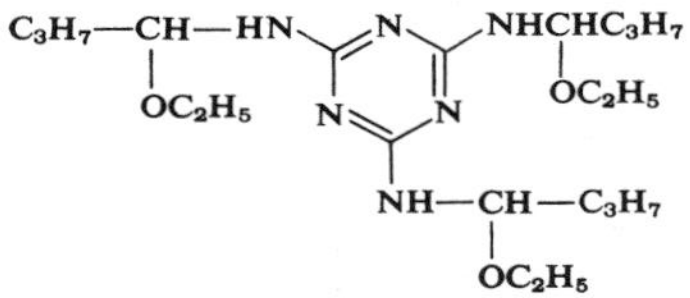

[e] Analyses in agreement with the following structure:

C₂H₅—CH—HN—N—NH—CH—C₂H₅
OCH₃   OCH₃
N   N
NH—CH—C₂H₅
OCH₃

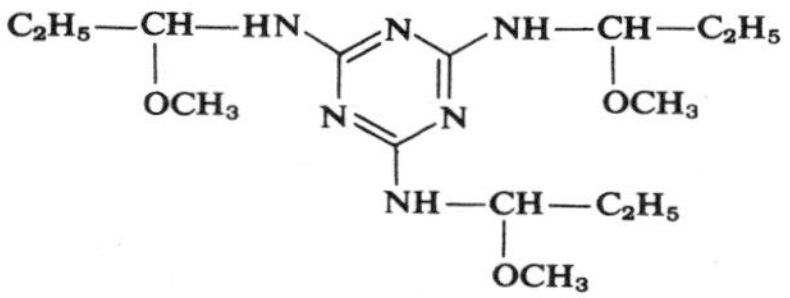

[f] Analyses in agreement with the following structure:

CH₃—CHHN—N—NHCH—CH₃
OC₂H₅   OC₂H₅
N   N
NH₂

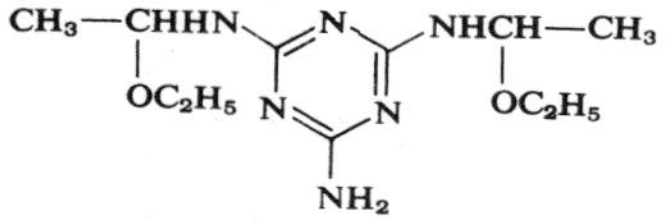

[g] Analyses in agreement with the following structure:

CH₃—CH—NH—N—NHCH—CH₃
OC₂H₅   OC₂H₅
N   N
NHCHCH₃
OC₂H₅

31

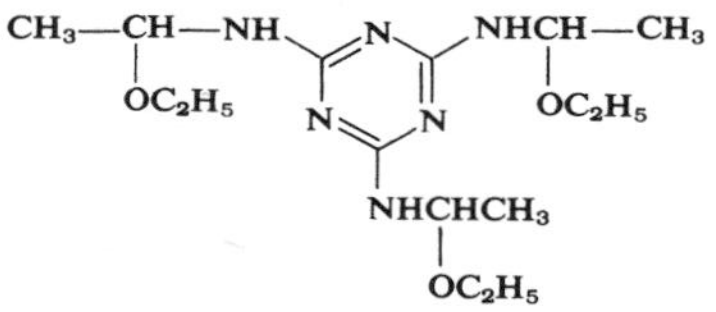

properties [45]. The addition of acetaldehyde to melamine–formaldehyde or melamine–urea–formaldehyde resins has been shown to improve the storage stability of the aqueous solutions of these resins [46]. Some typical examples of these types of condensations are described in Preparations 3-6, 3-7, and 3-8.

3-6. *Preparation of a Melamine–Urea–Formaldehyde Resin* [45]

$$
\begin{array}{c}
\text{H}_2\text{N}\diagdown\!\!\diagup\text{N}\diagdown\!\!\diagup\text{NH}_2 \\
\quad\quad\quad\quad + \text{H}_2\text{N—C—NH}_2 + \text{CH}_2{=}\text{O} \longrightarrow \text{resin} \quad (31) \\
\text{N}\diagup\diagdown\text{N} \\
\text{NH}_2
\end{array}
$$

To a resin flask equipped with a mechanical stirrer and condenser is added 63 gm (0.5 mole) of melamine and 120 gm (2.0 moles) or urea. Then 435 gm (31% aq; 4.5 moles) of neutral formaldehyde is added with stirring and heated by means of a water bath to dissolve the solids (approximately 0.5 hr). If a sample is cooled at this point, the resin will precipitate. The solution is now blended with 120 gm of cellulose and the entire mass is dried and ground. The resulting molding powder flows well when heated for 3 min at 145°C. Thiourea may be used in place of urea to give similar resins.

3-7. *Stable Aqueous Melamine–Urea–Formaldehyde Resins* [46]

$$
\begin{array}{c}
\text{H}_2\text{N}\diagdown\!\!\diagup\text{N}\diagdown\!\!\diagup\text{NH}_2 \\
\quad\quad\quad\quad + \text{H}_2\text{NCNH}_2 + \text{CH}_2{=}\text{O} + \text{CH}_3\text{CHO} \longrightarrow \text{stable aq. resin} \\
\text{N}\diagup\diagdown\text{N} \\
\text{NH}_2
\end{array}
$$

$$(32)$$

To a resin flask equipped with a mechanical stirrer and condenser are added 259.2 gm (3.2 moles) of 37% formaldehyde (inhibited with 13.2% methanol) and 4.4 gm (0.1 mole) of acetaldehyde. The pH is adjusted to 8.5 with aqueous sodium hydroxide and then simultaneously while stirring are added 50.4 gm (0.4 mole) of melamine and 60 gm (1.0 mole) of urea. The reaction mixture becomes warm (40°–50°C) and the temperature of the reaction is maintained at 50°C for about $\frac{1}{2}$ hr after the melamine and urea go into solution. The pH of the resulting condensation is adjusted to about 10.5 and the solution is stable without precipitation for 32 days (250 cP at 25°C).

Some additional preparations using this procedure are outlined in Table XIII.

TABLE XIII

PREPARATION OF STABLE AQUEOUS MELAMINE–UREA–FORMALDEHYDE RESINS [46]^a

Melamine	Urea	Formaldehyde	Acetaldehyde	Glyoxal	Stability (days)	Viscosity (25°C, cP)
0.4	1.0	3.2	0	—	<1	ppt
0.4	1.0	3.2	0.1	—	32	250
0.5	1.0	3.5	1.0	—	29	300
0.2	1.0	2.6	0.02	—	2	ppt
0.4	1.0	3.2	0.4	—	22	260
0.4	1.0	3.2	0.8	—	19	1000
0.5	0.5	3.0	—	2.0	50	80
0.5	0.5	3.0	—	1.0	48	240
0.8	0.8	3.0	—	2.0	43	160

^{a} Conditions: same as Preparation 3-7. Data given in moles.

3-8. *Preparation of a Melamine–Phenol–Formaldehyde Resin* [45]

$$ \text{(33)} $$

To a resin flask equipped as previously described are added 50 gm (0.4 mole) of melamine, 37.3 gm (0.4 mole) of phenol, and 119 gm of 40.3% (1.6 moles) of formaldehyde. The pH is adjusted to 3–6, and the solution is heated to 95°C and kept at that temperature for $\frac{1}{2}$ hr. At this point the solution forms two layers. The entire reaction mixture is blended with 60 gm of cellulose, dried, and ground up to a fine powder. The resulting material is suitable for use as a molding resin and thermosets at 150°C to an infusible solid.

4. BENZOGUANAMINE–ALDEHYDE CONDENSATIONS

Benzoguanamine (2,6-diamino-1-phenyl-1,3,5-triazine) reacts with form-aldehyde and other aldehydes to give hydroxyalkanol derivatives which can be converted to resinous products. Alone or as cocondensations with mel-amine, urea, or phenol, these hydroxyalkanol derivatives are used to give resins useful in adhesives, molding compounds, and varnishes, and resins with improved water resistance and flow properties [58]. Benzoguanamine resins offer better solubility and compatability in alkyd resins and the resulting alkyd varnishes give coatings with higher gloss [59].

Benzoguanamine (m.p. 227°C) is prepared by the reaction of dicyanodiamide with benzonitrile.

A. Formaldehyde Condensations

Benzoguanamine reacts with formaldehyde under conditions similar to melamine to give either di- [60] or tetramethylol derivatives [60,61]. The preparation of several alkoxymethyl derivatives of benzoguanamine has recently been reported [61], and the results are given in Tables XIV and XV.

4-1. Preparation of Tetramethylolbenzoguanamine [48a,60]*

$$
\underset{\substack{\text{N} \diagdown \text{N} \\ \text{C}_6\text{H}_5}}{\text{H}_2\text{N} \diagdown \text{N} \diagup \text{NH}_2} + 4\text{CH}_2=\text{O} \longrightarrow \underset{\substack{\text{N} \diagdown \text{N} \\ \text{C}_6\text{H}_5}}{(\text{HOCH}_2)_2\text{N} \diagdown \text{N} \diagup \text{N}(\text{CH}_2\text{OH})_2} \tag{34}
$$

To a flask are added 187 gm (1.0 mole) of benzoguanamine and 519 ml of 37% formaldehyde solution (6.39 moles) at pH 7.9. The mixture is warmed at 71°C for 80 min and then allowed to cool slowly to room temperature. The white solid is filtered, washed with methanol, and dried in a vacuum oven at 35°C to yield 126.8 gm, m.p. 130°–131°C and 118 gm, m.p. 128°–130°C (total yield = 79.5%).

4-2. Preparation of Tetramethoxymethylbenzoguanamine [48a]*

$$
\underset{\substack{\text{N} \diagdown \text{N} \\ \text{C}_6\text{H}_5}}{(\text{HOCH}_2)_2\text{N} \diagdown \text{N} \diagup \text{N}(\text{CH}_2\text{OH})_2} + 4\text{CH}_3\text{OH} \xrightarrow{\text{HCl}}
$$

$$
\underset{\substack{\text{N} \diagdown \text{N} \\ \text{C}_6\text{H}_5}}{(\text{CH}_2\text{OCH}_2)_2\text{N} \diagdown \text{N} \diagup \text{N}(\text{CH}_2\text{OCH}_3)_2} + 4\text{H}_2\text{O} \tag{35}
$$

TABLE XIV
SUMMARY OF SYNTHESIZED BENZOGUANAMINE COMPOUNDS[a]

$$(ROCH_2)_nN\text{---}N\text{---}N(CH_2OR)_n$$

$$C_6H_5$$

R	n	M.p. (°C)	Calc. (%)			Found (%)		
			C	H	N	C	H	N
H	2	130–131	50.80	5.54	22.84	50.72	6.11	22.30
CH_3	2	89–90	56.30	6.88	19.30	56.62	6.78	18.70
C_2H_5	2	48–49	60.10	7.87	16.70	60.97	7.97	16.80
C_3H_7	2	liq.[b]	63.10	8.65	—	62.30	8.44	—
C_4H_9	2	liq.	65.60	9.24	13.20	65.22	9.28	13.70
$C_6H_5CH_2$	2	74–75	73.70	6.16	10.50	74.00	6.22	10.53

[a] Reprinted from S. R. Sandler, *J. Appl. Polym. Sci.* **13**, 555 (1969). Copyright 1969 by the Journal of Applied Polymer Science. Reprinted by permission of the copyright owner.

[b] Mol. wt. calc., 475; found, 480 (vapor pressure osmometry).

To a flask are added 27.65 gm (0.09 mole) of tetramethylolbenzoguanamine, 150 gm (5.62 moles) of methanol, and 9 ml of concentrated hydrochloric acid. The mixture is stirred and the solution becomes clear. In a few minutes, a crystalline precipitate forms which is filtered and dried to yield 22.2 gm, m.p. 91°–92°C and 0.9 gm, m.p. 88°–89°C. Recrystallization of 2 gm of the first fraction from methanol yields a material of m.p. 89°–90°C.

4-3. *Preparation of Benzoguanamine–Formaldehyde Baking Enamel* [60]

To 200 gm (1.1 moles) of benzoguanamine is added 325 gm of 37% (4.0 moles) formaldehyde. The pH is adjusted to 8.0 with dilute sodium hydroxide and the temperature is raised to 70°C. The mixture is agitated and an equal volume of water is added 20 min after the solution becomes clear. The solution is cooled and filtered, and the precipitated tetrabis(hydroxymethyl)benzoguanamine is dried. Approximately 20 gm of this material is added with agitation to a solution of 100 gm of intercondensate of 25 gm phthalic anhydride–40 gm glycerol–35 gm soybean fatty acids in 200 gm of xylene at 50°C. This product gives a strong, clear, durable film when baked for 20 min at 150°C.

$(ROCH_2)_nN$⎯⎯⎯⎯⎯⎯$N(CH_2OR)_n$

C_6H_5

R	n	Major infrared bands (μm)[b]
H	2	2.95 (m), 3.40 (w), 6.26 (w), 6.47 (s), 7.20 (s), 9.85 (m), 10.07 (m), 11.50 (w), 12.12 (w), 12.78 (w), 14.25 (w)
CH_3	2	3.40 (m), 6.27 (m), 6.45 (s), 6.55 (s), 6.75 (s), 6.89 (m), 7.22 (s), 7.55 (s), 9.11 (s), 9.36 (m), 9.95 (s), 11.00 (m), 11.50 (s), 11.70 (m), 14.20 (m)
C_2H_5	2	3.35 (w), 6.09 (s), 6.20 (s), 6.50 (s), 7.24 (m), 7.55 (m), 9.15 (s), 9.53 (m), 9.85 (w), 10.00 (w), 11.22 (w), 11.50 (w), 11.85 (w), 12.90 (m), 14.35 (w)
C_3H_7	2	3.37 (w), 6.05 (s), 6.17 (s), 6.50 (s), 7.24 (m), 7.55 (s), 9.10 (s), 9.50 (m), 9.95 (w), 12.90 (m), 14.35 (w)
C_4H_9	2	3.37 (m), 3.46 (m), 6.25 (m), 6.50 (s), 6.73 (m), 6.87 (m), 7.25 (s), 7.55 (m), 9.20 (s), 12.09 (m), 14.22 (m)
$C_6H_5CH_2$	2	3.30 (w), 3.50 (w), 6.26 (w), 6.45 (s), 6.52 (s), 7.18 (m), 7.25 (m), 9.18 (m), 9.35 (m), 9.75 (m), 13.55 (m), 14.35 (m)

[a] Reprinted from S. R. Sandler, *J. Appl. Polym. Sci.* 13, 555 (1969). Copyright 1969 by the Journal of Applied Science. Reprinted by permission of the copyright owner.

[b] Relative intensities of bands are denoted by s = strong; m = medium; and w = weak.

4-4. Preparation of a Butylated Urea–Benzoguanamine Resin for Coatings [62]

To a mixture of 9 gm (0.05 mole) of benzoguanamine, 60 gm (1.0 mole) of urea, and 260 gm of 37% (3.2 moles) formaldehyde is added 10% aqueous sodium carbonate to bring the pH to 8.0. Then 370 gm (5.0 moles) of *n*-butanol is added and the solution is heated to 80°C for 40 min. Fifteen grams of concentrated hydrochloric acid is added and the mixture is heated to 93°C over 30 min with removal of water. The mixture is kept at 93°C for 2 hr, evaporated to 66.5% solid and diluted with xylene to 50% resin solids. The resin (3 gm) is blended with 7.0 gm of soybean oil-modified alkyd resin, coated, and baked at 130°C for 30 min to give a heat-stable coating, resistant to bending.

4-5. *Preparation of a Benzoguanamine–Urea–Formaldehyde Molding Powder* [63]

To a resin flask equipped with a mechanical stirrer and condenser is added 26.3 gm (10.14 moles) of benzoguanamine, 3.7 gm (0.026 mole) of hexamine, and 79 gm (0.98 mole) of formaldehyde. The mixture is heated to 80°–85°C and the pH adjusted with ammonium hydroxide. Then a solution containing 740 gm of 37% (9.0 moles) formaldehyde and 370 gm (6.2 moles) of urea is added and the mixture stirred and heated for 40 min at 75°–80°C and pH 7.5–8.0. The resulting reaction mixture is treated with 0.4 gm of ammonium chloride at 60°C and 250 gm of wood pulp is added. The mixture is agitated 20 min, dried, and pulverized to give a molding powder.

Other Aldehyde Condensations

Benzoguanamine has been found to react with substituted aldehydes in the presence of alcohols to give dialkoxyalkylbenzoguanamine as shown in Eq. (36) and Table XVI [64].

$$
\begin{array}{c}
H_2N \\
\end{array}
\;
\begin{array}{c}
N \\
\end{array}
\;
\begin{array}{c}
NH_2 \\
\end{array}
\quad + \; RCH{=}O \; + \; R'OH \; \longrightarrow
$$
(excess)   (excess)

$$
\begin{array}{c}
RCHHN \\
| \\
OR'
\end{array}
\;
\begin{array}{c}
N \\
\end{array}
\;
\begin{array}{c}
NHCHR \\
| \\
OR'
\end{array}
\qquad C_6H_5 \qquad (36)
$$

These results are similar to those found for melamine, as earlier described.

4-6. *Preparation of Di-2-ethoxyethylbenzoguanamine* [*2,4-Di(2-ethoxyethyl)-6-phenyl-s-triazine*] [65]

$$
\begin{array}{c}
H_2N \\
\end{array}
\;
\begin{array}{c}
N \\
\end{array}
\;
\begin{array}{c}
NH_2 \\
\end{array}
\quad + \; CH_3CH{=}O \; + \; C_2H_5OH \; \longrightarrow
\quad
\begin{array}{c}
CH_3CHHN \\
| \\
OC_2H_5
\end{array}
\;
\begin{array}{c}
N \\
\end{array}
\;
\begin{array}{c}
NHCHCH_3 \\
| \\
OC_2H_5
\end{array}
$$

C₆H₅ ... C₆H₅

(37)

To a three-necked round-bottom flask equipped with a mechanical stirrer and Dry Ice condenser is added 18.7 gm (0.12 mole) of benzoguanamine, 200 ml of ethanol (100%), and 35.2 gm (0.8 mole) of acetaldehyde. The mixture is heated to reflux for 40 min, whereupon it becomes a clear solution. Acetaldehyde loss is determined by connecting a Dry Ice trap to the condenser and acetaldehyde is replaced if necessary. The solution is heated an additional 45 min and then cooled to room temperature. Cooling the solution overnight in the refrigerator affords a crystalline precipitate of white crystals, m.p. 101°–103°C.

TABLE XVI

REACTION OF BENZOGUANAMINE (2,4-DIAMINO-6-PHENYL-s-TRIAZINE) WITH ALDEHYDES AND ALCOHOLS [64][a]

| Benzoguanamine (moles) | Aldehyde (RCH=O) | | Alcohol (R'OH) | | Yield (%) | M.p. (°C) | Analysis (%) | | | | | |
| | R= | Mole | R'= | ml | | | Calculated | | | Found | | |
							C	H	N	C	H	N
0.1	CH_3	0.8	C_2H_5	200	95	101–103	61.63	7.75	21.15	61.38	7.78	21.42
0.1	CH_3	0.8	CH_3	200	75	151–154	59.30	6.93	23.20	58.76	7.07	22.52
0.1	C_2H_5	0.8	CH_3	100	53	108–110	61.00	7.55	21.15	62.78	8.38	21.02
0.2	C_2H_5	1.0	C_2H_5	200	77	118–120	63.50	8.08	19.50	62.80	7.88	19.19
0.1	C_3H_7	0.8	C_2H_5	200	72	108–110	65.10	8.54	18.10	64.27	8.27	17.99
0.1	CCl_3	0.8	H_2O	125	81	189–191	32.37	2.28	14.54	32.91	2.48	14.51

[a] A typical preparation illustrating the reaction conditions is described in Preparation 4-6.

5. MISCELLANEOUS PREPARATIONS

1. Reaction of guanidine with formaldehyde to give either di- or trimethylolguanidines [66].

2. Transetherification of an alkoxyalkyl aminotriazine–aldehyde with a monohydroxy carboxylic acid [67].

3. Condensates of phosphate polyols with hexamethoxymethylmelamine useful as intumescent coatings [68].

4. Urea–formaldehyde foams [69].

5. Reaction of piperazine with aldehydes [70].

6. Cyclic urea derivatives and their cross-linking with cotton [71].

7. Resins based on *p*-toluenesulfonamides [72].

8. Benzoguanamine and acetoxyguanamine–aldehyde resins [73].

9. Reaction of dehydrated melamine–aldehyde condensate with fatty acid in the presence of a lower molecular weight alcohol [74].

10. Condensation of 6-phenylalkylguanamines with aldehydes in the presence of alcohols [75].

11. Self-hardening novolac aminoplast resins [76].

12. Melamine paper laminates formulation [77].

13. Methyl derivatives of ureido polyamines [78].

14. Preparation of melamine-N,N',N''-tris(methanesulfonic acid sodium salt) [79].

15. Hydroxymethylated 2-substituted 4,6-diamino-*s*-triazines [80].

16. Acrylamide–formaldehyde resins [81].

17. Acrylamide–glyoxal resins [82].

18. Aniline–formaldehyde polymers [83].

19. Aminoalkylamino-*s*-triazine–epoxy resins [84].

REFERENCES

1. S. Sjallena and L. Seekles, *Rec. Trav. Chim. Pays-Bas* **44**, 827 (1925); R. L. Wayland, Jr., U.S. Patent 2,825,732 (1958); A. L. Wilson, U.S. Patent 2,517,750 (1950); C. F. H. Allen, C. O. Edens, J. Van Allan, *Org. Synth., Collect. Vol.* **3**, 394 (1955); H. B. Goldstein and M. A. Silvestri, U.S. Patent 3,049,446 (1962).

2. J. F. Blais, "Amino Resins." Van Nostrand-Reinhold, Princeton, New Jersey, 1959; B. Bann and S. A. Miller, *Chem. Rev.* **58**, 131 (1958); E. M. Smolin and L. Rapoport, "The Chemistry of Heterocyclic Compounds, s-Triazine and Derivatives," Wiley (Interscience), New York, 1959; H. P. Wohnsiedler, *Kirk-Othmer Encycl. Chem. Technol. 2nd Ed.* **2**, 225 (1963).

2a. G. Lidmer, *Encycl. Polym. Sci. Technol.* **2**, 1 (1965).

3. *Nat. Saf. News*, pp. 95–104 (1974).

4. H. Jahn, U.S. Patent 1,355,834 (1920).

5. G. Widmer, *Encycl. Polym. Sci. Technol.* **2**, 1 (1965).
6. H. Holzer, *Ber. Dtsch. Chem. Ges.* **17**, 659 (1884).
7. E. Ludy, *J. Chem. Soc.* **56**, 1059 (1889).
8. C. Goldschmidt, *Ber. Dtsch. Chem. Ges.* **29**, 2438 (1896).
9. A. Einhorn and A. Hamburger, *Ber. Dtsch. Chem. Ges.* **41**, 24 (1908).
10. H. Scheibler, F. Trosler, and E. Scholz, *Z. Angew. Chem.* **41**, 1305 (1928).
11. H. Kadowaki, *Bull. Chem. Soc. Jpn.* **11**, 248 (1936).
12. J. I. DeJong and J. DeJong, *Recl. Trav. Chim. Pays-Bas* **72**, 139 and 1027 (1953).
13. L. E. Smythe, *J. Am. Chem. Soc.* **73**, 2735 (1951).
14. B. R. Glutz and H. Zollinger, *Helv. Chim. Acta* **52**, 1976 (1969).
15. J. I. DeJong and J. DeJong, *Recl. Trav. Chim. Pays-Bas* (**72**, 1027 (1953); H. Kadowaki, *Bull. Chem. Soc. Jpn.* **11**, 248 (1936).
16. A. Ilicento, *Ann. Chim.* (*Rome*) **43**, 625 (1953).
17. H. Staudinger and G. Niessen, *Makromol. Chem.* **15**, 75 (1955).
18. P. S. Hewett, U.S. Patent 2,456,191 (1948).
19. W. G. Simons, U.S. Patent 2,518,388 (1950).
20. T. S. Hodgins and A. G. Hovey, U.S. Patent 2,226,518 (1940); *Ind. Eng. Chem.* **30**, 1021 (1938).
21. E. E. Novotny and W. W. Johnson, U.S. Patent 1,827,824 (1931); G. Barsky and H. P. Wohnsiedler, U.S. Patent 1,896,276 (1933); L. Amaral, W. A. Sandstrom, and E. H. Cordes, *J. Am. Chem. Soc.* **88**, 225 (1966); K. K. Ibrakhimov, K. R. Rustamov, and R. V. Usmanov, *Dokl. Akad. Nauk Uzb. SSR* **27**, 29 (1970); K. K. Ibrakhimov and K. R. Rustamov, *Zh. Fiz. Khim.* **44**, 1563 (1970); *Uzb. Khim. Zh.* **13**, 49 (1969); A. Kawaski and Y. Ogata, *Mem. Fac. Eng., Nagoya Univ.* **19**, 1 (1967); Y. Ogata, A. Kawasaki, and N. Okumura, *Tetrahedron* **22**, 1731 (1966).
22. S. L. Vail, R. H. Barker, and P. G. Mennit, *J. Org. Chem.* **30**, 2179 (1965).
23. H. Feichtinger, German Patent 1,468,541 (1969).
24. K. Mihara and K. Takaoka, Japanese Patent 15,635 (1967).
25. F. Bihari, A. Madarasz, R. Veroczei, and O. Szatala, Hungarian Patent 156,805 (1969).
26. B. Reibnitz, U.S. Patent 2,731,472 (1956); H. Pauly and H. Sauter, *Ber. Dtsch. Chem. Ges. B* **63**, 2063 (1930).
27. H. Biltz, *Ber. Dtsch. Chem. Ges.* **40**, 4806 (1907).
28. F. Kohler and J. Brandeis, German Patent 968,904 (1958); A. Woerner and W. Rümens, German Patent 962,795 (1957); Badische Anilin & Soda Fabrik A. G., British Patent 783,051 (1957).
29. H. B. Goldstein and M. A. Silvestri, U.S. Patent 3,049,446 (1962); Tootal Broadhurst Lee Co., Ltd., British Patent 831,297 (1960).
30. Anonymous, *Ind. Eng. Chem.* **57**, 13 (1965).
31. J. Liebig, *Ann. Chem.* **10**, 17 (1834).
32. W. Fisch, U.S. Patent 2,164,705 (1939); I. G. Farbenindustrie A.-G., French Patent 817,895 (1937); G. Widmer, W. Fisch, and J. Jakl, U.S. Patent 2,161,940 (1939); D. Jayne, U.S. Patent 2,180,295 (1939); Gesellschaft für Chemische Industrie a Basel, French Patents 811,804 and 814,761 (1937).
33. Soc. pour l'ind. chim. a' Bale, British Patent 468,677 (1936).
34. G. Widmer and W. Fisch, U.S. Patent 2,197,357 (1940).
35. E. Friedheim, U.S. Patent 2,430,462 (1947); J. Dudley and E. Lyon, *Symp. Fibrous Proteins, Soc. Dyers Color.* p. 215 (1948); A. Gams, G. Widmer, and W. Fisch, *Helv. Chim. Acta* **24**E, 302 (1941); *Br. Plast. Moulded Prod. Trader* **14**, 508 (1943); A. G. Hovey, *Acta* **24**E, 302 (1941); *Br. Plast. Moulded Prod. Trader* **14**, 508 (1943);

T. Hodgins, A. G. Hovey, S. Hewett, W. R. Barrett, and C. M. Meeske, *Ind. Eng. Chem.* **33**, 769 (1941); E. M. Smolin and L. Rapoport, "The Chemistry of Heterocyclic Compounds, *s*-Triazine and Derivatives," pp. 338–386. Wiley (Interscience), New York, 1959; E. Werner, *J. Chem. Soc.* **107**, 721 (1915).

36. P. J. Flory, *Chem. Rev.* **39**, 137 (1946); "Principles of Polymer Chemistry." Cornell Univ. Press, Ithaca, New York, 1953.
37. F. W. Billmeyer, Jr., "Textbook of Polymer Science." Wiley (Interscience), New York, 1971.
38. H. P. Wohnsiedler, *Ind. Eng. Chem.* **45**, 2307 (1953).
39. M. Okano and Y. Ogata, *J. Am. Chem. Soc.* **74**, 5728 (1952).
40. J. Koeda, *J. Chem. Soc. Jpn., Pure Chem. Sect.* **75**, 571 (1954).
41. K. Sata, *Bull. Chem. Soc. Jpn.* **40**, 724 (1967); *Chem. Abstr.* **67**, 53270c (1967).
42. H. P. Wohnsiedler, *Ind. Eng. Chem.* **44**, 2679 (1952).
42a. H. P. Wohnsiedler and W. M. Thomas, U.S. Patents 2,345,543 (1944); 2,485,079 and 2,485,080 (1949).
43. C. G. Landes and C. S. Maxwell, U.S. Patent 2,559,220 (1945).
44. H. P. Wohnsiedler, *Kirk-Othmer, Encycl. Chem. Technol., 2nd Ed.* **2**, 225 (1963).
45. G. Widmer and W. Fisch, U.S. Patent 2,328,592 (1943).
46. S. R. Sandler, U.S. Patent 3,793,280 (1974).
47. W. N. Oldham, U.S. Patent 2,378,724 (1945); *Chem. Abstr.* **39**, 5088 (1945).
48. G. Widmer and W. Fisch, U.S. Patent 2,328,592 (1943).
48a. S. R. Sandler, *J. Appl. Polym. Sci.* **13**, 555 (1969).
49. M. M. Donaldson, U.S. Patent 3,488,350 (1970).
50. Technical Bulletin, Cymel 303/Cymel 370. American Cyanamid Co., Wayne, New Jersey, 1974.
51. Technical Bulletin, CRT 21. American Cyanamid Co., Wallingford, Connecticut, 1964.
52. L. Rudelburgher, *Chem. Zentralbl.* **17**, 2110 (1913).
53. Henkel & Co., British Patent 455,008 (1936).
54. Gesellschaft für Chem. Industrie, Basel, German Patent 671,724 (1939).
55. E. R. Atkinson and A. H. Bump, *J. Am. Chem. Soc.* **72**, 629 (1950).
56. Soc. pour l'ind. Chim. a' Bale, British Patent 468,746 (1937).
57. W. Talbot, U.S. Patent 2,260,239 (1941).
58. Ciba, Ltd., Swiss Patent 308,280 (1955); A. A. Varela and H. T. Bangs, U.S. Patent 2,781,553 (1957); A. A. Varela and R. J. Schupp, German Patent 942,168 (1956); K. Hiratsuka, S. Kono, and S. Omura, Japanese Patents 15,246 and 16,791 (1963).
59. A. Coutras, U.S. Patent 2,867,600 (1959); F. P. Grimshaw, *J. Oil Colour Chem. Assoc.* **40**, 1060 (1957); F. R. Spencer, U.S. Patent 2,579,980 (1951); W. J. Kissel, U.S. Patent 3,075,945 (1963).
60. H. Rider, T. Anas, and G. L. Fraser, U.S. Patent 2,816,865 (1957).
61. S. R. Sandler, *J. Appl. Polym. Sci.* **13**, 555 (1969); J. R. Stephens, U.S. Patent 3,091,612 (1963).
62. M. Moroi and Y. Nakamura, Japanese Patent 6822,872 (1968).
63. B. Ogaki, French Patent 1,347,975 (1964); *Chem. Abstr.* **61**, 8488 (1964).
64. S. R. Sandler using procedure 4-6; see also Sandler [48a].
65. S. R. Sandler using procedure described in ref. [48a].
66. Anonymous, *Plast. Molded Prod.* **7**, 401 (1931); Technical Brochure "The Chemistry of Guanidine," Tech. Brochure, No. 1C-1365-11-1. American Cyanamid Co., Wayne, New Jersey, 1963.
67. C. E. Coats and J. D. Nordstrom, U.S. Patent 3,519,627 (1970).

68. F. H. Thomas, H. M. Hendrick, and E. L. Schulz, U.S. Patent 3,422,046 (1969).
69. Allied Chem.-Corp., British Patent 1,021,248 (1966); D. S. Shriver, R. R. Mac-Gregor, and W. P. Moore, Jr., U.S. Patent 3,256,067 (1966); F. Brasco and P. R. Temple, U.S. Patent 3,284,379 (1966); S. Coppick and R. L. Beal, U.S. Patent 3,189,479 (1965).
70. S. R. Sandler and M. L. Delgado, *J. Polym. Sci., Polym. Chem. Ed.* **7**, 1323 (1969).
71. H. Z. Jung, R. R. Benerito, E. J. Gonzales, and R. J. Berni, *J. Appl. Polym. Sci.* **13**, 1949 (1969).
72. Switzer Brothers, Inc., British Patent 1,067,666 (1967).
73. W. Yasutake, U.S. Patent 3,523,051 (1970).
74. Manuf. de Produits Chim. Protex., French Patent 1,547,691 (1969).
75. Y. Nomura, K. Honda, T. Yoshida, T. Nogachi, and F. Kukarai, Japanese Patent 71/21,744 (1931); *Chem. Abstr.* **75**, 15348 (1972).
76. H. Matsuda, T. Iwaki, and T. Tanaka, Japanese Patent 68/28,479 (1968); *Chem. Abstr.* **70**, 78785 (1969).
77. Henkel & Cie, German Patent 1,282,975 (1969).
78. R. S. Yost and R. W. Anten, U.S. Patent 2,616,874 (1952).
79. H. Schoenenberger and A. Adam, *Arzneim.-Forsch.* **15**, 30 (1965).
80. T. Tashiro, *J. Appl. Polym. Sci.* **18**, 2903 (1974).
81. H. Kamogawa, Japanese Patent 69/14,938 (1969); *Chem. Abstr.* **73**, 99633i (1970).
82. American Cyanamid Co., Netherlands Patent Appl. 6,609,764 (1967); *Chem. Abstr* **67**, 12758z (1967).
83. G. Kunath and K. Hanschild, German (East) Patent 62,447 (1968); *Chem. Abstr.* **70**, 78782w (1969); M. F. Brooks and V. Kerrigan, German Offen. 1,959,168 (1970); *Chem. Abstr.* **70**, 77895k (1970); J. A. Nicholas and H. J. Twitchett, British Patent 1,192,121 (1970); *Chem. Abstr.* **70**, 26213d (1970).
84. H. Feichtinger and W. Raudenbusch, U.S. Patent 3,413,268 (1968); *Chem. Abstr.* **70**, 78783x (1969).

PHENOL–ALDEHYDE CONDENSATIONS

I. INTRODUCTION

This chapter is concerned with the practical aspects of reaction of formaldehyde with phenols to give both methylolated and cross-linked products. The reaction of other aldehydes to give resinous products will be briefly mentioned. The mechanism of these reactions will not be discussed.

A. Phenol–Aldehyde Condensates

In 1843, Pira [1] reported that phenol alcohols are converted to resins (called saliretins) on heating. Baeyer [2] in 1872 reported that the reaction of phenols with acetaldehyde in the presence of acid catalysts also gives resinous products. Kleeberg [3] in 1891 reported that formaldehyde undergoes similar reactions. However, Dianin [4,4a] found that acetone reacts with phenol to give a crystalline bisphenol (now known as bisphenol A). In 1874 Lederer [5] and Manasse [6] independently synthesized *o*-hydroxybenzyl alcohol (saligenin) by the low-temperature alkaline-catalyzed formaldehyde reaction.

Baekeland [7,7a] was granted a patent in 1909 describing his alkaline-catalyzed Bakelite resins ("resoles") and also the acid-cured "novolak" product. He conceived the idea of using counterpressure during hot cure to prevent bubbles and foaming from heat and was able to produce strong cured resinous products.

In connection with nomenclature, *Chemical Abstracts* uses the terms

TABLE I

PHENOLICS: PATTERN OF CONSUMPTION[a]

	1000 metric tons	
Market	1972	1973
Bonding and adhesive resin for:		
Coated and bonded abrasives	9.1	11.2
Fibrous and granulated wood	40.0	42.0
Friction materials	13.4	14.7
Foundry and shell moldings	43.6	50.0
Insulation materials	107.0	112.0
Laminating		
Building	26.1	26.2
Electrical/electronics	7.3	7.3
Furniture	16.0	17.0
Other	2.7	2.7
Plywood	163.3	125.0
Molding compounds	155.6	173.1
Protective coatings	9.6	10.1
Exports	13.1	13.6
Other	44.0	48.6
Total	650.8	653.5

[a] Reprinted from *Mod. Plast.* **51**, 40 (1974). Copyright 1974 by Modern Plastics. Reprinted by permission of the copyright owner.

TABLE II
PHENOLICS: MOLDING POWDER MARKETS[a]

| | 1000 metric tons | |
Market	1972	1973
Appliances	31.8	41.4
Business machines	6.1	6.8
Closures	4.5	4.1
Electrical/electronics		
Controls and switches	56.0	61.0
Telephone and communications	9.3	9.5
Wiring devices	15.9	16.3
Housewares		
Utensils and handles	14.3	14.7
Other	4.7	5.3
Machine parts, etc.	4.8	5.1
Other	8.2	8.9
Total	155.6	173.1

[a] Reprinted from *Mod. Plast.* **51**, 40 (1974). Copyright 1974 by Modern Plastics. Reprinted by permission of the copyright owner.

novolak and resol. Following more popular usage we shall use the terms novolac and resole in this chapter.

Some historical reviews and monographs on phenolic resins are worthwhile sources for additional details [7,8–8c].

Today phenolic resins are a large-volume item achieving sales of approximately 1.3 billion lb in 1973, up from 450 million lb in 1956. The pattern of consumption and sales growth are shown in Tables I and II. As a result of their stability [9] and other useful properties, these resins continue to find widespread use.

Much information is being published on the phenolic resins as evidenced by the proliferation of references in a recent *Chemical Abstracts* issue (8th Collect. Vol. CA 66-75) which lists 22 pages of subject matter under "phenol condensation products."

Some additional references describing the use and versatility of resins based on phenolics are also worthwhile to consult for more details [8a,10].

CAUTION (*Toxicity*): Phenols, formaldehyde, and other aldehydes are toxic and should be handled with adequate ventilation and skin protection [11]. Where possible the use of hydrogen chloride in the presence of formaldehyde or formaldehyde sources (hexamethylenetetramine, etc.) should be avoided. Recent reports have indicated that formaldehyde and hydrogen chloride can spontaneously react to give the known carcinogen bis(chloromethyl) ether [12].

2. CONDENSATION OF PHENOL WITH ALDEHYDES

A. Phenol–Formaldehyde Condensations

Freeman and Lewis [13] published one of the first complete kinetic investigations of phenol with formaldehyde. The hydroxymethylation at 30°C was followed by analysis using quantitative paper chromatography with detection of five products (2-hydroxymethylphenol, 4-hydroxymethylphenol, 2,4-dihydroxymethylphenol, 2,6-dihydroxymethylphenol, and 2,4,6-trihydroxymethylphenol). More recently, Zavitsas and Beaulieu [14] used GLC to investigate the kinetics of the phenol–formaldehyde reaction using only catalytic amounts of base and at pH ranges where the second-order rate expression was shown to be valid.

$$\text{Rate} = k[\text{P}^-][\text{F}] \tag{1}$$

where P^- is the anion of a phenolic component and F is the concentration of unreacted formaldehyde (hydrate).

This expression is inadequate for concentrated systems where the formaldehyde concentration is greater than 1 M (3 wt.%) [14].

It is interesting to note from the data of Zavitsas [15] that at 30°C, with a mole ratio of formaldehyde to phenol of 1.95 and 1.99% of the phenolic component in the anion form, there are 5 hydroxymethyls produced as shown in Scheme 1. The rate of their production at 30°C and 57°C is shown in Figs. 1 and 2.

Gel permeation and gas–liquid chromatography of phenolic resins have been reported [16,16a]. The technique allows one to study the molecular weight distribution (average molecular weights) of novolac and offers a means to monitor effects of process changes on product composition. The results of these techniques are shown in Figs. 3 and 4.

Proton NMR, differential thermal analysis (DTA), thermogravimetric analysis (TGA) and capacitance versus cure time have also been reviewed in connection with analyses of phenolic resins [16a].

For an average novolac resin with 10 phenol rings chance contact at the three positions (ortho and para) on phenol by formaldehyde could lead to 13,203 linear isomers [17]. This complexity gives rise to amorphous-type

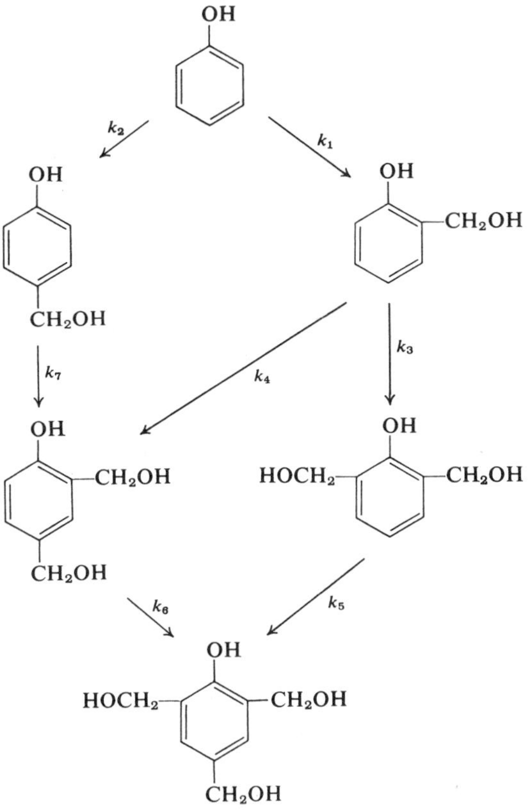

SCHEME 1. Possible reactions of phenol and formaldehyde (excess) under alkaline conditions. [Reprinted from A. A. Zavitsas, *Am. Chem. Soc., Div. Org. Coat. Plast. Chem., Pap.* **26**, 93 (1966). Copyright 1966 by the American Chemical Society. Reprinted by permission of the copyright owner.]

products and makes isolation of pure compounds difficult. Isomerism studies on a series of resoles has recently been reported [18].

Phenol and formalin at pH 3.0–3.1 do not react to any appreciable degree within days or weeks. Adjustment to either pH 0.5 to 1.5 or pH 4–7 or 7–11 gives an accelerated rate of reaction. Equimolar amounts of phenol and formaldehyde (or excess formaldehyde) at alkaline pH values give controllable one-step reactions (yielding resoles) whereas acid pH values give an uncontrollable reaction [8,19]. At one mole of phenol and less than one mole of formaldehyde, acid pH values give controllable phenolic resins (novolacs) whereas alkaline pH gives a highly ortho-substituted novolac as described in Scheme 2.

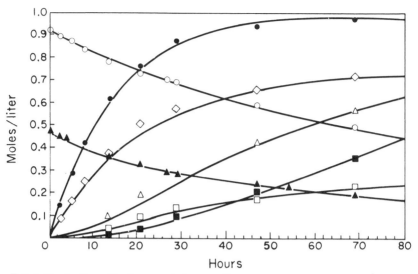

FIG. 1. Production of hydroxymethyls according to Scheme 1. ▲ 1:10 phenol; ○ 1:10 formaldehyde; ● 2-HMP; ◇ 4-HMP; □ 2,6-DHMP; △ 2,4-DHMP; ■ 2,4,6-THMP. Temperature = 30°C. [Reprinted from A. A. Zavitsas, *Am. Chem. Soc., Div. Org. Coat. Plast. Chem., Pap.* **26**, 93 (1966). Copyright 1966 by the American Chemical Society. Reprinted by permission of the copyright owner.]

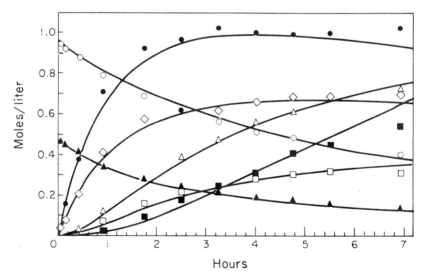

FIG. 2. Production of hydroxymethyls. Symbols have the same meaning as in Fig. 1. Temperature = 57°C. [Reprinted from A. A. Zavitsas, *Am. Chem. Soc., Div. Org. Coat. Plast. Chem., Pap.* **26**, 93 (1966). Copyright 1966 by the American Chemical Society. Reprinted by permission of the copyright owner.]

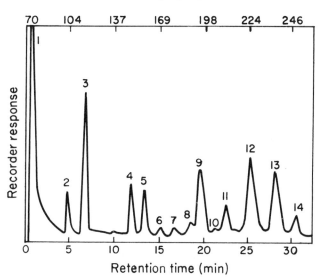

FIG. 3. Gas–liquid chromatographic analysis of phenolic resole. 1, Solvent, acetone; 2, phenyl acetate; 3, benzyl acetate standard; 4, 2-HMP diacetate; 5, 4-HMP diacetate; 6, 2-HMP monohemiformal diacetate; 7, 4-HMP monohemiformal diacetate; 8, 2,6-DHMP triacetate; 9, 2,4-DHMP triacetate; 10, 2,6-DHMP monohemiformal triacetate; 11, 2,4-DHMP monohemiformal triacetate; 12, 2,4,6-THMP tetraacetate; 13, 2,4,6-THMP monohemiformal tetraacetate; 14, 2,4,6-THMP dihemiformal tetraacetate. [Reprinted from M. F. Drumm, *Am. Chem. Soc., Div. Org. Coat. Plast. Chem., Pap.* **26**, 87 (1966). Copyright by the American Chemical Society. Reprinted by permission of the copyright owner.]

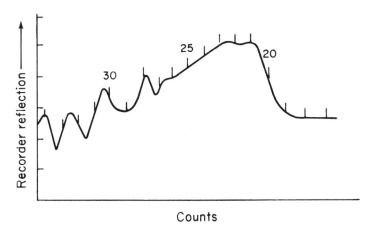

FIG. 4. Gel permeation chromatography: phenolic novolacs—5 ml per count, 5 min per count, solvent: THF. [Reprinted from M. F. Drumm, *Am. Chem. Soc., Div. Org. Coat. Plast. Chem., Pap.* **26**, 89 (1966). Copyright 1966 by the American Chemical Society. Reprinted by permission of the copyright owner.]

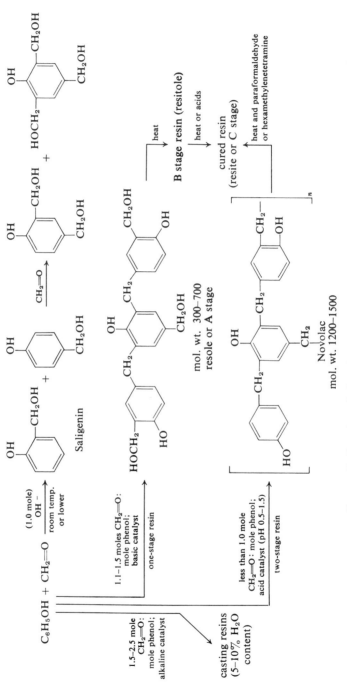

SCHEME 2. Reaction of phenol with formaldehyde at various pH values.

A-Stage resins or resoles are soluble in water and in conventional solvents. If the polymerization continues then one obtains higher molecular weight resins at the B or resitol stage which are insoluble in water or alkali solutions but are still fusible and soluble in conventional organic solvents such as acetone. Further condensation of the B-stage resin affords an extensively cross-linked, fully cured, insoluble, infusible resite or C-stage phenolic resin.

The mean molecular weights of novolacs are usually about 1000 or less. Novolac can react further with formaldehyde [from paraformaldehyde or hexamethylenetetramine (hexa)] to give products that are considerably cross-linked and are said to be "cured," thermoset, or thermohardened.

As seen above, the mole ratio of phenol to formaldehyde determines the type of resin and degree of cross-linking obtained. This particularly affects the physical properties and resulting strength of the casting resins [20]. Using a basic catalyst, Sprung found that the rate of formaldehyde disappearance follows the relationship in Eq. (2) [21]:

$$K = 6.9 \times 10^{-3} \times C \tag{2}$$

where K = rate of formaldehyde disappearance and C = molar phenol: formaldehyde ratio. This relationship indicates that the K value increases as the proportion of phenol to formaldehyde increases [21].

The reaction of various phenols with various mole ratios of formaldehyde (paraformaldehyde) at 98°C leads to different average chain lengths of products as described in Table III.

B. Other Phenols

The reaction of formaldehyde with unsubstituted phenols leads to either soluble or cross-linked resins since condensation occurs at either ortho or para positions. Monosubstituted [22] (ortho or para) phenols give cross-linking with difficulty but phenols doubly substituted in ortho or para positions yield only low molecular weight products. If only one ortho or para position is available on the phenol then the phenol cannot produce resins and reacts with difficulty with aldehydes [21]. Sometimes cresols and phenol are blended together to obtain fully cured resins. In addition to phenol, the other important phenols that are used to give phenolic resins are *o*-cresol, mixed cresols, *p-tert*-butylphenol (from isobutylene and phenol), *p*-phenylphenol (by-product from phenol manufacture), resorcinol, and cardanol (from cashew nutshell liquid).

The reactions of alkylphenols (containing substituents having at least three carbon atoms) with formaldehyde afford resins known as alkylphenolic resins [22]. Examples of suitable phenols include *tert*-butylphenol, octylphenol and phenylphenol.

TABLE III

Average Chain Length of Phenol–Paraformaldehyde Reaction Products[a]

Phenol	Moles of CH_2O per mole phenol		Reaction time, at 98°C (min)	Apparent reactive positions		n (av.)	Remarks
	Taken	Reacted		Initial	Final		
Phenol	0.87	0.85	291	3.00	2.03	2.6	—
Phenol	1.05	1.04	300	3.00	1.90	2.3	—
Phenol	1.25	1.20	285	3.00	1.89	2.0	—
o-Cresol[b]	0.87	0.78	545	2.57[b]	2.13	1.4	—
p-Cresol	0.87	0.81	420	2.27	1.96	1.3	—
m-Cresol	0.87	0.86	90	2.98	1.71	2.1	98°C
m-Cresol	0.87	0.84	101	2.98	2.43	1.6	88°C
m-Cresol	0.87	0.83	150	2.98	2.57	1.4	78°C
m-Cresol	0.87	0.83	300	2.98	2.68	1.3	68°C
3,5-Xylenol	0.87	0.85	30	2.95	1.91	1.3	—
2,6-Xylenol	1.05	0.53	855	2.56	2.58	2.0	Dimer formed
2,3,5-Trimethylphenol	0.87	0.77	155	2.15	1.94	1.2	—
3,4-Xylenol	0.87	0.77	120	2.26	1.90	1.3	—
2.5-Xylenol[b]	0.87	0.74	120	2.67[b]	2.14	1.6	—

[a] Reprinted from M. M. Sprung, *J. Am. Chem. Soc.* **63**, 334 (1941). Copyright 1941 by the American Chemical Society. Reprinted by permission of the copyright owner.

[b] o-Cresol and 2,5-xylenol, structurally related to o-cresol, show an appreciably higher absorption of bromine in the presence of formaldehyde than in its absence. Slightly high initial bromine values were shown by other phenols in the presence of formaldehyde, but in general the discrepancies were within the experimental error. The peculiarity of these two phenol–paraform systems in this regard is not at present understood.

The alkyl-substituted resins show good compatibility with oils, neutral resins, and rubbers.

Reactive alkylphenolic resins are obtained from bismethylol derivatives.

$$\text{HOCH}_2-\!\!\left[\begin{array}{c}\text{OH}\\\\\text{R}\end{array}\right.-\text{CH}_2-\left.\begin{array}{c}\text{OH}\\\\\text{R}\end{array}\right]_n-\text{CH}_2\text{OH} \xrightarrow[-\text{H}_2\text{O}]{} \begin{array}{c}\text{reactive}\\\text{alkylphenolics}\end{array} \quad (3)$$

The alkylphenolic resins are of more interest than the corresponding novolacs since they find applications as modifying and cross-linking agents for oil varnishes and in the production of resins for coatings and printing inks based on rosin. Alkylphenolic resins are also used as antioxidants and stabilizers.

Resorcinol reacts rapidly with formaldehyde in the absence of catalysts, but both acid and alkaline catalysts have been used [23]. Resorcinol–formaldehyde resins are used to produce low-temperature curing resins alone or in combination with phenol–formaldehyde resin. Some combinations permit curing at room temperature under neutral conditions. This property of low-temperature cure is important in the manufacture of laminates and adhesives. Base resins for adhesives are made by reacting one mole of resorcinol with less than one mole of formaldehyde, and the condensate is usually dissolved in a solvent such as aqueous alcohol. The resin base at pH 7.0 has a long shelf-life and is hardened by the addition of formaldehyde or paraformaldehyde [24].

Trihydric phenols have only played a minor role because of their high price. Bisphenol A [2,2'-bis(hydroxyphenyl)propane] is still being used to a small extent in phenolic resin production. However, there is renewed interest in some specialty phenols such as chlorinated and brominated types as a result of legislation regarding flame retardancy.

C. Reaction Conditions

Reaction uniformity in phenol–aldehyde condensations is obtained by careful control of reaction conditions such as temperature of condensation [25], catalyst concentration, and pH. At the early stages the reaction is exothermic and the temperature must be controlled to prevent a runaway reaction [25a]. Gel permeation chromatography offers a means to monitor product uniformity. Specific catalysts (usually 1–6%) are used to obtain either a resole (alkaline catalysts or basic salts) or novolac (acid catalysts).

The basic catalysts are usually $NaOH$, Na_2CO_3, KOH, K_2CO_3, $Bu(OH)_2$, $R_4N^+OH^-$, NH_3, RNH_2, and R_2NH. The acidic catalysts commonly used are $HCOOH$, HX, $HOCO-COOH$, H_3PO_4, H_2SO_4, NH_2SO_3H, $CH_3C_6H_4SO_3H$, and CCl_3COOH. Carrying out the alkali-catalyzed reaction in nonpolar solvents increases the relative amount of ortho substitution [8–8c]. In addition the hydroxides of Zn, Mg, or Al or the divalent electroposition metal and organic acid salts are used to obtain high yields of ortho-methylolated novolacs which are more active in curing [26,26a].

Recently Partansky [25] reinvestigated Bender's [26] and Fraser's [26a] work and reported, as did Bender, that the best ortho-directing catalysts were small amounts of mildly alkaline oxides of zinc and magnesium. The acetate of zinc and magnesium also gave similar results, affording 85–95% ortho substitution. Following these metals were the tertiary amines which gave 70–80% ortho substitution. Strong alkali gave 70–75% ortho substitution with high molecular weight products being formed. Most acids gave close to a 50:50 ortho:para ratio except when 5% acetic acid was used in alcohol or toluene, as shown in Table IV.

The last entry in Table IV indicates that in order to prepare about 89% 4,4′-di(hydroxyphenyl)methane one should run the reaction for about 27 days at room temperature using a mixture of 100 parts of phenol to 7 parts 91% paraformaldehyde and about 14–20 parts concentrated hydrochloric acid dissolved in 100–150 parts 50% ethanol.

The most common grade of formaldehyde is known as formalin (37% formaldehyde). Polymer precipitation is usually inhibited by the addition of 7–15% methanol. The methanol does not interefere in resin production but is removed in the stripping operation. For economic reasons 44–50% uninhibited formaldehyde is used in commercial practice to save costs in stripping and to allow larger batch operations to be carried out. Occasionally paraformaldehyde is also used. The latter contains 91% formaldehyde and 9% water.

2-1. *Preparation of a Phenol–Formaldehyde Resole* [27,28]

(4)

TABLE IV

EFFECT OF CATALYSTS ON THE ORTHO:PARA RATIO IN PHENOL–FORMALDEHYDE CONDENSATIONS[a]

CH_2=O:C_6H_5OH (mole ratio)[b]	Catalyst	Parts catalyst	Solvent	Parts solvent[c]	Conditions		Product (mol. wt.)	Ortho:para ratio
					Temp. (°C)	Time (hr)		
0.1	KOH	1.0	H_2O	0.5	100	4	260	66:34
0.2	NaOH	2.0	H_2O	100	96	2	312	65:35
0.2	$Mg(OH)_2$	2.0	H_2O	100	96	2	360	69:31
0.2	MgO	2.0	Toluene	100	98	2	420	95:5
0.15	ZnO	0.4	H_2O	10	115	3	235	83:17
0.2	Et_3N	2.0	H_2O	100	96	2	270	55:45
0.2	Et_3N	1.5	H_2O	100	150	2	280	87:13
0.2	$Zn(OAc)_2$	1.0	C_2H_5OH	100	150	2	High	88:12
0.2	$Zn(OAc)_2$ } AcOH	1.0 / 0.5	C_2H_5OH	100	175	2	238	100
0.2	$Mg(OAc)_2$ } MgO	2.0 / 0.3	Toluene	100	150	2	297	100
0.2	AcOH	5	C_2H_5OH	100	70	24	315	82:18
0.2	Oxalic acid	2	H_2O	20	96	2	215	50:50
0.2	Oxalic acid	2	H_2O	100	96	2	215	42:58
0.2	HCl	0.5	C_2H_5OH	100	70	24	203	42:58
0.2	HCl	1.0	H_2O	10	25	1 day	216	44:56
0.2	HCl	5.0	C_2H_5OH	63.2	25	27 days	216	11:89

[a] Data taken from A. M. Partansky, *Am. Chem. Soc., Div. Org. Coat.* **27**, 115 (1967).

[b] Paraformaldehyde used (91% CH_2=O, 9% H_2O).

[c] H_2O is present in all solvent systems due to paraformaldehyde content.

Method A [27]: To a resin kettle are added 800 gm (8.5 moles) of phenol, 80.0 gm of water, 940.5 gm (37%) of formaldehyde (11.6 moles), and 40 gm of barium hydroxide octahydrate. The reaction mixture is stirred and maintained at 70°C for 2 hr. Then sufficient oxalic acid is added to bring the pH to 6–7. The water is removed from the resin at 30–50 mm Hg at a temperature no higher than 70°C. Samples (1–2 ml) are withdrawn every 15 min to check the extent of condensation. The end point is taken when sample of resin placed on a hot plate at 160°C gels in less than 10 sec or when the cooled resin is brittle and nontacky at room temperature. The resin at this point is termed resole or A stage. Further heating will convert it into a B-stage and finally to a C-stage resin as earlier described. The A-stage resin is used for adhesives, laminates, varnishes, and molding powders and is converted to the C stage by heating to 100°C for various times.

Method B [28]: To a flask equipped with a reflux condenser and mechanical stirrer are added 47 gm (0.5 mole) of phenol, 80 ml of 37% aqueous formaldehyde (1.0 mole), and 100 ml of 4 *N* sodium hydroxide. The reaction mixture is stirred at room temperature for 16 hr and then heated on a steam bath for 1 hr. The reaction mixture is cooled and the pH adjusted to 7.0. The aqueous layer is decanted from the viscous brown liquid product and the wet organic phase is taken up in 500 ml of acetone and dried over anhydrous $MgSO_4$ followed by molecular sieving. The dried acetone product solution is filtered and evaporated to yield a water-free light brown syrup. The IR spectrum of the uncured resole resin dried at room temperature for 3 days is shown in Fig. 5. The same resole resin cured for 3 hr at 120°C in air is shown in Fig. 6.

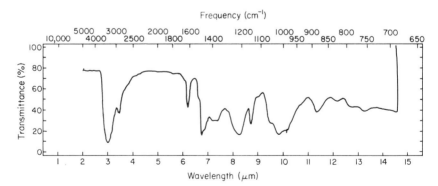

FIG. 5. Infrared spectrum of uncured resole resin—3 days at room temperature. [Reprinted from W. M. Jackson and R. T. Conley, *J. Appl. Polym. Sci.* **8**, 2163 (1964). Copyright 1964 by the Journal of Applied Polymer Science. Reprinted by permission of the copyright owner.]

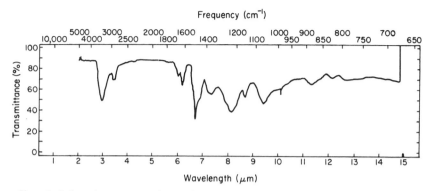

FIG. 6. Infrared spectrum of a typical phenolic resin air-cured for 3 hr at 120°C. [Reprinted from W. M. Jackson and R. T. Conley, *J. Appl. Polym. Sci.* **8**, 2163 (1964). Copyright 1964 by the Journal of Applied Polymer Science. Reprinted by permission of the copyright owner.]

Aniline is sometimes added to resole resin preparations to slow the condensation reaction and thereby give products with improved rheological properties which are of value in molding operations [29].

2-2. *Preparation of a Phenol–Formaldehyde Casting Resin* [30,31]

Method A [30]: To a resin kettle are added 100 gm (1.06 moles) of phenol, 200 gm of 37% formaldehyde (2.47 moles), and 12.0 gm of a 25% aqueous sodium hydroxide solution. The mixture is preheated for 25 min, refluxed for 17 min, neutralized, and then acidified to about pH 4.5 by the addition of approximately 19.5 gm of 51% aqueous lactic acid solution. The reaction mixture is dehydrated at 30–50 mm Hg at 80°C in order to remove the excess water. The dehydration takes 1.5–2.5 hr depending on the vacuum available. The final resin has about 14% water content. During the dehydration the resin is checked as in Method A, Preparation 2-1, to determine when the product has reached the A stage. This resin, when poured into a suitable mold and heated at 75°–95°C in an oven, gives a clear, hard casting [32].

Method B [31]: To a 1-liter resin kettle equipped with a mechanical stirrer, condenser, and thermometer are added 100 gm (1.06 moles) of phenol, 203 gm of 37% aqueous formaldehyde (2.5 moles), and 15 gm of 20% aqueous sodium hydroxide (0.075 mole). The reaction mixture is heated to 75°–80°C for about $2\frac{1}{2}$–$3\frac{1}{2}$ hr. It is then vacuum stripped of water at about 30 mm Hg while heating until the pot temperature gradually rises to about 64°C. At this point 7.5 gm (90% grade) of lactic acid (0.075 mole) is added to neutralize the starting alkali, followed by the addition of 15–20 gm of 98% glycerol. The vacuum removal of water is continued at low heat until the resin is almost completely dehydrated. When the resin is dehydrated sufficiently, a sample of the resin when dropped into a beaker of water of about 11°–13°C should congeal and solidify. The resin should also be hard enough only barely to yield when pressed between the fingers. The resin is now ready to be poured or cast into open molds and maintained at temperatures below the boiling point of water until completely set to a stable solid. Usually a period of 100–200-hr is required while heating at temperatures no higher than 78°–82°C.

Casting-type resins are generally prepared using a large excess of formaldehyde (2–3:1) with an alkali metal hydroxide as a catalyst. After reaction the pH is adjusted to about 4.5–6 with a weak organic acid. A polyol such as glycerol or ethylene glycol is sometimes added to improve the clarity of the product. The resin is cured by heating in open molds at 75°–95°C for extended periods of time. The casting is clear if sufficient water is removed during the resin preparation step.

Adhesives [33] can be obtained from the casting or "one-step" resin by the addition of acids such as sulfamic acid. The pH is usually lowered to 3 to obtain the most rapid hardening rate. For example [34]: 76.4 gm of the resin is mixed with 11.5 gm of a mixture of $\frac{2}{3}$ sulfamic acid and $\frac{1}{3}$ resorcinol; 114.4 gm of sand is then added to give a cement adhesive. The adhesive sets at room temperature in 48 hr.

Casting resins give varying degrees of light transparency depending on the choice of acidifying agent. For example, methoxyacetic acid or lactic acid give opaque cast resins but phenoxyacetic acid gives transparent resins. Tables listing the effect of type of acidifying agents on the properties of cast resins can be found in Harris and Neville [32].

Several types of alkaline catalysts may be used in the resole preparation stage. These are LiOH, NaOH, KOH, Ca(OH)$_2$, Ba(OH)$_2$, MgO, ZnO, NH$_2$OH, and organic alkyl or alkanol amines. The rate of the reaction varies with each catalyst and the resin properties may be somewhat different [29].

Paraformaldehyde or trioxane can be used to give a nonaqueous condensation, thus eliminating the need for a dehydration step. However, some water is still present in the resin since paraformaldehyde usually contains 9% water. If completely water-soluble resins are desired the reflux time is shortened and the dehydration step eliminated.

In place of phenol alkylated, halogenated, or acylated phenols or mixtures (cresols, xylenols, alkoxyphenols, phenylphenols, chlorophenols) can be used but the reaction conditions (rates of reaction) will vary with each. As an example, a cresol resole is described in Preparation 2-3.

2-3. Preparation of a Cresol–Formaldehyde Resole [27]

$$(5)$$

To a resin kettle are added 1500 gm (13.9 moles) of cresol (containing at least 38% *m*-cresol), 1300 gm of 37% formaldehyde (16.0 moles) and 75 gm of 25% ammonia (sp. gr. 0.90). The mixture is refluxed for $1\frac{1}{2}$–2 hr and then water is removed under reduced pressure (50–60 mm Hg) by raising the temperature from 45°C to 75°–78°C (at 30 mm Hg) and heating until the resin is stiff enough to slow the mechanical stirrer.

Both ortho- and para-substituted phenols yield oil-soluble phenolic resins. The para-substituted products discolor to a lesser extent under weathering conditions. Acid-catalyzed condensations lead to typical permanently fusible novolacs, whereas alkaline-catalyzed preparations afford heat hardenable products [35].

Oil-soluble *p*-butylated (*tert*-butyl)phenol–formaldehyde resins are offered commercially by Union Carbide for coating and varnish applications [36].

2-4. Preparation of a Phenol–Formaldehyde Novolac [37,38]

$$(6)$$

Method A [37]: To a 5-liter, 3-necked resin kettle are added 2400 gm (25.6 moles) of phenol, 1645 gm of 37% formaldehyde (20.3 moles), and

30 gm (0.33 mole) of oxalic acid (see Note a). The mixture is stirred and refluxed until the distillate is free of formaldehyde (1–3 hr). Water is distilled off until the resin temperature reaches 154°C. An aliquot of resin at this point has a viscosity of 105 sec at 150°C inclined plate flow (see Note b). The pressure is slowly reduced while a slow current of nitrogen is bubbled through the resin and the mixture is heated to 175°C at 6 mm Hg. Approximately 6% phenol is recovered and then the resin (see Note c) is poured into an aluminum dish and cooled. The resin has a melt viscosity of 510 sec at 150°C.

NOTES: (a) Formaldehyde:phenol ratio = 0.794. (b) The melt viscosity was checked by placing a 1-gm sample on a grooved polished hot plate which is inclined to 10° from the horizontal. The time in seconds to flow 40 mm at 150°C is recorded. For the higher viscosity novolacs one uses the same procedure but heats to 170°C. (c) Residual phenol should be no higher than 4–5%.

Method B [38]: To a flask equipped with a dropping funnel, reflux condenser, and mechanical stirrer are added 188 gm (2.0 moles) of phenol and 2.0 ml of concentrated sulfuric acid. The flask is heated to 80°C and while stirring 138 ml of 37% formaldehyde (1.7 moles) is added dropwise at such a rate to maintain this temperature. After the addition, the reaction mixture is refluxed for $1\frac{1}{2}$ hr and the water then removed by vacuum distillation to afford the crude resin. The resin is washed with fresh water to remove unreacted phenol. The solid is dissolved in acetone and dried over anhydrous magnesium sulfate. The absence of free phenol is checked by using the infrared spectral band at 14.6 μm. The resin is isolated as a fine white powder by evaporation of the acetone solution.

The recovered phenol layer may also contain 1,3-benzodioxane [39].

b.p. 101°C at 20 mm Hg

$n_D^{20.2°C}$ 1.5478

This material can also act as a source of phenol and formaldehyde in future novolac preparations. This product may possibly be derived from benzyl hemiformal a product known to exist during the early stages of resinification [40]. Recently Rohm and Haas have developed a process in which a new polymeric adsorbent selectively removes phenol and most phenolic wastes from waste streams. The phenol can be recovered for recycling in the production of resins [41].

The use of sulfuric acid or phosphoric acid leads to thermally unstable novolacs since these acids will remain in the resin whereas oxalic and hydrochloric acids are both volatile enough to be removed on vacuum stripping [42]. A practical application embodying these concepts has led to the preparation of self-curing phenol–formaldehyde resins [43].

Novolacs with some branched molecules [44] are produced by the reaction of 1.0 mole of phenol and 3.0 moles of formaldehyde (37% aqueous solution) using 0.06 mole of calcium hydroxide and heating at 50°C until 2.4 moles of formaldehyde have reacted. Then 2.6 moles of phenol are added and the mixture is neutralized with dilute sulfuric acid. Then more acid is added and a novolac is prepared in the usual manner, as described above. This novolac cures faster with hexamethylenetetramine (Hexa) and does not produce a foamy mass [44].

Molding powders [45] using novolac may be prepared by blending approximately 50 parts novolac with 50 parts of a finely ground filler such as woodflour and then adding 6–7 parts Hexa along with 2–2½ parts magnesium oxide. The addition of 1–1½ parts magnesium or calcium stearate is also used as a mold release aid. The mixture is blended together on a ball mill and then placed in a mold which is heated to 160°C while 2000 psi is applied for 5 min. The resulting resin is a hard, cured solid.

2-5. *Preparation of Salts of Tris(hydroxymethyl)phenate* [46,47]

$$\text{(ArONa)} + 3CH_2{=}O \longrightarrow \text{(product)} \qquad (7)$$

Method A: Sodium salt [46]. To a flask containing 88 gm (1.1 moles) of 50% sodium hydroxide solution is added 94 gm (1.0 mole) of phenol. The solution solidifies when cooled to 25°–30°C, and then 270 ml of 40% formalin (3.6 moles) is added. The resulting solution is cooled to keep the temperature below 40°C. The reaction mixture is kept at 40°C for 24 hr or at room temperature for several days. The reaction mixture is then poured into 3 liters of ethanol to give a white granular precipitate. The product is filtered after 3 hr, rinsed with acetone, then ether, and finally dried in a vacuum desiccator to give 165 gm (80%).

m-Cresol reacts under similar conditions to give a tris product but 3,5-xylenol gives a bis(hydroxymethyl) derivative.

Method B: Lithium salt [47]. To a flask equipped with a mechanical stirrer, reflux condenser, and thermometer are added 94 gm (1.0 mole) of phenol, 42.0 gm (1.0 mole) of lithium hydroxide monohydrate and 100–120 ml of distilled water. Formaldehyde as a 40% solution (225 ml or 3.0 moles) is added. The slurry dissolves slowly on warming to give a pale yellow solution and at about 45°C an exothermic reaction commences. The temperature is kept below 55°C and the mixture is allowed to stand overnight at room temperature. Isopropanol (800 ml) is added with stirring to give a voluminous precipitate of fine white granules. After $\frac{1}{2}$ hr the precipitate is filtered and washed with 200 ml isopropanol followed by 400 ml of acetone. The product is filtered (do not allow much air to pass through the product), air-dried for 15 min, and then placed in a vacuum desiccator for 4 days to afford 95–124 gm (50–65%). The product is soluble in water to the extent of 0.10 gm/ml at 23°C and 0.19 gm/ml at 40°C.

At pH greater than 8.5 and at 40°C, two moles of trimethylolphenol condense to form tetramethylol dihydroxydiphenylmethane [47].

The commercial method of preparation of trimethylbutylphenol [48,49] or its calcium salt [50] has been reported. The process involves reacting slightly over 3.0 moles of formaldehyde with 1.0 mole of sodium phenate at low temperature. The salt-free product is obtained by neutralization with phosphoric acid and decantation from the salt solution. The product is mainly free of resinous material and has found use in laminates, adhesives [51], and foundry sand molds [52].

2-6. Condensation of p-Chlorophenol with Formaldehyde [53]

$$\text{OH} \quad + \text{CH}_2\!\!=\!\!\text{O} \longrightarrow \left[\text{OH} \quad \text{—CH}_2\!\!—\! \right]_n + \text{H}_2\text{O} \qquad (8)$$

Method A: To a resin kettle equipped with a mechanical stirrer and dropping funnel is added a solution of 64.3 gm (0.5 mole) of *p*-chlorophenol, 15.0 gm (0.5 mole of formaldehyde) of *s*-trioxane, and 1.0 gm of *p*-toluenesulfonic acid monohydrate in 30 ml of bis(2-ethoxyethyl) ether. The solution is heated under reflux at 110°C for 3 hr. At this time 100 ml of benzene is added and water (8.98 gm, 0.5 mole) is removed by azeotropic distillation over a 5 hr period using a Dean & Stark trap. Then the resin is heated for 12 hr at 110°C and for 6 hr at 135°C. The product is now diluted with 225 ml of

N,N-dimethylformamide (DMF), and poured into a large volume of water to give a gray solid, n_{sp} 0.040.

Method B: To a Carius tube is added 38.6 gm (0.3 mole) *p*-chlorophenol, 9.0 gm (0.3 mole of formaldehyde) of *s*-trioxane, 0.6 gm of toluenesulfonic acid, and 45 ml of bis(2-ethoxyethyl) ether. The shielded tube is sealed and heated at 150°C for 24 hr. The tube is cooled to room temperature, wrapped in a towel, and cautiously opened. The polymer is dissolved in 150 ml acetone, filtered, and then added with rapid stirring to 1600 ml of water containing 15 ml of concentrated hydrochloric acid. The light tan polymer is washed thoroughly with distilled water and dried. The polymer is isolated in 95–100% yield (soluble in acetone) n_{sp} 0.059, softening point 90°–100°C, chars at 120°C, and decomposes at 200°C or more on a melting point block. If placed immediately on a preheated hot bar, the polymer melts at about 145°–155°C.

Dehalogenation of *p*-chlorophenol–formaldehyde polymers is effected by means of sodium in liquid ammonia and gives a linear phenol–formaldehyde polymer free of chlorine as described below [54].

2-7. *Dehalogenation of p-Chlorophenol–Formaldehyde Polymers* [54]

$$\left[\begin{array}{c} OH \\ \underset{Cl}{\bigcirc}-CH_2- \end{array} \right]_n \xrightarrow[\text{liq. NH}_3]{Na} \left[\begin{array}{c} OH \\ \bigcirc-CH_2- \end{array} \right]_n \tag{9}$$

To a flask containing 2 liters of liquid ammonia cooled in a methanol–Dry Ice bath is dissolved 10 gm (0.07 mole) of *p*-chlorophenol–formaldehyde polymer. Then 6.0 gm sodium metal (0.26 gm atom) is added portionwise with stirring and a blue color forms. The reaction mixture is cooled and stirred for 24 hr and the excess sodium is then destroyed by the cautious addition of ammonium chloride. The liquid ammonia is removed by evaporation in a hood, followed by addition of 150 ml of 10% aqueous hydrochloric acid to the residue. The resulting solid is filtered and then washed with distilled water. The solid is dissolved in 75 ml of acetone and reprecipitated by adding with vigorous stirring to 500 ml of distilled water containing 5 ml of hydrochloric acid. The light tan polymer is filtered, washed with distilled water, air dried, and then dried further over refluxing toluene for 24 hr under reduced pressure (approximately 0.5 mm Hg).

The product is acetylated by adding 2.0 gm of it to 20 ml of pyridine followed by addition of 9.0 ml of acetic anhydride. After 6 days at room temperature the reaction mixture is added with stirring to 200 ml of water containing 5 ml of concentrated hydrochloric acid. The white product is removed by filtration, washed with water, and dried. The molecular weight is determined cryoscopically in benzil to be 2980 [degree of polymerization (DP) 20.3].

D. Other Aldehydes

The structure of the aldehyde used for condensation with phenolic compounds can vary from formaldehyde to others [55] as shown below:

$$R—CH=O$$

$$R = H—, CH_3—, C_3H_7—, CHO, \quad \text{(furyl)}, \quad COONa, CCl_3, C_6H_5$$

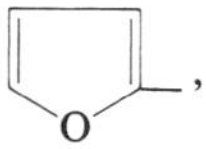

These aldehydes react on acid condensation with phenols to give novolac-type products. Base-catalyzed condensation is not practical with acetaldehyde since it undergoes rapid aldol condensation and self-resinification reactions. Acid condensations involving acetaldehyde, or its trimer paraldehyde, and phenol give soluble and permanently fusible resins, comparable to the novolacs. Aldehyde with no α hydrogens react in a manner similar to formaldehyde:

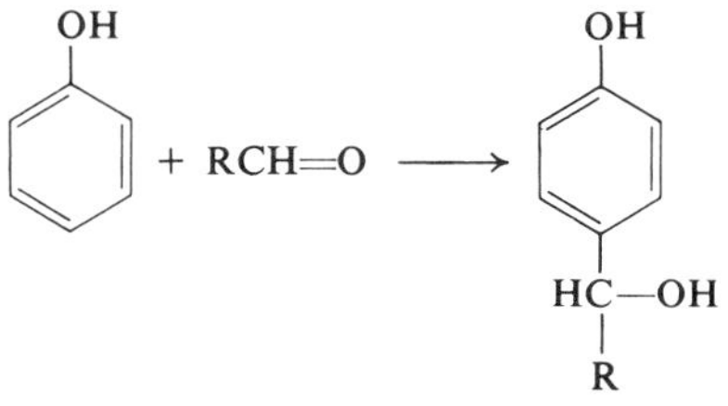

$$(10)$$

Para-substituted phenols lead to only ortho-substituted phenol alcohols.

Increasing the ratio of furfural to phenol raises the melting point and speeds the stroke cure. None of these resins are true novolacs since they are not permanently fusible (although they are made with less than one mole of aldehyde per mole of phenol). Novolacs can be prepared by lowering the furfural to phenol ratio and recovering the excess phenol by vacuum distillation. The data on a series of furfural novolac resins is shown in Table V [55a].

The furfural–phenol novolacs are stable under neutral or alkaline conditions but readily undergo a thermoset reaction on the acid side to give a black resin [56]. It is possible that the furfuryl ring system also takes part in the curing reaction to give in addition furan-type polymers [57].

TABLE V

SERIES OF FURFURAL–PHENOL NOVOLAC RESINS[a]

Ratio of furfural to phenol in charge	Ratio of furfural to phenol reacted	Melting point (°C)	Stroke cure (min)	
			300°F	330°F
0.5	0.875	105	3.5	1.5
0.45	0.84	96	4	1.5
0.35	0.795	92	4	1.75
0.25	0.675	59	4.5	2.25
0.20	0.60	52	5.25	2.25

[a] Reprinted from L. H. Brown, *Ind. Eng. Chem.* **44,** 2673 (1952). Copyright 1952 by the American Chemical Society. Reprinted by permission of the copyright owner.

Mikes recently reported on the preparation of mixed furfural–formaldehyde phenolic resins [58].

2-8. *Preparation of Furfural–Phenol Resin* [59]

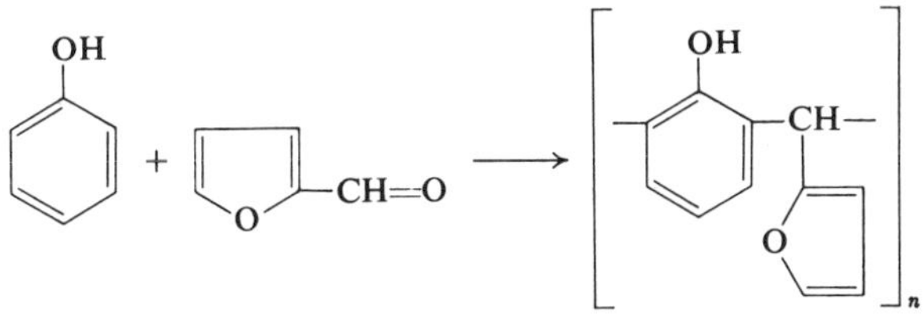

$$(11)$$

To a resin kettle equipped with a mechanical stirrer, thermometer, and reflux condenser are added 8.0 gm of potassium carbonate and 400 gm (4.06 moles) of phenol. The mixture is heated to 135°C to form a solution and then 300 gm (3.1 moles) of furfural is added dropwise over a 30 min period while the temperature is maintained at 135°C. The mixture is refluxed for 3–3½ hr and the temperature drops to 116°C. The separated resin has a melting point of 82°C and is obtained in 106% yield based on phenol (resin contains water). The stroke cure (see Note below) is taken after adding 2% calcium oxide and 10% hexamethylenetetramine and is approximately 5 min.

NOTE: Stroke cure is obtained by placing 0.5 gm of the catalyzed resin mixture on a hot plate set at 150°C surface temperature and stroking it with a spatula until the sol–gel transition occurs.

E. Stability

Resoles are not as stable as novolacs and tend to react further on standing. Storage at temperatures below room temperature is recommended. For example, storage of a resole at 90°F causes it to gel in about 20–24 hr whereas storage at −50°F gives a storage stability of 100–150 days. At the gel point the resin is too stiff to work properly but is not yet finally cured.

Novolacs are extremely stable (1 or more years) provided they contain no residual strong acids (e.g., sulfuric acid).

F. Curing Conditions for Phenolic Resins

One-stage resins or resoles contain an adequate number of methylol groups to be cured either thermally (160°–200°C) or by acids. Novolac resins (two-stage resins) which are prepared with an acidic catalyst and less than one mole of formaldehyde per mole of phenol are permanently soluble and fusible and cure only upon the addition of a curing agent (heat and paraformaldehyde or more preferably hexamethylenetetramine) [8c,60]. Usually hexamethylene-tetramine (Hexa) is used at 10% of the novolac resin. Both resin and Hexa are ground together before curing the resin. These mixes are stable in the dry state. The physical properties of the cured resole or novolac reach a maximum degree of cross-linking at or near approximately 1.5:1.0 ratio of formaldehyde to phenol.

Baekeland [7,61] and Lebach [62] defined three stages which the phenolic resins go through to become insoluble and infusible. The first is the A stage, in which the resin or resole is both soluble and fusible. The second stage is the B stage or resitol stage, in which the resin is still fusible but is no longer soluble in alkalies and is only partially soluble in solvents such as acetone. In the final C stage or resite stage the resin is insoluble in all solvents and does not melt below the decomposition point.

Some other reported curing agents are quinone [63], chloranil [63], anhydroformaldehyde aniline [64], methylolureas and melamines [65], ethylenediamine–formaldehyde products, and paraformaldehyde. Resorcinol novolac resins can be cured with paraformaldehyde [66] and the mildly acid nature of resorcinol helps to catalyze the curing reaction [66]. Ammonium salts have also been reported as accelerators for the curing of phenolic resins [67].

Cure rates of phenolic resins have been studied by differential thermal analysis and infrared spectroscopy [68].

The thermal hardening of one-stage resins involves simultaneous formation of methylene and ether linkages [69]. Further thermal reactions involve loss of formaldehyde from dibenzyl ether linkages [70]. The latter reaction is catalyzed by either acids or bases.

The major curing agent for two-stage resins is hexamethylenetetramine (hexa) [71]. The hexa reacts to form bis- and tris(hydroxybenzyl)amine and/or polynuclear nitrogen derivatives. At elevated temperatures the nitrogen products are unstable and may decompose into other products such as azomethines.

Bender [72] in a study of the hexa curing reaction of novolacs used model compounds to study the cure rate. The data show that novolacs with ortho methylene linkages give the fastest cure rates whereas the para is the slowest. This information was used to develop a synthesis for a new set of phenolic resins which have tailor-made curing speeds with hexa [44,73]. Selective catalysts such as alkaline salts of zinc, aluminum, or magnesium force the linkages in the ortho position, producing fast curing resins with an ortho-(2,2') content of more than 50% [44,73]. Further work [74] has defined these catalysts as electropositive bivalent metal ions with the most effective being Mn, Cd, Zn, and Co, and those of lesser effectiveness being Mg and Pb. Recently [75] it has been reported that the hydroxides of the transition metals Cu, Cr, Ni, and Co are the most effective in directing the ortho hydroxy-methylation reaction. Acid catalysts favor p,p' isomers where ammonia or sodium hydroxide clause o,p formation.

Saligenin, the parent of the ortho-substituted hydroxymethyl phenols, is prepared by the low-temperature reaction of equimolar amounts of phenol and formaldehyde using calcium hydroxide catalyst [76]. Zinc acetate can be used with excess phenol to give a 53% yield of saligenin, m.p. 86°C, after precipitation from carbon tetrachloride [77]. The product is not stable for long periods of time at room temperature and must be kept refrigerated. Exposure of saligenin to air will cause the sample to resinify, possibly because of contamination by acidic impurities.

Some reactions of phenol with aldehydes leading to nonresinous products such as bisphenols are described below. The bisphenols are used in epoxy resin formation [78], antioxidants [79,79a], pharmaceuticals [80], pesticides [81], fungicides [82], and bactericides [79a,83].

G. Compounding with Phenolic Resins

Fillers or extenders are added to phenolic resins to improve various physical properties of phenolic plastics [84]. The modifying agents are combined chemically with the resins and include such components as woodflour, cashew nutshell liquid, tannins, natural resins, glycerol, fatty acids, certain alkyl resins, and polyvinyl resins [for example, poly(vinyl acetate) or co-polymers].

H. Dihydroxydiphenylalkanes (Bisphenols)

Bisphenols are mainly prepared by one of the five methods (Scheme 3) described below:

(a) Acid-catalyzed condensation reaction of aldehydes and ketones with phenol [85].

(b) Base-catalyzed reaction of formaldehyde with phenols [86].

(c) Condensation reaction of phenol alcohols or hydroxybenzyl alcohols with phenols [87].

(d) Self-condensation reaction of phenol alcohols with loss of formaldehyde [88].

(e) Loss of formaldehyde from dihydroxydibenzyl ethers [89].

Para-substituted phenols favor the 2,2 product bisphenol. Unsubstituted phenol gives mixtures of the 4,4′ and 2,4′ products.

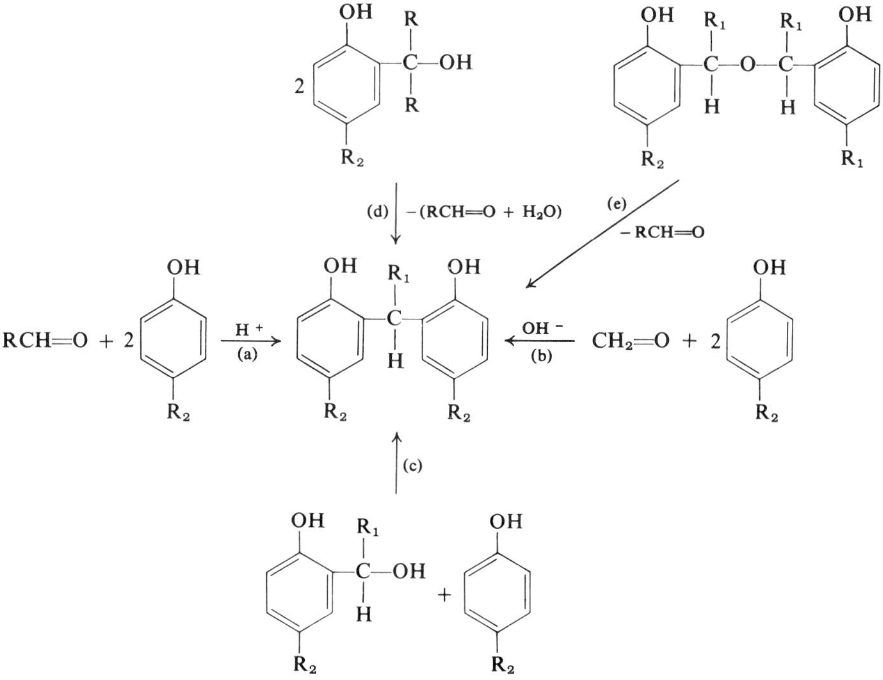

Scheme 3. Preparation of bisphenols.

In the structure of RCH=O above, R may be H, CH_3, CCl_3, [furan structure], or C_6H_5 [90].

Ketones such as acetone are used to produce the large volume dihydroxydiphenylalkane known as bisphenol A by reaction with phenol [4a,91]. If the para-position on phenol contains a substituent then the bisphenol is mainly the 2,2' isomer. Mercaptans and mercapto acids have a beneficial effect on the yield and purity probably by allowing the ketone to react in the thioacetal or thiohemiacetal forms which are less susceptible to side reactions [4a,91].

$$2\; \text{C}_6\text{H}_4\text{—OH} + \text{R}_1\text{R}_2\text{C}{=}\text{O} \longrightarrow \text{HO—}\underset{R_2}{\overset{R_1}{\text{C}_6\text{H}_4\text{—C—C}_6\text{H}_4}}\text{—OH} \qquad (12)$$

$R_1, R_2 = H$ or alkyl

3. MISCELLANEOUS PREPARATIONS AND APPLICATIONS

1. Catechol–formaldehyde resins [92].
2. Preparation of catechol–methylol derivatives [93].
3. Preparation of phenolic resin foams [94].
4. Phenol–formaldehyde poly(vinyl acetate) adhesives [95].
5. Compounding rubber with phenolic resins [96].
6. Phenolic detergents by reaction of C_8–C_{18} alkyl phenols with formaldehyde and then reaction with alkylene oxide [97].
7. Textile treatments and applications [98].
8. Phenolic resins for ion-exchange resins [99].
9. Phenolic resins as antioxidants [100].
10. Phenol–formaldehyde fibers (Kynol from Carborundum Co.) [101].
11. Xylene–formaldehyde–phenol resins [102].
12. Phenol–formaldehyde condensation products and their polyalkylene ether alkanol derivatives [103].
13. Blending phenol–aldehyde resins with poly(vinyl butyral crotonal) or other poly(vinyl acetals) [104].
14. Preparation of flame-retardant phenolic resins [105].
15. Nitration of linear phenol–formaldehyde polymers [106].
16. Preparation of inorganic esters of novolacs [107].
17. Preparation of linear 4-chloro-3,5-dimethylphenol–formaldehyde polymers [108].
18. Synthesis of dimeric and trimeric phenol–formaldehyde polymer model compounds [109].
19. Epichlorohydrin reaction products with phenolic resins [110].
20. Novolac resins from shale tars [111].
21. Phenolic latex adhesive [112].

22. Injection moldable phenoplasts [113].

23. Phenolic resin–rubber adhesive [114].

24. Continuous manufacture of water-soluble phenol–formaldehyde resins [115].

25. Sulfonation of A-stage resins using inorganic sulfites [116].

REFERENCES

1. R. Pira, *Ann. Chem. Pharm.* **48**, 751 (1843).
2. A. Baeyer, *Ber. Dtsch. Chem. Ges.* **5** (1), 25–26, 280–282, and **5** (2), 1094–1100 (1872).
3. A. Kleeberg, *Justus Liebigs Chem. Ann.* **263**, 283 (1891).
4. A. Dianin, *J. Russ. Phys-Chem. Ges.* **1**, 488, 523, and 601 (1891).
4a. A. Dianin, *Ber. Dtsch. Chem. Ges.* **25**, 334R (1892).
5. L. Lederer, *J. Prakt. Chem.* [N.S.] **50**, 223 (1894).
6. O. Manasse, *Ber. Dtsch. Chem. Ges.* **27**, 2409 (1894).
7. L. H. Baekeland, *J. Ind. Eng. Chem.* **1**, 149 (1909).
7a. L. H. Baekeland, U.S. Patents 939,966 and 942,852 (1909); *J. Ind. Eng. Chem.* **6**, 506 (1913).
8. W. A. Keutgen, *Encycl. Polym. Sci. Technol.* **10**, 1 (1969); R. S. Daniels, U.S. Patent 2,398,361 (1946).
8a. D. F. Gould, "Phenolic Resins." Van Nostrand-Reinhold, Princeton, New Jersey, 1959.
8b. T. S. Carswell, "Phenoplasts," Vol. VII, High Polym. Ser. Wiley (Interscience), New York, 1947; N. J. L. Megson, "Phenolic Resin Chemistry." Academic Press, New York, 1958; R. W. Martin, "The Chemistry of Phenolic Resins." Wiley, New York, 1956; H. Kaemmerer, *Kunststoffe* **56**, 154 (1966); L. V. Redman and A. V. H. Mory, *Ind. Eng. Chem.* **23**, 595 (1931).
8c. K. C. Frisch, *Kirk-Othmer Encycl. Chem. Technol., 1st Ed. Vol.* **10**, p. 335 (1953).
9. R. T. Conley, *Am. Chem. Soc., Div. Org. Coat. Plast. Chem., Pap.* **26**, 138 (1966).
10. E. F. Borro, Sr., *Mod. Plast.* **42**, 102 (1965); W. H. Boyd and C. N. Merriam, *Am. Chem. Soc., Div. Org. Coat. Plast. Chem., Pap.* **34**, 22 (1974).
11. *Nat. Saf. News* **109**, 95–104 (1974).
12. R. Goodson, *Chem. & Eng. News* **52** (42), 5 (1974); J. M. Stellman, *Chem. & Eng. News* **52** (25), 3 (1974); E. M. Beavers, *Chem. & Eng. News* **52** (29), 3 (1974); Rohm & Haas Co., *Catalyst (Philadelphia)* **59**, 196 (1974).
13. J. H. Freeman and C. W. Lewis, *J. Am. Chem. Soc.* **76**, 2080 (1954).
14. A. A. Zavitas and R. D. Beaulieu, *Am. Chem. Soc., Div. Org. Coat. Plast. Chem., Pap.* **27**, 100 (1967).
15. A. A. Zavitsas, *Am. Chem. Soc., Div. Org. Coat. Plast. Chem., Pap.* **26**, 93 (1966); **27**, 100 (1967).
16. J. J. Gardikes and F. W. Konrad, *Am. Chem. Soc., Div. Org. Coat. Chem., Pap.* **26**, 131 (1966).
16a. M. F. Drumm, *Am. Chem. Soc., Div. Org. Coat. Plast. Chem., Pap.* **26**, 85 (1966).
17. S. H. Hollingdale and N. J. L. Megson, *J. Appl. Chem.* **5**, 616 (1955).
18. M. I. Siling, *Vysokomol. Soedin., Ser. B* **13**, 527 (1971).
19. G. K. Vogelsang, U.S. Patent 2,634,249 (1953).

20. W. R. Thompson, *Chem. Ind. (N.Y.)* **48**, 450 (1941).
21. M. M. Sprung, *J. Am. Chem. Soc.* **63**, 334 (1941).
22. K. Hultzsch, *Am. Chem. Soc., Div. Org. Coat. Plast. Chem., Pap.* **26**, 121 (1966).
23. A. J. Norton, U.S. Patents 2,385,370 (1946); 2,414,417 (1947); P. H. Rhodes, U.S. Patents 2,385,372 (1946); 2,414,416 (1947); R. A. V. Raff and B. H. Silverman, *Ind. Eng. Chem.* **43**, 1423 (1951).
24. P. H. Rhodes, *Mod. Plast.* **24**, 145 (1947).
25. A. M. Partansky, *Am. Chem. Soc., Div. Org. Coat. Plast. Chem., Pap.* **27**, 115 (1967).
25a. Société industrielle de recherches et de fabrications and J. Viguier, French Patent 1,145,603 (1957); *Chem. Abstr.* **53**, 23110 (1959).
26. H. L. Bender, A. G. Farnham, and J. W. Guyer, U.S. Patents 2,464,207 and 2,475,587 (1949).
26a. D. A. T. Fraser, *J. Appl. Chem.* **7**, 676 (1957).
27. *U.S. Dep. Commer., Off. Tech. Serv., PB Rep.* **25,642** (1945).
28. R. T. Conley and J. F. Bieron, *J. Appl. Polym. Sci.* **7**, 171 (1963).
29. J. Nelson, *Macromol. Synth.* **3**, 1 (1966).
30. R. W. Bentz and H. A. Neville, *J. Polym. Sci.* **4**, 673 (1949).
31. O. Pantke, U.S. Patent 1,909,786 (1933).
32. T. G. Harris and H. A. Neville, *J. Polym. Sci.* **10**, 19 (1953).
33. D. A. Fraser, R. W. Hall, and A. L. J. Raum, British Patent 773,611 (1957).
34. R. S. Daniels, U.S. Patent 2,398,361 (1946).
35. V. H. Turkington and I. Allen, *Ind. Eng. Chem.* **33**, 966 (1941).
36. "Bakelite Resins for Coatings and Adhesives," Tech. Brochure No. J-2295-C. Union Carbide, New York, New York, 1974; "Coating Materials," Tech. Brochure No. F4-2504, 106-9-30M. Union Carbide, New York, New York, 1974.
37. H. E. Hoyt, H. W. Keuchel, and R. B. Dean, *Paint Varn. Prod.* **31**, 33 (1958).
38. R. T. Conley and J. F. Bieron, *J. Appl. Polym. Sci.* **7**, 103 (1963).
39. W. Baker, *J. Chem. Soc.* p. 1765 (1931).
40. J. C. Woodbrey, H. P. Higginbottom, and H. J. Culbertson, *J. Polym. Sci., Part A* **3**, 1079 (1965).
41. E. H. Brook, J. J. McNulty, and R. P. McDonnell, *Am. Chem. Soc., Div. Org. Coat. Plast. Chem., Pap.* **34**, 30 (1974).
42. H. E. Hoyt, H. W. Keuchel, and R. B. Dean, *Paint Varn. Prod.* **31**, 331 (1958).
43. K. Iwamura and S. Ogaki, Japanese Patent 71/09,478 (1971).
44. R. Dijkstra and J. DeJonge, *Am. Chem. Soc., Div. Org. Coat. Plast. Chem., Pap.* **26**, 107 (1966).
45. O. Pantke, U.S. Patent 1,909,786 (1933).
46. R. W. Martin, *J. Am. Chem. Soc.* **73**, 3952 (1951).
47. J. H. Freeman, *Am. Chem. Soc., Div. Org. Coat. Plast. Chem., Pap.* **27**, 84 (1967).
48. C. Y. Meyers, U.S. Patent 2,889,374 (1959).
49. G. A. Senior, U.S. Patent 3,109,033 (1963).
50. G. A. Senior, U.S. Patent 3,023,252 (1962).
51. B. P. Barth, U.S. Patent 3,046,103 (1962).
52. B. P. Barth, U.S. Patent 2,999,283 (1961).
53. W. J. Burke, W. E. Craven, A. Rosenthal, S. H. Ruetman, C. W. Stephens, and C. Weatherbee, *J. Polym. Sci.* **20**, 75 (1956).
54. W. J. Burke, S. H. Ruetman, C. W. Stephens, and A. Rosenthal, *J. Polym. Sci.* **22**, 477 (1956).
55. L. H. Brown, *Ind. Eng. Chem.* **44**, 2673 (1952); L. N. Ferguson, *Chem. Rev.* **38**, 231 (1946); G. C. Helintzer and B. Rodnevich *J. Appl. Chem. USSR (Engl. Transl.)*

8, 909 (1935); D. R. Stevens and A. C. Dubbs, U.S. Patent 2,515,909 (1950); H. Pauly and A. Schantz, *Ber. Dtsch. Chem. Ges.* **56**, 979 (1923); M. Ikawa and K. P. Link, *J. Am. Chem. Soc.* **72**, 4373 (1950); F. G. Pope and H. Howard, *J. Chem. Soc.* **97**, 78 (1910); S. Krishna and F. G. Pope, *J. Chem. Soc.* **119**, 286 (1921).

56. E. E. Novotny and D. S. Kendall, U.S. Patents 1,398,142–1,398,149 (1921); 1,705,493–1,705,496 (1929).
57. K. J. Siegfried, *Encycl. Polym. Sci. Technol.* **7**, 432 (1967).
58. J. A. Mikes, *J. Polym. Sci.* **53**, 1 (1961).
59. L. H. Brown, *Ind. Eng. Chem.* **44**, 2673 (1952).
60. N. B. Sunshine, in "Flame Retardancy of Polymeric Materials" (W. C. Kuryla and A. J. Papa, eds.), Vol. 2, p. 201. Dekker, New York, 1973.
61. L. H. Baekeland, *J. Ind. Eng. Chem.* **4**, 737 (1912).
62. H. Lebach, *Z. Angew. Chem.* **22**, 1598 (1909); *J. Soc. Chem. Ind.* **32**, 559 (1913).
63. F. Seebach, U.S. Patent 1,883,145 (1932).
64. L. H. Baekeland, U.S. Patent 1,216,265 (1917).
65. F. J. Groten, U.S. Patent 2,287,256 (1943); H. von Diesbach, German Patent 611,210 (1930); H. von Diesbach, O. Wanger, and A. von Stockalper, *Helv. Chim. Acta* **14**, 355 (1941).
66. E. E. Novotny, U.S. Patents 1,767,696 (1930); 1,802,390 (1931); 1,849,109 (1932); G. E. Little and K. W. Pepper, *Br. Plast.* **19**, 430 (1947).
67. E. Bock, German Patent 833,121 (1952); *Chem. Abstr.* **51**, 11764 (1957).
68. Z. Katović, *J. Appl. Polym. Sci.* **11**, 85 (1967); R. H. White and T. Rust, *ibid.* **9**, 777 (1965).
69. H. von Euler, E. Adler, G. Eklund, and O. Torngren, *Ark. Kemi, Mineral. Geol.* B **15**, 1 (1942).
70. E. Ziegler and K. Lercher, *Ber. Dtsch. Chem. Ges.* **74** (1), 841 (1941); E. Ziegler and R. Kohlhauser, *Monatsh. Chem.* **79**, 92 (1928); A. Zinke and E. Ziegler, *Ber. Dtsch. Chem. Ges.* **74** (1), 541 (1941).
71. J. W. Aylsworth, U.S. Patent 1,102,594 (1912); A. Zinke, G. Zigeuner, G. Weisz, and W. Leopold-Lowenthal, *Monatsh. Chem.* **81**, 1098 (1950).
72. H. L. Bender, *Mod. Plast.* **30**, 136 (1953); **31**, 115 (1954).
73. H. L. Bender and A. G. Farnham, U.S. Patent 2,475,587 (1949); H. L. Bender, A. G. Farnham, and J. W. Guyer, U.S. Patent 2,464,207 (1949).
74. D. A. Fraser, R. W. Hall, and A. L. J. Raum, *J. Appl. Chem.* **7**, 676 (1957).
75. H. G. Peer, *Recl. Trav. Chim. Pays-Bas* **79**, 825 (1960).
76. G. C. H. Clark, E. G. Peppiatit, A. Poole, and R. J. Wieker, British Patent 751,845 (1956).
77. A. L. J. Raum, British Patent 774,696 (1957).
78. S. O. Greenlee, U.S. Patent 2,494,295 (1950); P. Castan, U.S. Patents 2,324,483 and 2,444,333 (1948).
79. A. J. Dietzler, U.S. Patent 2,482,748 (1950).
79a. D. J. Beaver and P. J. Stoffel, *J. Am. Chem. Soc.* **74**, 3410 (1952).
80. R. Baltzly, J. S. Buck, E. J. DeBeer, and F. J. Webb, *J. Am. Chem. Soc.* **71**, 1301 (1949).
81. L. Haskelberg and D. Lavie, *J. Org. Chem.* **14**, 498 (1949).
82. P. B. Marsh and M. L. Butler, *Ind. Eng. Chem.* **38**, 701 (1946).
83. H. E. Faith, *J. Am. Chem. Soc.* **72**, 837 (1950).
84. M. M. Sprung, *J. Am. Chem. Soc.* **63**, 334 (1941).
85. J. B. Fishman, *J. Am. Chem. Soc.* **42**, 2297 (1920); S. R. Finn and J. W. G. Musty, *J. Soc. Chem. Ind., London* **69**, 549 (1950).

86. R. W. Martin, *J. Am. Chem. Soc.* **73**, 3952 (1951).
87. J. B. Niederl and I. W. Ruderman, *J. Am. Chem. Soc.* **67**, 1176 (1945).
88. J. H. Freeman and C. W. Lewis, *J. Am. Chem. Soc.* **76**, 2080 (1954).
89. K. Hultzsch, "Chemie der Phenolhärze." Springer-Verlag, Berlin and New York, 1950.
90. T. Zincke, *Justus Liebigs Chem. Ann.* **363**, 246 (1908); I. P. Losev and M. S. Akutin, *J. Appl. Chem. USSR (Engl. Transl.)* **13**, 916 (1940); J. B. Niederl and J. S. McCoy, *J. Am. Chem. Soc.* **63**, 1731 (1941); A. S. Briggs and J. Haworth, U.S. Patent 2,559,932 (1951).
91. J. E. Jansen, U.S. Patent 2,468,982 (1949); A. J. Dietzler, U.S. Patent 2,526,545 (1950).
92. W. E. Hillis and G. Urbach, *J. Appl. Chem.* **9**, 474 (1959); Dow Chem. Co., Netherlands Patent Appl. 6,506,110 (1965); *Chem. Abstr.* **64**, 10691b (1965); "Applications For Catechol," Tech. Brochure. Crown Zellerbach Corp., Camas, Washington, 1973.
93. "Reactions of Catechol," Tech. Brochure. Crown Zellerbach Corp., 1974; F. Langmaier, *Kozarstvi* **16**, 15 (1966); *Chem. Abstr.* **65**, 18701e (1966).
94. H. Scherr, A. Gottfurcht, and R. W. Stenzel, *Plastics (N.Y., 1944–49)* **9**, 8 (1949); R. G. B. Mitchell and D. Smith, *Plastics (London)* **24**, 44, 85 (1959); Compagnie de Saint Gobain, French Patent 1,480,362 (1957); Y. Oto, S. Suzuki, and K. Udagawa, German Offen. 2,037,930 (1971).
95. S. R. Sandler, U.S. Patents 3,547,771 (1970), 3,619,346 (1971), and 3,677,883 (1972); J. Teppema, U.S. Patents 2,902,458 and 2,902,459 (1959); W. B. Armour, U.S. Patents 3,433,701 and 3,444,037 (1969).
96. A. N. Iknayan, A. C. Peterson, and H. J. Butts, U.S. Patent 2,702,286 (1955).
97. L. H. Bock and J. L. Rainey, U.S. Patent 2,454,541 (1941).
98. E. I. Du Pont de Nemours, Canadian Patent 523,167 (1956); H. A. Pohl, *Text. Res. J.* **28**, 473 (1958).
99. S. L. S. Thomas and J. I. Jones, British Patent 762,085 (1956).
100. H. Palfreeman, British Patent 788,794 (1958).
101. A. A. Wosilait and J. Economy, *in* "Proceedings of the 1974 Symposium on Textile Flammability April 17–18, 1974 at Charlotte, N.C." (R. B. Le Blanc, ed.), p. 165. LeBlanc Research Corp., East Greenwich, Rhode Island, 1974.
102. E. C. Winegartner and R. D. Wesselhoft, U.S. Patent 3,394,203 (1968).
103. J. W. Cornforth, P. M. D. Hart, G. A. Nicholls, and R. J. W. Rees, British Patent 774,282 (1957); W. T. Romanov and K. I. Tarasov, U.S.S.R. Patent 105,277 (1957).
104. G. S. Bridskii, R. S. Derevyanko, I. V. Kamenskii, V. I. Kandela, E. T. Margunova, L. D. Radchik, R. A. Shereleva, and M. A. Chervinskaya, U.S.S.R. Patent 286,223 (1970).
105. N. B. Sunshine, *in* "Flame Retardancy of Polymeric Materials" (W. C. Kuryla and A. J. Papa, eds.), Vol. 2, p. 212. Dekker, New York 1973; H. K. Nason, M. L. Nielson, and J. E. Malowan, U.S. Patents 2,661,341 and 2,661,342 (1954).
106. W. J. Burke and H. P. Higginbottom, *Am. Chem. Soc., Div. Org. Coat. Plast. Chem., Pap.* **26**, 112 (1966).
107. B. F. Daniels and A. F. Shepard, *Am. Chem. Soc., Div. Org. Coat. Plast. Chem., Pap.* **27**, 125 (1967).
108. R. H. Peterson, C. R. Guilbert, and W. J. Burke, *Am. Chem. Soc., Div. Org. Coat. Plast. Chem., Pap.* **27**, 152 (1967).
109. R. H. Peterson, W. J. Burke, and W. C. Nickles, *Am. Chem. Soc., Div. Org. Coat. Plast. Chem., Pap.* **27**, 164 (1967).

110. H. Lee and K. Neville, "Handbook of Epoxy Resins." McGraw-Hill, New York, 1967.
111. L. Kalde and E. F. Petukhov, *Dobycha Pererab. Goryuch. Slantsev* No. 15, p. 315 (1966); *Chem. Abstr.* **68**, 51585 (1968).
112. T. Tamaki, Japan Kokai 74/52, 236 (1974); *Chem. Abstr.* **81**, 122108 (1974).
113. M. Chevalier, *Plast. Mod. Elastomeres* **26**, 75 (1974); *Chem. Abstr.* **81**, 122014 (1974).
114. K. Hultzsch, German Patent 1,002,489 (1957); P. B. Barth, U.S. Patent 3,817,922 (1974).
115. J. M. Colombani, German Offen. 2,029,934 (1971); *Chem. Abstr.* **74**, 77020 (1971).
116. W. S. Niederhauser and M. W. Miller, U.S. Patent 2,357,798 (1944).

EPOXY RESINS

I. INTRODUCTION

Epoxy resins are usually prepared from compounds (or polymers) containing two or more epoxy groups which have been reacted with amines, anhydrides, or other groups capable of opening the epoxy ring and forming thermosetting products. Polymers from monoepoxy compounds have already been described in the previous volume of this series [1].

Schlack [2] and Castan [3,3a] are credited with the earliest U.S. patents describing epoxy resin technology. Greenlee [4] further emphasized the use of bisphenols and their reaction with epichlorohydrin to yield diepoxides capable of reaction with crude tall oil resin acids to yield resins useful for coatings. The use of diepoxide resins that are cured with amines was reported by Whittier and Lawn [5] in a U.S. patent in 1956.

The introduction of epoxidation techniques for polyunsaturated natural oils by Swern [6,6a] led to industrial interest in the preparation of epoxy compounds useful for resin production [7,7a].

Epoxy resin technology has been reviewed and a number of relevant references are available [6,8,8a].

A summary of epoxy resin uses and growth is summarized in Fig. 1. Note that 44% of the epoxy resin outlet is in the area of coatings used for appliances, automobiles, and cans. These resins have the added advantage of being solventless systems, which helps avoid air pollution problems in plants. The pattern of consumption of epoxy resins is described in Table I.

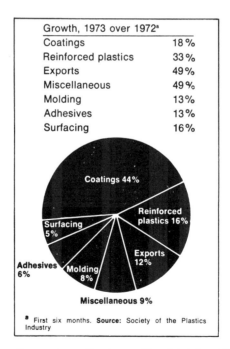

Growth, 1973 over 1972[a]	
Coatings	18%
Reinforced plastics	33%
Exports	49%
Miscellaneous	49%
Molding	13%
Adhesives	13%
Surfacing	16%

Coatings 44%

Reinforced plastics 16%

Surfacing 5%

Exports 12%

Adhesives 6%

Molding 8%

Miscellaneous 9%

[a] First six months. **Source:** Society of the Plastics Industry

FIG. 1. Epoxy growth extends to all end uses. [Reprinted from W. F. Fallwell, *Chem. Eng. News* **51** (40), 10 (1973). Copyright 1973 by the American Chemical Society. Reprinted by permission of the copyright owner.]

The approximate epoxy resin production capacity as of 1973 is more than 300 million pounds per year. Major producers and approximate capacities are tabulated below.

Producers	Approximate production capacity for 1973 in million pounds per year[a]
Shell Chemical Company	100
Dow Chemical Company	70
Ciba-Geigy Co.	70
Resyn Corp. Celanese (Devoe and Reynolds) Union Carbide Corp. Wilmington Chemical Co. Reichhold Chemical Co.	60

[a] Approximately 0 in 1940 when Devoe and Reynolds in the U.S.A. and Ciba in Europe started the commercial epoxy business.

The growth in production of epoxies is shown in Fig. 2 and is seen to be up about 20% for the first half of 1973. The estimated epoxy sales will reach 220 million pounds in 1973 up from 183 million pounds in 1972 [9].

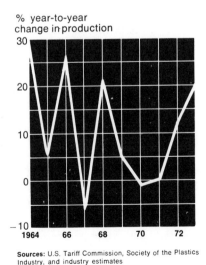

Sources: U.S. Tariff Commission, Society of the Plastics Industry, and industry estimates

Fig. 2. Epoxies hit old growth stride. [Reprinted from W. F. Fallwell, *Chem. Eng. News* **51** (40), 8 (1973). Copyright 1973 by the American Chemical Society. Reprinted by permission of the copyright owners.]

TABLE I
EPOXY: PATTERN OF CONSUMPTION [a]

Market	1000 metric tons	
	1972	1973
Bonding and adhesives [b]	5.1	5.4
Flooring, paving, and aggregates	4.6	5.4
Protective coatings		
Appliance finishes	2.9	3.8
Auto primers	5.6	6.7
Can and drum coatings	9.5	10.9
Pipe coatings	2.3	2.2
Plant maintenance	7.4	8.7
Other (including trade sales)	9.6	12.1
Reinforced plastics [c]		
Electrical laminates	6.2	9.1
Filament winding	3.4	3.8
Other	2.9	3.0
Tooling, casting, and molding	7.8	7.9
Exports	9.1	12.2
Other	6.6	7.9
Total	83.0	99.1

[a] Reprinted from *Mod. Plast.* **51**, 39 (1974). Copyright 1974 by Modern Plastics. Reprinted with permission of the copyright owner.

[b] Includes flooring, road coating, and TV resins.

[c] Does not include reinforcements.

a. Safety Precautions

Epoxy resins and their curing agents are considered primary skin irritants. Contact with epoxy resins should only be made using gloves and face shields and while working in hoods or well-ventilated areas [10]. Some individuals on prolonged contact with epoxy resins may develop a skin sensitization evidenced by blisters or other dermatitis conditions. Other individuals may develop an asthma-like condition. Contaminated gloves and work clothes should be changed immediately and either laundered or discarded. Contaminated shoes should be discarded. Frequent washing of hands is advisable and strict personal hygiene must be practiced. Note that some aromatic amine curing agents may be carcinogenic. A properly cured epoxy resin system usually presents no health problems relating to skin irritation. Individuals who show sensitivity should discontinue handling epoxy compounds. Otherwise hypersensitivity develops, making it difficult even to come close to those materials without developing dermatitis or other reactions.

b. Analysis of Epoxy Resins

Epoxy resins are analyzed for epoxy or oxirane content, which is reported as the epoxy or oxirane equivalent or epoxy equivalent weight, i.e., the weight of resin in grams which contains one gram equivalent of an epoxy group. The "epoxy value" designates the fractional number of epoxy groups per 100 grams of epoxy resin. Percent oxirane oxygen is used for epoxidized oils and dienes.

Analytically, epoxy groups are determined by the reaction with hydrogen halide and back titration with standard base. Other functional groups present may cause interference problems and result in poor end points. Pyridinium chloride–pyridine is a recommended reagent for the analysis of bisphenol-diglycidyl ether resins [11,11a].

Recently improved analytical procedures allow direct titration of the epoxy group [11a]. This is achieved by the use of hydrogen bromide dissolved in an anhydrous protic solvent such as glacial acetic acid [12]. The methods developed by Jay [13] and Dijkstra and Dahmen [14] using quaternary ammonium bromide or iodide in acetic and perchloric acid solutions for titration are considered the best general techniques for a wide variety of epoxides (hindered and unhindered alicyclic epoxides). The use of the halogen acid procedure fails for epoxides that undergo intramolecular rearrangement to aldehydes or ketones [15].

More recently Eggers and Humphrey have reported on the application of gel permeation chromatography to monitor epoxy resin molecular distributions and curing [16].

2. CONDENSATION–ELIMINATION REACTIONS

A. Epoxy Compounds via Epichlorohydrin

Epoxy resins are generally prepared by the reaction of epichlorohydrin with active hydrogen-bearing compounds:

$$Cl-CH_2CH-CH_2 + R-XH \xrightarrow[-HCl]{base} RX-CH_2-CH-CH_2 \tag{1}$$

X = O, S
R = aliphatic or aromatic

The reaction involves a chlorohydrin intermediate which is then treated with base to give the resulting epoxy compound.

$$Cl-CH_2-\overset{O}{\overset{\diagup\diagdown}{CH}}-CH_2 + RXH \xrightarrow[\text{base}]{\textit{Step 1}} RX-CH_2-\overset{OH}{\underset{|}{CH}}-CH_2-Cl \xrightarrow[-HCl]{\textit{Step 2}}$$

$$RX-CH_2-\overset{O}{\overset{\diagup\diagdown}{CH}}-CH_2 \quad (2)$$

Step 1 has also been reported to be catalyzed by Friedel–Crafts-type catalysts such as $ZnCl_2$ or BF_3 [17].

Bisphenol A is reacted with excess epichlorohydrin in the presence of aqueous sodium hydroxide to obtain the diglycidyl ether derivative [18,19]. Instead of sodium hydroxide one can use an anhydrous system with either a quaternary ammonium salt [20] or lithium hydroxide monohydrate [21,22].

Other aromatic polyhydroxy [23] and polynuclear [24] compounds such as resorcinol, hydroquinone, catechol, phloroglucinol, and tetra(p-hydroxyphenyl)ethane have been reported to react with epichlorohydrin to give glycidyl ether derivatives.

Aliphatic polyols such as glycerol [25], 2,5-bis(hydroxymethyl)tetrahydrofuran [26], and chlorinated butenediol [27,28] have been reported to be used to give epoxy compounds.

2-1. Preparation of the Diglycidyl Ether of 2,2',6,6'-Tetrabromobisphenol A [2,2-Di(3,5-dibromo-4-hydroxyphenyl)propane] [29]

$$(3)$$

To a 1-liter resin kettle equipped with a stirrer, condenser, and thermometer are added 136 gm (0.25 mole) of tetrabromobisphenol A (Michigan Chemical Corp.), 462.5 gm of epichlorohydrin (5.0 moles), and 2.5 ml of water. The mixture is stirred until the solids dissolve and then 6 gm (0.15 mole) of sodium hydroxide pellets is added. The temperature is raised to 100°C and then lowered to 95°C while more sodium hydroxide (35 gm, 0.87 mole) is added in 6-gm batches. After the last addition the reaction mixture becomes yellow and opaque. Stirring is continued until no further exotherm is observed. The excess epichlorohydrin is distilled off using a water aspirator while the reaction mixture is kept below 150°C. Benzene (25 ml) is added to precipitate sodium chloride and the mixture is filtered through a Büchner funnel. The salts are washed with another 25 ml of benzene and the washings together with the filtrate are concentrated under reduced pressure to yield 117 gm (71%) of a dark brown syrup having an epoxy equivalent weight of 331 (calculated 328).

The condensation reactions of (2,3-epoxy-1-chloropropane)epichlorohydrin have assumed great commercial significance in the preparation of the epoxides for polymer applications in plastics, adhesives, and coatings. For example, the diepoxides of bisphenol A are prepared by condensing two moles of epichlorohydrin with a basic solution of bisphenol A [Eq. (4)] [30–32].

$$2ClCH_2-CH-CH_2 + HO-\underset{O}{\bigcirc}-\overset{CH_3}{\underset{CH_3}{C}}-\bigcirc-OH \xrightarrow{NaOH}$$

$$\left(CH_2-CHCH_2O-\bigcirc-\overset{CH_3}{\underset{CH_3}{C}}- \right)_2 \qquad (4)$$

Free-flowing crystals (m.p. 43.5°C) of 2,2-bis(2,3-epoxypropoxyphenyl)-propane are obtained by carrying out the reaction of bisphenol A with epichlorohydrin which contains at least about 90% by weight of the diglycidyl ether, to which is added 8–15% by weight of *n*-butyl glycidyl ether as a solvent [33]. The synthesis of monoglycidyl aryl ethers has also been reported [34].

A review of the ether-forming reactions of epichlorohydrin (2,3-epoxy-1-chloropropane) has been published [35].

The literature on the synthesis of epoxy resins is extensive, and several good sources are available [8,36]. Using Preparation 2-1, additional epoxy resins have been prepared as shown in Tables II and III.

Glycidyl amines have been prepared by the reaction of epichlorohydrin with aliphatic or aromatic amines followed by dehydrohalogenation [22]. The aromatic glycidyl amines are much more stable on storage than the aliphatic types and can be cured with acid or anhydride curing agents to give resins.

TABLE II

PREPARATION OF EPOXY COMPOUNDS FROM PHENOLIC COMPOUNDS AND EPICHLOROHYDRIN

Hydroxy compound (moles)	Epichlorohydrin (moles)	Base	Reaction conditions			Yield (%)	Physical properties		Ref.
			Solvent	Temp. (°C)	Time (hr)		M.p. (°C)	Approx. eq. wt.	
p-Aminophenol 0.44	4.0	0.65 gm LiOH·H_2O in 6.0 gm CH_2O 66.5 gm 50% aq. NaOH	84 gm 95% C_2H_5OH —	25 55–60	137 3½	— 88	— —	— 104 gm/ gm mole	a
Bisphenol A 1.5	15.0	40% aq. NaOH	—	112	3½	509 g (97%)	—	0.519 equiv./ 100 gm	b,d
Di-α-naphthol 1.0	15.0	22 gm NaOH	—	95	1½	100	196°– 200°	0.32 equiv./ 100 gm (mol. wt. 412)	c
Resorcinol 1.0	2.0	2.0 gm 20% NaOH	—	75	1	—	—	—	d,e
Bis(hydroxyphenyl)- [1,4,5,6,7-hexachlorobi- cyclo(2.2.1)-5-heptene-2]- methane 1.0	20.0	1.0 gm NaOH	400 gm CH_3OH	60–100	1	91	—	0.313 equiv./ 100 gm	f

[a] N. H. Reinking, B. P. Barth, and F. J. Castner, U.S. Patent 2,951,825 (1960).
[b] J. M. Goppel, U.S. Patent 2,801,227 (1957).
[c] V. C. E. Burhop, J. Polym. Sci. 12, 699 (1968).
[d] P. Castan, U.S. Patent 2,324,483 (1943).
[e] S. O. Greenlee, U.S. Patent 2,493,486 (1950).
[f] C. G. Schwarzer, U.S. Patent 3,385,908 (1968).

TABLE III
PREPARATION OF EPOXY RESINS[a]

Starting material	Product	Yield (%)
1,1,3-Tri(4-hydroxyphenyl)propane	Amber viscous liquid	86
1,1-Bis(4-hydroxyphenyl)cyclohexane	Tan soft solid	85
1,1-Di(4-hydroxyphenyl)-4-methylcyclohexane	Yellow viscous liquid	76
Tetrachlorobisphenol A	Amber liquid	93
Tetrabromophenol A	Brown syrup	71
2,2-Methylenebis(4-ethyl-6-*tert*-butyl)phenol	Brown soft solid	39
2,2-Bis(4-hydroxy-3-methylphenyl)butane	Amber viscous liquid	91
2,2-Bis(4-hydroxy-3,5-dimethylphenyl)propane	Amber viscous liquid	100
2,2-Bis(4-hydroxy-3-methylphenyl)propane	Amber viscous liquid	100
Bisphenol A	Amber viscous liquid	89
Bis(4-hydroxyphenyl) sulfone	Brown solid	41
Bis(4-hydroxyphenyl) ether	Tan solid	100
2,2-Bis(4-hydroxy-3,5-dibromophenyl)-hexafluoropropane	Solid, m.p. 28°–29°C	65
2,2-Bis(4-hydroxyphenyl)-*sym*-tetrafluoro-dichloropropane	Amber syrup	83
$\left(HO-\bigcirc-\right)_2\!\!\overset{\displaystyle CH_3}{\underset{\displaystyle |}{C}}(CH_2)_8CH_3$	Amber viscous liquid	64
$\left(HO-\bigcirc-\right)_2\!\!\overset{\displaystyle CH_3}{\underset{\displaystyle |}{C}}(CH_2)_{14}CH_3$	Amber viscous liquid	74
$\left(HO-\bigcirc-\right)_2\!\!\overset{\displaystyle CH_3}{\underset{\displaystyle |}{C}}(CH_2)_6CH_3$	Amber viscous liquid	87

[a] Reprinted in part from S. R. Sandler and F. R. Berg, *J. Appl. Polym. Sci.* **9**, 3707 (1965). Copyright 1965 by the Journal of Applied Polymer Science. Reprinted by permission of the copyright owner.

2-2. *Preparation and Polymerization of the Glycidyl Derivative of p-Aminophenol* [22]

$$HO-\bigcirc-NH_2 + 3CH_2\!\!-\!\!CH\!-\!\!CH_2\!-\!Cl \longrightarrow$$

$$CH_2\!\!-\!\!CH\!-\!\!CH_2\!-\!O-\bigcirc-N(CH_2\!\!-\!\!CH\!-\!\!CH_2)_2 \quad (5)$$

To a flask containing 48.5 gm (0.444 mole) of p-aminophenol, 370 gm (4.0 moles) of epichlorohydrin, and 84 gm of 95% ethyl alcohol is added 0.65 gm of lithium hydroxide monohydrate dissolved in 6 ml of water. The mixture is stirred at room temperature for 137 hr and then heated at 55°–60°C while 66.5 gm (1.66 moles) of 40% aqueous sodium hydroxide is added over $3\frac{1}{2}$ hr.

Excess water and epichlorohydrin are removed by distillation under reduced pressure (30 mm Hg) until the temperature in the flask reaches 65°C. The residue is dissolved in toluene and washed with water to remove salts and residual caustic. The toluene solution is then stripped at 30 mm Hg to 120°C to yield 104 gm (88% yield) of the triglycidyl derivative.

B. Epoxy Esters via Glycidol

Oxirane compounds such as 3,4-epoxy-1-butene [37], (2,3-epoxy-1-propanol) glycidol [38], and 2,3-epoxybutanoic acid [39] can be used to react with other functional groups while keeping the epoxy groups intact. Recently several pure diglycidyl esters have been prepared in 25–94% yields by the simultaneous addition of an acid chloride and triethylamine to glycidol at 0°–5°C [40].

$$CH_2\!\!-\!\!CH\!\!-\!\!CH_2OH + R(COCl)_2 + 2Et_3N \longrightarrow (CH_2\!\!-\!\!CH\!\!-\!\!CH_2\!\!-\!\!O\overset{\overset{\textstyle O}{\|}}{C})_2R$$

$$+ 2Et_3N\cdot HCl \quad (6)$$

Glycidyl esters of aromatic acids appear to have been relatively unexplored compared to those of the aliphatic acids [41] except for descriptions in some patents [41–51]. The patents have described the preparation of these materials rather crudely, except for one case where carbon and hydrogen analyses have been reported for crystalline diglycidyl terephthalate [43,47].*

Diglycidyl phthalate and diglycidyl terephthalate are described in detail in a procedure [43,50] which recommends addition of the acid chloride to a mixture of glycidol in the presence of triethylamine at 0°–5°C in benzene or toluene solution. This procedure suffers from the disadvantage that glycidol itself polymerizes exothermically in the presence of basic materials and thus contaminates the products, which are subsequently isolated in low yields. The products are easily polymerized by the application of heat in the presence of basic catalysts such as triethylamine. Some investigators [43,47] have ne-

* Reprinted from S. R. Sandler and F. R. Berg, *J. Chem. Eng. Data* **11**, 447 (1966). Copyright 1966 by The American Chemical Society. Reprinted by permission of the copyright owner.

glected this fact and give procedures wherein the products are heated to 160°C at 10 mm Hg pressure in vessels which originally contained basic catalysts. Therefore, to prepare purer products with acceptable analyses, a modified procedure was developed which utilized low temperatures and separate addition of triethylamine. Many of the crystalline products obtained have been reported by earlier investigators as resins. The melting points and analyses for the glycidyl esters prepared by this method appear in Table IV. The general procedure is illustrated below, utilizing the preparation of diglycidyl isophthalate as an example. The properties of diglycidyl hexahydroterephthalate are shown for comparison, although this is not an aromatic system.*

An additional method that is useful for the preparation of glycidyl esters is the reaction of the allyl esters of aromatic acids with *m*-chloroperbenzoic acid (see Section 3,A of this chapter).

2-3. *Preparation of Diglycidyl Isophthalate* [40]*

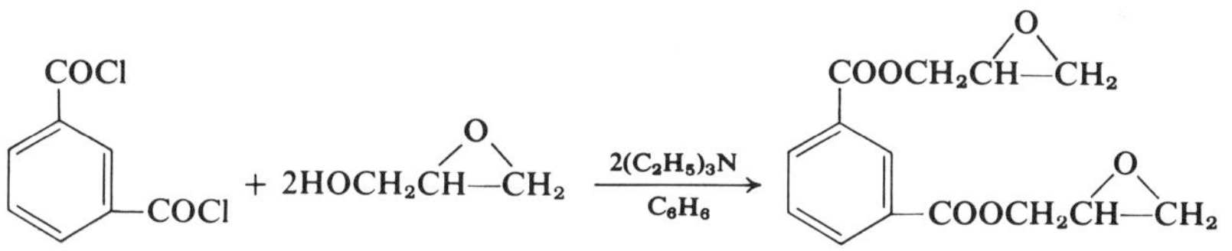

$$\text{(7)}$$

A 1-liter resin kettle is fitted with a stirrer, thermometer, condenser with drying tube, and two 300-ml dropping funnels. All the equipment is carefully dried and flushed with nitrogen for 10 min. Glycidol (74.1 gm, 1.0 mole) and 200 ml of benzene are placed in the flask and cooled, with stirring, to 0°C in an ice water–methanol bath. A solution of isophthaloyl chloride (101.5 gm, 0.5 mole) in 150 ml of benzene is placed in one dropping funnel and a solution of triethylamine (101 gm, 1.0 mole) in 150 ml of benzene in the second funnel. The dropwise addition of the acid chloride is begun first, followed by the dropwise addition of the triethylamine solution. The rates are controlled so that the pot temperature does not exceed 5°C, and the acid chloride addition is slightly faster than that of the triethylamine. Complete addition requires 3 hr. Stirring is continued for 3 hr longer while the pot reaches room temperature. The solids (triethylamine hydrochloride) are filtered, rinsed with 50 ml of benzene, and dried. The weight of triethylamine hydrochloride is 123 gm (theoretical weight 137.5 gm).

The filtrates are washed in a separatory funnel with 200 ml of saturated

* Reprinted from S. R. Sandler and F. R. Berg, *J. Chem. Eng. Data* 11, 447 (1966). Copyright 1966 by the American Chemical Society. Reprinted by permission of the copyright owner.

TABLE IV

Preparation of Polyglycidyl Esters of Polycarboxylic Acids According to Eq. (6)[a]

Product	Yield (%)[b]	Analysis				M.p. (°C)[c]	Ref. to previously reported m.p. (°C)	Characteristic infrared absorption bands (μm)[d]
		Calc.		Found				
		%C	%H	%C	%H			
Diglycidyl isophthalate	78	60.43	5.07	60.42	5.13	60–63	Viscous liquid[g]	3.40(w), 5.80(s), 6.20(w), 7.65(s), 8.20(s), 9.40(m), 11.04(m), 13.68(s), 14.05(m)
Diglycidyl terephthalate	67	60.43	5.07	60.37	5.19	94–96[e]	108–109[h]	3.40(w), 5.80(s), 6.12(w), 7.85(s), 8.95(s), 9.08(s), 9.35(m), 11.80(m), 11.80(m), 13.27(m), 13.70(s)
Diglycidyl hexahydroterephthalate	25	59.14	7.09	58.67	6.86	88–90	—	3.40(w), 3.50(vw), 5.77(s), 7.26(w), 7.58(m), 8.05(m), 8.55(s), 8.65(s), 9.57(m), 11.00(w), 13.26(w)
Diglycidyl phthalate	94	60.43	5.07	60.04	5.39	Viscous liquid	Viscous liquid[h]	3.40 (m), 5.75(vs), 6.23(m), 6.30(m), 7.85(vs), 8.95(s), 8.40(s), 11.05(m), 13.50(m), 14.25(m)
Diglycidyl 3,5-dichloroterephthalate	53	48.44	3.45	48.66	3.68	113–114	—	3.40 (vw), 5.76(vs), 7.45(s), 7.80(vs), 8.10(vs), 8.94(vs), 9.25(s), 9.33(vs), 11.15(m), 12.83(m), 13.18(m)

Diglycidyl diphenoate	85	67.80	68.04	5.08	5.10	Viscous liquid	—	3.40(vw), 5.80(vs), 6.25(m), 6.33(w), 7.77(vs), 8.00(vs), 8.30(m), 8.60(w), 8.90(s), 9.25(s), 11.03(m), 13.25(s), 14.17(m)
Triglycidyl trimesoate[g]	58	57.44	57.18	4.80	4.86	58–63	—	3.40(m), 5.75(s), 6.20(w), 6.92(m), 7.43(m), 7.57(m), 8.10(s), 9.05(m), 9.34(w), 11.00(m), 13.55(s), 14.02(m), 14.56(m)
Triglycidyl trimellitate[f]	52–63	57.14	57.98	4.80	5.35	Liquid	—	3.40(m), 5.78(s), 6.12(w), 6.18(w), 7.43(m), 7.80(s), 8.10(s), 9.00(s), 9.40(m), 11.03(m), 11.75(m), 13.30(m)
Tetraglycidyl pyromellitate	30	55.23	55.18	4.64	4.81	113–123	—	3.40(vw), 5.77(vs), 6.15(vw), 8.0(vs), 8.85(s), 9.02(m), 9.25(w), 11.05(m), 12.10(m), 13.30(m), 13.42(m)

[a] Reprinted from S. R. Sandler and F. R. Berg, *J. Chem. Eng. Data* **11**, 447 (1966). Copyright 1966 by the American Chemical Society. Reprinted by permission of the copyright owner.
[b] For crude reaction products before recrystallization.
[c] Solid products recrystallized from petroleum ether–benzene.
[d] Intensity of absorptions: (s) strong, (vs) very strong, (m) medium, (w) weak, and (vw) very weak.
[e] After preparing this compound several times and recrystallizing from petroleum ether–benzene or benzene alone melting point was not raised to 108°–109°C as previously reported by Henkel and Cie, British Patent 735,001 (1955).
[f] Obtained as an oil and could not be crystallized. Attempted distillations under high vacuum gave a resinous polymer. Triglycidyl trimesoate was obtained as an oil in 46% yield when triethylamine was present with glycidol and trimesoyl chloride was added at 0°–5°C. If a procedure similar to that for diglycidyl isophthalate was followed, a crystalline solid (m.p. 50°C) was obtained. Three recrystallizations from petroleum ether–benzene gave a solid (m.p. 58°–63°C).
[g] G. Schade, German Patent 1,176,122 (1964).
[h] Henkel and Cie, British Patent 735,001 (1955).

aqueous sodium chloride, twice with 200-ml portions of distilled water, and dried over anhydrous calcium chloride. The salt is removed by gravity filtration and the benzene is stripped from the filtrate by means of a vacuum pump and a warm water (40°–45°C) bath. The residue, a white solid, is mixed with petroleum ether and filtered. The crude product weighs 111 gm (theoretical yield 139 gm) and melts at 48°–53°C. The product is dissolved in 700 ml of petroleum ether–benzene (1:1) solution, stirred with about 5 gm of activated charcoal, filtered, and cooled. About 25 ml of petroleum ether is added. The recrystallized material is filtered and dried in a vacuum oven at room temperature. Final yield is 36 gm (m.p. 60°–63°C).

2-4. Preparation of Glycidyl Benzoate [52]

$$C_6H_5COCl + HOCH_2CH\overset{O}{\overset{\diagup\diagdown}{-}}CH_2 \xrightarrow{Et_3N} C_6H_5COOCH_2CH\overset{O}{\overset{\diagup\diagdown}{-}}CH_2 + Et_3N\cdot HCl \quad (8)$$

To a 2-liter flask are added 51.9 gm (0.7 mole) of glycidol and 400 ml of benzene. The flask is cooled to 0°C and a solution of 81.2 ml (0.7 mole) of benzoyl chloride in 60 ml of benzene is placed in one dropping funnel while in another is placed a solution of 97.0 ml (0.7 mole) of triethylamine in 40 ml of benzene. The solutions are added dropwise simultaneously over a 2-hr period and then stirred for an additional 2 hr. The solids are filtered and rinsed with two 50-ml portions of benzene. The filtrate is shaken twice with dilute hydrochloric acid, washed with water until neutral, and dried over sodium sulfate. The benzene is then removed using a water aspirator and the residue is vacuum distilled to yield 90 gm (71%) of product, b.p. 97.5°–98.5°C (0.8–0.9 mm Hg). The reported [47] boiling point is 103°C (1 mm Hg).

Epichlorohydrin can also be reacted with carboxylic acids using tertiary amines or benzyltrimethylammonium chloride to give the chlorohydrin ester. The latter can be treated with base to give epoxy esters [53].

C. Epoxy Compounds via the Darzens Glycidic Ester Synthesis

The Darzens synthesis [54–56] involves the condensation of aldehydes or ketones with ethyl chloroacetate in the presence of sodium amide or ethoxide to give α,β-epoxy esters in one step. Aromatic ketones and aldehydes as well as aliphatic ketones give good yields. Yields from aliphatic aldehydes, however, are poor.

$$R_2C{=}O + Cl{-}CH_2{-}COOC_2H_5 \xrightarrow{base} R\overset{R}{\underset{\underset{O}{\diagdown\diagup}}{-}}C{-}CHCOOC_2H_5 \quad (9)$$

Ethyl dichloroacetate also condenses with aldehydes and ketones with the aid of magnesium amalgam to give α-chloro-β-hydroxy esters which, upon treatment with sodium ethoxide, give glycidic esters.

In place of chloroacetic esters other halogenated compounds have been found to give good yields of epoxides, e.g., α-chloro ketones [57] with benzyl [58] or benzal halides [59].

$$\text{RCHO} + \text{ClCH}_2\text{COR}' \xrightarrow{\text{C}_2\text{H}_5\text{ONa}} \underset{\displaystyle \text{RCH-CH-COR}'}{\overset{\displaystyle O}{\triangle}} \qquad (10)$$

$$\text{RCHO} + \text{R}'\text{CH}_2\text{Cl} \xrightarrow{\text{CH}_3\text{OK}} \underset{\displaystyle \text{RCH-CHR}'}{\overset{\displaystyle O}{\triangle}} \qquad (11)$$

2-5. *Preparation of Phenylmethylglycidic Ester* [60]

$$\text{C}_6\text{H}_5\text{COCH}_3 + \text{ClCH}_2\text{COOC}_2\text{H}_5 + \text{NaNH}_2 \longrightarrow \underset{\displaystyle O}{\overset{\displaystyle \text{CH}_3}{\text{C}_6\text{H}_5\text{C-CHCOOC}_2\text{H}_5}} + \text{NH}_3$$
$$+ \text{NaCl} \quad (12)$$

To a flask equipped with a stirrer, condenser, and thermometer are added 120 gm (1 mole) of acetophenone, 120 gm (1 mole) of ethyl chloroacetate, and 200 ml of dry benzene.

NOTE: The reaction should be carried out in a hood since a large volume of ammonia gas is given off.

During a period of 2 hr, 47.2 gm (1.2 moles) of finely powdered sodium amide is added (strongly exothermic) while the temperature is kept at 15°–20°C. Following the addition, the reaction mixture is stirred for 2 hr at room temperature and then the red mixture is slowly poured into a beaker containing 700 gm of cracked ice while stirring by hand. The organic layer is separated and the aqueous layer is extracted with 200 ml of benzene. The combined organic layers are washed with three 300-ml portions of water, the last one containing 10 ml of acetic acid. The benzene solution is dried over sodium sulfate, filtered, and concentrated, using an aspirator, to give a residue which upon fractionation under reduced pressure yields 128–132 gm (62–64%) of product, b.p. 111°–113°C (3 mm Hg).

Glycidic thio esters have recently been reported for the first time to be prepared by the reaction of α-bromothiol esters with aldehydes as shown in Eq. (13). See data tabulated below.*

$$\underset{C_6H_5CH}{\overset{O}{\|}} + \underset{\underset{Br}{|}}{\overset{O}{\|}}RCHCX\text{-}t\text{-Bu} \longrightarrow \underset{II}{C_6H_5CH\overset{O}{\diagup}\underset{\underset{R}{|}}{\overset{O}{\underset{\|}{C}}}CX\text{-}t\text{-Bu}} \qquad (13)$$

X = O or S

I				II			
X =	R =	Base	Solvent	% cis	% trans	% yield	
S	H	LiN[Si(CH$_3$)$_3$]$_2$,	THF	10	90	59	
		NaH	DMF	70	30	67	
S	CH$_3$	NaH	CMF	10	90	58	
O	CH$_3$	NaH	DMF	55	45	61	

3. OXIDATION REACTIONS

Epoxy resins can be prepared by oxidation of suitable unsaturated compounds with hydrogen peroxide, air, peracetic acid, or other organic peroxy acids such as m-chloroperbenzoic acid [40].

Unsaturated polymers, prepared from butadiene or isoprene, can be epoxidized with peroxyacids. The peroxyacids used can either be preformed or prepared in situ by reacting hydrogen peroxide with lower aliphatic carboxylic acids. The epoxidized polymers can be reacted with diamines or dianhydrides to give a cross-linked resin useful for adhesive and coating applications.

Although peracetic acid [42] has been reported to peroxidize diallyl esters, the epoxy esters decompose in the presence of acetic acid. m-Chloroperbenzoic acid gives a clean product and is separated from the by-product m-chlorobenzoic acid by washing the organic layer with dilute sodium carbonate or sodium hydroxide. Recently other investigators [61,62] have also found m-chloroperbenzoic acid useful for preparing volatile epoxides. The procedure for preparing diglycidyl terephthalate from diallyl terephthalate is given below.

* Data reprinted from D. J. Dagli and J. Wemple, J. Org. Chem. 39, 2938 (1974). Copyright by the American Chemical Society. Reprinted by permission of the copyright owner.

3-1. Preparation of Diglycidyl Terephthalate [40]*

$$COCl \quad + 2HOCH_2CH{=}CH_2 \xrightarrow[\text{(2) } m\text{-chloro-}\atop \text{perbenzoic acid}]{\text{(1) } 2(C_2H_5)_3N,C_6H_6} \quad COOCH_2{-}CH{-}CH_2 \qquad (14)$$

The reaction of diallyl terephthalate with *m*-chloroperbenzoic acid consists of cooling to 0°–3°C a solution of the acid (0.4 mole) in chloroform (700 ml) and adding the diester (0.2 mole) in small portions with gentle agitation over a period of 4 hr. The stoppered reaction flask is stored at 3°–5°C for 3 days and then brought to room temperature. The solids are removed by filtration and the filtrate is washed with a solution of sodium carbonate (0.2 mole) in 250 ml of water, then several times with water, and finally dried over calcium sulfate. The washings are acidified and the solids collected. The weight of the combined recovered solids is equal to the theoretical weight of *m*-chlorobenzoic acid (62.6 gm). The dried chloroform filtrate is concentrated by water suction, and residual chloroform then removed by the vacuum pump. The product is obtained as a pale yellow liquid from which solids separate. Recrystallization from petroleum ether–benzene yields the pure diglycidyl terephthalate (28% yield). Similarly obtained are diglycidyl isophthalate (87% yield) and liquid glycidyl phthalate (100% yield). The infrared spectra of the above products are identical to those of the epoxy esters prepared using glycidol and the acid chlorides.

Butadiene–styrene and butadiene–acrylonitrile copolymer latexes have also been epoxidized with peracetic acid [63].

Unsaturated polyesters and partially unsaturated PVC also have been used as starting materials to prepare epoxidized resins [7a,64,65].

4. FREE RADICAL REACTIONS

Unsaturated epoxy compounds such as allyl glycidyl ethers and glycidyl methacrylate may be homo- or copolymerized with other monomers to give a polymer with epoxy groups [66]. The resulting polymer or copolymer is then cured in the conventional manner to give the epoxy resin.

---

* Reprinted from S. R. Sandler and F. R. Berg, *J. Chem. Eng. Data* **11**, 447 (1966). Copyright 1966 by the American Chemical Society. Reprinted by permission of the copyright owner.

TABLE V

REACTIVITY RATIOS OF GMA WITH VARIOUS MONOMERS

Comonomer	r_2	r_1	Temp. (°C)	Ref.
Methacrylic acid	0.38	1.90	—	a
Methyl methacrylate	0.82	0.75	60	b
	1.05	0.80		
Vinyl chloride	8.84	0.04	50	c
Vinyl acetate	32.2	0.028	—	a
Acrylonitrile	1.32	0.14	60	d
Styrene	0.53	0.44	60	d

^{a} Data sheet on glycidyl methacrylate from Blemmer Chemical Corp. No. 7D11-1000A, p. 3 (1974).

^{b} K. Iwakura, *Makromol. Chem.* **97**, 128 (1966); M. S. Gluckman, *J. Polym. Sci.* **37**, 411 (1959).

^{c} K. Imoto, *Kobunshi Kagaku* **18**, 747 (1961).

^{d} K. Iwakura, *Kobunshi Kagaku* **17**, 187 (1960).

4-1. Preparation of an Epoxy Resin by the Free-Radical-Initiated Polymerization of Allyl Glycidyl Ether [66]

$$(15)$$

Allyl glycidyl ether is refluxed at 155°C while air is bubbled through it. After 48 hr the refractive index of the solution is found to increase to 1.4990 at 20°C (an increase of n_D^{20} 0.0412) which corresponds to a 48% conversion to polymer with a viscosity of 27.0 poises at 25°C. (Color 9-10 on Gardner scale.)

The reactivity ratios of glycidyl methacrylate (GMA) with other monomers are given in Table V.

4-2. Solution Copolymerization of Glycidyl Methacrylate with Methyl Methacrylate [67]

$$(16)$$

To a sealable vial 15 mm in diameter and 250 mm in length is added 104.3 gm of a mixture (1:1 by weight) of methyl ethyl ketone (MEK) and xylene. Then 14.2 gm of glycidyl methacrylate, 90.11 gm of methyl methacrylate, and 2.42 gm of benzoyl peroxide are added. The tube is purged five times with nitrogen, cooled, evacuated with 3 mm Hg, and sealed by melting its upper end with a burner. The sealed tube is placed in a shaker and heated at 80°C for 8 hr. Then the tube is cooled, cautiously opened, diluted with MEK to 5% solids concentration, and added to 5 times its weight of methanol to precipitate the polymer. The polymer is filtered and dried under reduced pressure to yield a white powder.

4-3. Emulsion Copolymerization of Glycidyl Methacrylate with Vinyl Acetate [68]

(17)

To a flask is added 492 gm of water and 20 gm of poly(vinyl alcohol) (100%). The mixture is stirred until the poly(vinyl alcohol) dissolves. Then 6 gm of Nonion E230 [polyoxyethylene (30) oleyl ether from Nippon Oils & Fats Co., Ltd.] and 2.0 gm of calcium carbonate are added and stirred in order to dissolve them and the solution then purged with nitrogen for 5 min. The flask is heated to 65°C and 40 gm of vinyl acetate and 40 gm of 2.5% potassium persulfate are added. The temperature is raised to 75° ± 1°C and the reaction allowed to proceed for 1 hr. Then 270 gm of vinyl acetate and 30 gm of 2.5% potassium persulfate (aqueous solution) are added to the flask separately (using two dropping funnels) over a 3-hr period at 75° ± 1°C. A solution of 70 gm vinyl acetate and 20 gm glycidyl methacrylate is then added dropwise while 11 parts 2.5% aqueous $K_2S_2O_8$ is added at the same time from another dropping funnel. The latter additions are done over a 1 hr period at 75° ± 1°C. The copolymerization reaction is continued for another hour at 75° ± 1°C.

Glycidyl methacrylate can also be reacted with various nucleophilic reagents to open the epoxy group and thus give compounds containing a polymerizable double bond. The latter may be useful as cross-linking agents, grafting monomers, and monomers to copolymerize with vinyl compounds to add

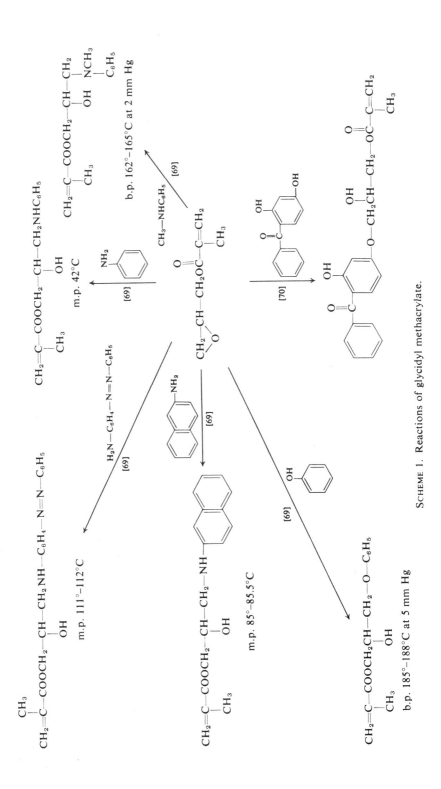

SCHEME 1. Reactions of glycidyl methacrylate.

additional properties related to the nucleophile used in opening the epoxy group (see Scheme 1) [69,70].

Alternately, glycidyl methacrylate can be polymerized as a free radical and the same types of reactions shown above can be effected on the polymer.

5. CURING–POLYMERIZATION REACTIONS OF EPOXY COMPOUNDS AND RESINS

Epoxy resins are cured [71] by reaction of the epoxy group with other functional groups to give linear, branched, or cross-linked products as described in Eq. (18) and in Tables VI–XIII.

$$Z + CH_2\text{—}CHR \longrightarrow Z^+\text{—}CH_2\text{—}CH\text{—}R \qquad (18)$$

$$Z = R_3N, ROH, RCOOH, (RCO)_2O, RNH_2, RCONH_2, RSH, \text{etc.}$$

The compounds Z are active hydrogen compounds such as amines, anhydrides, and acids [72]. The curing reaction can also involve homopolymerization catalyzed by Lewis acids or tertiary amines. In most cases the reactions catalyzed by Lewis acids are too fast for practical systems and the reactions catalyzed by tertiary amines too slow and too highly temperature and concentration dependent to make them reliable as the sole curing agent.

It is interesting to note the activity of tertiary amines toward epoxide does not always correlate with their basicity but depends also strongly on steric factors. For example, triethylamine appears more reactive than tributylamine or benzyldiethylamine [73]. On the other hand, all epoxy groups are of equal reactivity. Electron-attracting groups increase the rate with nucleophilic agents (amines, inorganic bases) whereas electron-donating groups (methylene, vinyl) improve reactivity with acidic-type (electrophilic) curing agents. For example, internal epoxy groups favor reaction with electrophilic curing agents [74].

Some of the typical curing reactions are illustrated in Scheme 2 [75–85].

The choice of the specific curing agent is based on end-use considerations with regard to desired properties of the cured resin. The amount of curing agent, if of the active hydrogen type, is calculated as one epoxy group for each active hydrogen of the reagent. Too much curing agent may cause embrittlement. Tertiary amines and Lewis acids are required only in catalytic amounts such as 5–15 phr.

CAUTION: It should be noted that some aromatic and cyclic amines used in curing epoxy resins may be carcinogenic. All amine curing agents or N-heterocyclic types should be suspect and handled with great care.

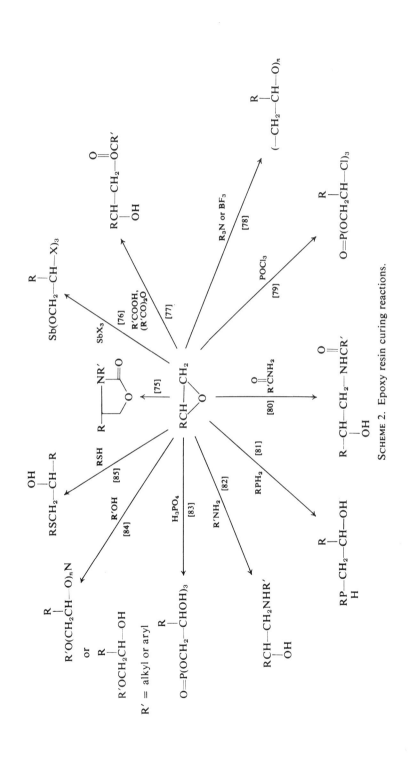

SCHEME 2. Epoxy resin curing reactions.

TABLE VI

REACTIVITY OF AMINE ACCELERATORS WITH ANHYDRIDE–EPOXY RESIN SYSTEMS[a,b]

Accelerator	Araldite 6010 plus 136 phr of dodecenylsuccinic anhydride and accelerator			Araldite 6010 plus 91 phr of Nadic methyl anhydride and accelerator		
	1.0% (hr)	2.0% (hr)	3.0% (hr)	1.0% (hr)	2.0% (hr)	3.0% (hr)
Aniline	—	NR	NR	NR	NR	NR
N-Methylaniline	—	NR	NR	NR	NR	NR
N,N-Dimethylaniline	—	20	—	—	—	—
p,p'-Methylenedianiline	—	6R	—	—	—	—
p,p'-Methylenebis(N,N-dimethylaniline)	14	4	—	—	6	2
α-Methylbenzyldimethylamine	NR	8	23	6	7	—
N-Methylcyclohexylamine	—	NR	—	—	6	—
N-Benzylmethylamine	—	NR	—	NR	NR	NR
Diethylenetriamine	—	NR	NR	NR	—	—
Triethylenetetramine	—	NR	NR	>30<48	>30<48	24
Pyridine	10	5	—	4	—	—
2-Aminopyridine	—	NR	24	—	—	8
2-Amino-3-methylpyridine	24	5	>24	17	3	4
2-Amino-4-methylpyridine	—	24	—	24	5	6
2-Amino-5-methylpyridine	NR	NR	—	6	4	—
2-Amino-6-methylpyridine	NR	NR	24	24	—	5
2,6-Diaminopyridine	—	NR	—	—	5	5
Diethylamine	—	NR	—	—	—	—
Triethylamine	16	8	18	>30<48	>30<48	24
Dibutylaminopropylamine	23	7	>24	6	—	3
Dimethylaminopropylamine	—	23	8	25	—	4
Diethylaminopropylamine	—	>24	24	—	—	4

[a] NR indicates no reaction after 48 hr.

[b] Reprinted from G. M. Kline, *Mod. Plast.* **42**, 152 (1964). Copyright April, 1964 by Modern Plastics. Reprinted by permission of the copyright owner.

The recent report that the bicyclic phosphorus compound 4-ethyl-1-phospha-2,6,7-trioxabicyclo[2,2,2]octane-1-oxide [86] is highly toxic should cause one to be careful in the handling of related compounds such as triethanol

$$O=P \left\langle \begin{array}{l} O-CH_2 \\ O-CH_2 \\ O-CH_2 \end{array} \right\rangle C_2H_5$$

TABLE VII

Reactivity of Amine Accelerators with Anhydride–Epoxy Resin Systems[b]

	Gel time at 65°C, 100 gm samples[a]		
Accelerator	A (hr)	B (hr)	C (hr)
Pyridine	1	$\frac{1}{2}$	$1\frac{1}{2}$
Diethylamine	—	9	42
Triethylamine	5	$2\frac{1}{2}$	$4\frac{1}{2}$
α-Methylbenzyldimethylamine	5	$2\frac{1}{2}$	$4\frac{1}{2}$
Dibutylaminopropylamine	5	$2\frac{1}{2}$	$1\frac{1}{2}$
Diethylaminopropylamine	5	$3\frac{1}{2}$	$3\frac{1}{2}$
Dibutylaminopropylamine acetate	4	4	$6\frac{1}{2}$
p,p′-Methylenebis(*N,N*-dimethylaniline)	1	1	1
2-Aminopyridine	3	$1\frac{1}{2}$	$2\frac{1}{2}$
2-Amino-3-methylpyridine	3	2	$3\frac{1}{2}$
2-Amino-4-methylpyridine	16	—	7
2-Amino-5-methylpyridine .	17	$2\frac{1}{2}$	$3\frac{1}{2}$
2-Amino-6-methylpyridine	4	$5\frac{1}{2}$	$8\frac{1}{2}$
2,6-Diaminopyridine	8	21	—

[a] Formulations: Araldite 6010 plus 2% of the indicated accelerator and (A) 102.4 phr hexahydrophthalic anhydride:HET anhydride 50:50, (B) 93.9 phr hexahydrophthalic anhydride:HET anhydride 60:40, and (C) 85.5 phr hexahydrophthalic anhydride:HET anhydride 70:30.

[b] Reprinted from G. M. Kline, *Mod. Plast.* **42**, 152 (1964). Copyright April, 1964 by Modern Plastics. Reprinted by permission of the copyright owner.

amine phosphate [87] recommended in catalyst preparations for epoxy resins.

$$O=P \begin{array}{c} O-(CH_2)_2 \\ O(CH_2)_2-N \\ O-(CH_2)_2 \end{array}$$

A wide variety of compounds can be used as curing agents and a typical sample as gleaned from a recent *Chemical Abstracts* subject index is listed below [88–90].

Triethylenetriamine	Cyanuric acid
Triethylenetetramine	N–Si compounds
Diethylaminopropylamine	Imidazoles
N-Aminoethylpiperazine	Amine oxides
Dicyandiamide	Betaines
m-Phenylenediamine	Biurea
Methylene dianiline	Borane derivatives

Diaminodiphenyl sulfone
BF$_3$–monoethylamine
Nadic methyl anhydride
Phthalic anhydride
Pyromellitic dianhydride
Chlorendic anhydride
Trimellitic anhydride
Dodecenylsuccinic anhydride
Hexahydrophthalic anhydride
Alkyl titanates and N bases
Aniline–formaldehyde resins and
   a dicarboxylic acid

Guanamine
Imines
Ketimines
N–P organic compounds
Polyamides
Polyamines
Polyisocyanates
Polymercaptans alone or with amines
   or anhydrides
Cyanuric chloride
Methoxyalkyl melamines
Sn(II) alcoholates and carboxylates

With diglycidyl derivative of bisphenol A, the aromatic amines such as 4,4'-methylene dianiline or diaminodiphenyl sulfone provide good thermal stability for the final cured resin. Although the aliphatic primary amines react more rapidly (triethylenetetramine cures the above epoxy resin based

TABLE VIII

GEL TIME EFFECT OF VARIOUS AMINES (1%)
ON THE REACTION OF SHELL EPOXY RESIN
E-824 (100 PARTS) AND HEXAHYDROPHTHALIC
ANHYDRIDE (60 PARTS)[a]

Amine catalyst	Gel time (min)
$(C_2H_5)_3N$	5
$(C_2H_5)_2NH$	20
$HOCH_2CH_2NH_2$	129
$(HOCH_2CH_2)_2NH$	38
$(HOCH_2CH_2)_3N$	7
$HOCH_2CH_2NCH_3$	20
$(HOCH_2CH_2)_2NCH_3$	3
$(HOXH_2CH_2)_2NCH \overset{\displaystyle CH_3}{\underset{\displaystyle C_6H_5}{\big<}}$	41
$C_6H_5CH_2N(CH_3)_2$	2
$C_6H_5CH{-}N(CH_3)_2$ $\quad\mid$ $\quad CH_3$	3
$(CH_3)_2CH_2NH$	43
$(CH_3)_2C{-}NH_2$ $\quad\mid$ $\quad OH$	164

[a] Data from Y. Tanaka and H. Kakiuchi, *J. Appl. Polym. Sci.* **7**, 1063 (1963).

TABLE IX

Temperature of Cure of Epoxy Resins[a] by Various Salts and
Adduct Compounds[b]

Compound	Amount (wt.%)	Cure temperature (°C)	Base pK_a
$C_6H_5NH_2BF_3$	2.0	55	4.69
$C_6H_5NH_3^+BF_4^-$	2.25–2.5	58	4.69
$2,6\text{-}Et_2C_6H_3NH_2 \cdot BF_3$	3.54	64	—
$2,6\text{-}Et_2C_6H_3NH_3^+BF_4^-$	3.86	63	—
$EtNH_2 \cdot BF_3$	1.96	140	10.63
$EtNH_3^+BF_4^-$	2.29	140	10.63
$sec\text{-}Bu_2NH \cdot BF_3$	3.15	160	11.05
$sec\text{-}Bu_2NH_2^+BF_4^-$	3.12	160	11.05
$i\text{-}Pr_2NH_2^+BF_4^-$	3.06	160	11.01
$Et_2NH \cdot BF_3$	3.48	210–215	11.09
$(C_6H_5)_3PH^+BF_4^-$	5.7	150	2.73
$(C_6H_5)_3P \cdot BF_3$	4.5	160	2.73
$C_6H_5NMe_2 \cdot BF_3$	4.4	180	5.18
$i\text{-}Pr_2EtNH^+BF_4^-$	3.85	180	(11.37)
Pyridine $\cdot BF_3$	2.0	225	5.23
Pyridine $H^+BF_4^-$	3.0	225	5.23
$Et_3NH^+BF_4^-$	3.36	241	10.65
$Et_3N \cdot BF_3$	3.0	243	10.65

[a] Diglycidyl ether of bisphenol A.

[b] Reprinted in part from J. J. Harris and S. C. Temin, *J. Appl. Polym.
Sci.* **10**, 523 (1966). Copyright 1966 by the Journal of Applied Polymer
Science. Reprinted by permission of the copyright owner.

on bisphenol A in 30 min at room temperature and causes it to exotherm up
to 200°C), they are more difficult to handle and offer poor thermal stability.

The anhydrides generally provide pot lives of days or months and cure
usually at 100°–180°C with very little exotherm. Tertiary amines accelerate
the time for gelation but still require the elevated temperature cure to obtain
optimum properties. Anhydrides usually give brittle products but the addition
of polyether flexibilizing groups yields more elastomeric products [91].

Some typical examples of amines known to catalyze the reaction of anhydrides with epoxy resin are shown in Tables VI to XIII.

Additionally, amine salts have been effectively used as curing agents, as
shown in Table IX.

Epichlorohydrin can be reacted with itself to give polyepichlorohydrin
[92].

Amines and amides react with epichlorohydrin to give either epoxy amines or amides which can be cured to give resins as shown in Table X [22,93].

Mercaptan and dimercapto compounds also react with epichlorohydrin to give diepoxy compounds [94].

Polymeric products [95] are obtained when the mole ratio is less than 2 for the epichlorohydrin–bisphenol A reaction and a molecular weight of 1420 is reported at a mole ratio of 1.2 [96]. High molecular weight resins (30,000 or more) are known as phenoxy resins, and these thermoplastic resins are used for coatings, adhesives, and various molding applications.

Infrared absorption spectroscopy is a fast and accurate means of determining the extent of cure of a given epoxy resin. One can follow either the decrease in the epoxy group absorption at approximately 11.0 μm or the increase in hydroxyl group absorption at approximately 2.9–3.0 μm [97,98]. Furthermore, depending on the temperature of the curing reaction, the degree of cure (cross-linking, etc.) will vary. The latter has a decided effect on the physical properties of the final cured resin.

NOTE: The type of curing agent will also greatly affect physical properties.

5-1. *Curing Reaction of Bisphenol A Diglycidyl Ether with Poly(adipic acid anhydride)* [3]

Three-hundred grams of the epoxy resin is mixed with 125 gm of poly(adipic acid anhydride) and heated to 130°C for 1 hr. The resulting resin hardens at 150°C in 1½ hr and is rather elastic. If the resin is heated at 150°–170°C for any length of time it becomes cross-linked and shows good adhesion to a wide variety of nonporous materials (glass, metal).

5-2. *Curing of the Triglycidyl Derivative of p-Aminophenol* [22]

$$CH_2\text{—}CH\text{—}CH_2O\text{—}\langle\ \rangle\text{—}N(CH_2CH\text{—}CH_2)_2$$

+ $\longrightarrow$ polymer   (19)

$$NH_2\text{—}\langle\ \rangle\text{—}CH_2\text{—}\langle\ \rangle\text{—}NH_2$$

To 100 gm of the above epoxy compound is added 15–33 gm of a curing agent consisting of a eutectic mixture of 60 parts *m*-phenylenediamine and 40 parts of 4,4′-methylenedianiline. The mixture is agitated at room temperature and cured for 1 hr at 75°C, 1 hr at 100°C and 16 hr at 185°C. With 15 parts of resin the heat distortion temperature is 200°C and the flexural strength at 25°C is 16,900 psi.

TABLE X

CURING REACTIONS OF VARIOUS EPOXY COMPOUNDS AND RESINS

Epoxy compound or resins

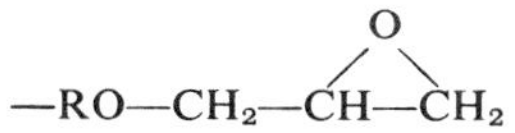

$$-\text{RO}-\text{CH}_2-\overset{\displaystyle O}{\overset{\displaystyle \diagup\!\diagdown}{\text{CH}}}-\text{CH}_2$$

R equals or is derived from

Curing agents	Temperature (°C)[a]	Ref.
Bisphenol A		
Triethylene tetraamine	—(234°F)	b
Ester dianhydride derived from trimellitic anhydride	100 (1 hr)	c
Bisphenol A		
4,4-Methylenedianiline	100 (2 hr)	c
Novolac resin		
4-p-Phenyl trimellitate anhydride	125	d
Epon 562		
Diethylenetetramine	105 (18 hr)	e
Dinaphthol (4,4'-dihydroxy-1,1'-dinaphthyl)		
Maleic anhydride + trace $(\text{CH}_3)_2\text{N}-\text{C}_6\text{H}_5$	160 (1 hr)	e
Phthalic anhydride	160 (80 hr)	e
Epikote 828		
Condensates of trimellitic acid anhydride with liquid hydrogenated polybutadiene or butadiene polymer containing ≥ 1 OH group + trace benzylmethylamine	90–130	f
Glycidyl ether epoxy resin		
Acid anhydride + trace of organic phosphonium salt	—	g
Bisphenol		
Polysebacic polyanhydride	90	h
Phthalic anhydride–triethylamine	—	i
BF_3-monoethylamine	—	j
Stannous octoate alone or with amines, anhydrides, polycarboxylic acids, polythiols, polyisocyanates, polyhydric phenols	—	k
Amides and polyamides	—	l
Ureas	—	m
Polyurethanes	—	n
Isocyanates	—	o
Glycidylthiomethyl ether		
Cellulose acetate + acetic acid + ZnCl_2	90	p
$\text{R}_3\text{Si}(\text{CH}_2)_n-\overset{\displaystyle \text{CH}_2}{\underset{\displaystyle \text{CH}_3}{\overset{\diagup O \diagdown}{\text{C}}}}$		
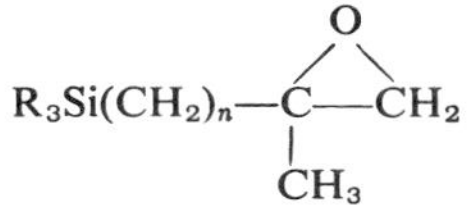 Amines, silanes, etc.	—	q
$\text{H}_2\text{C}-\overset{\displaystyle }{\underset{\displaystyle O}{\text{CH}}}-(\text{CH}_2)_n\text{OH}$		
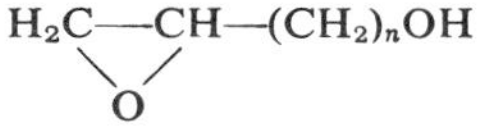 Tertiary amines, KOH, NaOCH_3, etc.	25	r

102

Glycidyl esters of aromatic acids	Amines, amides	—	s
2-Vinyl oxyethyl glycidyl ether	Maleic anhydride	—	t
1,2-Diisobutylene oxide	Dimethylhydrazine hydrochloride	—	u
Vinyl acetate–4-vinylcyclohexene monoepoxide copolymer	15% Solution of orthophosphonic acid	—	v
Triglycidyl derivative of p-aminophenol	4,4'-Methylenedianiline	75–185	w
Allyl glycidyl ether	Air, alkyl hydroperoxide	155	x
Bis[4-(1-methyl-1,2-epoxyethyl)-phenyl] ether	BF$_3$, PTSA	—	y

[a] Times given in parentheses indicate total time at a given temperature required to fully cure the epoxy resin.
[b] T. Ramos, U.S. Patent 2,894,920 (1959).
[c] S. R. Sandler and F. R. Berg, U.S. Patent 3,437,671 (1969).
[d] D. F. Loncrini, U.S. Patent 3,140,299 (1964).
[e] V. C. E. Burnop, J. Appl. Polym. Sci. 12, 699 (1968).
[f] A. Fukami and T. Moriwaki, Japan Kokai 73/88,198 (1973).
[g] J. D. B. Smith, U.S. Patent 3,784,583 (1974).
[h] R. G. Black, J. J. Seiwert, and J. B. Boylan, Plast. Technol. 10, 37 (1964).
[i] K. Cressy and J. Delmonte, 136th Am. Chem. Soc. Meet., Atlantic City, N.J., 1959 Paper 30, p. 100 (1959).
[j] S. O. Greenlee, U.S. Patent 2,521,912 (1950).
[k] W. R. Proops and G. W. Fowler, U.S. Patent 3,117,099 (1964).
[l] S. O. Greenlee, U.S. Patent 2,589,245 (1952).
[m] S. O. Greenlee, U.S. Patent 2,713,569 (1955).
[n] Olin Matheson Corp., Netherlands Patent Appl. 6,411,810 (1965).
[o] S. R. Sandler, J. Polym. Sci., Polym. Chem. Ed. 5, 1481 (1967).
[p] P. Schlack, U.S. Patent 2,131,120 (1938).
[q] L. V. Nozdrina, V. I. Midlin, and K. A. Andrianov, Russ. Chem. Rev. (Engl. Transl.) 42, 509 (1973).
[r] S. R. Sandler and F. R. Berg, J. Polym. Sci., Part A-1 4, 1253 (1966); U.S. Patent 3,425,960 (1969).
[s] S. R. Sandler, F. R. Berg, and G. Kitazawa, U.S. Patent 3,442,752 (1969).
[t] G. C. Murdoch and H. J. Schneider, U.S. Patent 3,000,690 (1961).
[u] B. Rudner, U.S. Patent 2,953,570 (1960).
[v] J. A. Robertson, U.S. Patent 2,687,404 (1954).
[w] N. H. Reinking, B. P. Barth, and F. J. Castner, U.S. Patent 2,951,825 (1960).
[x] T. W. Evans and E. C. Shokal, U.S. Patent 2,599,817 (1952).
[y] R. G. Neville and J. W. Mahoney, J. Appl. Polym. Sci. 11, 2029 (1967).

TABLE XI

CURING REACTIONS OF THE DIGLYCIDYL ETHER OF BISPHENOL A

Curing agents	Temp. (°C)	Ref.
Phthalic anhydride	120–170	a
Polyadipic acid anhydride	130	a
Coal tar pitch and diethylene triamine	—	b
Tall oil plus cobalt naphthenate	—	c
Organosiloxanes containing Si–OH groups, plus catalysts [ferric naphthenate, ferric bromide, zinc acetate, zinc benzoate, zinc stearate]	—	d
Triethylenetetramine plus molasses	—	e

a P. Castan, U.S. Patent 2,324,483 (1943).
b F. Whittier and R. J. Lawn, U.S. Patent 2,765,288 (1956).
c S. O. Greenlee, U.S. Patent 2,493,486 (1950).
d C. L. Frye and W. M. McLean, U.S. Patent 3,055,858 (1962).
e T. Ramos, U.S. Patent 2,894,920 (1959).

5-3. Curing (Polymerization) of 2-Vinyloxyethyl Glycidyl Ether with BF$_3$·Etherate [99]

$$CH_2\!-\!CH\!-\!CH_2OCH_2CH_2\!-\!O\!-\!CH\!=\!CH_2 \xrightarrow[\text{ether}]{BF_3} (-CH\!-\!CH_2\!-\!)_n$$

$$\underset{O}{\diagdown\diagup}$$

with the pendant structure:
$$\begin{array}{l} O \\ | \\ CH_2 \\ | \\ CH_2 \\ | \\ O\!-\!CH_2\!-\!(CH\!-\!CH_2O\!-\!)_n \end{array} \qquad (20)$$

To a 300 ml flask containing 7.8 gm diethyl ether is added 20 drops of 45% BF$_3$ etherate. The mixture is cooled in an ice bath and a solution of 2-vinyloxyethyl glycidyl ether (15.2 gm, 0.11 mole) in 31.2 gm diethyl ether is added dropwise with stirring. After the addition and 4 hr of stirring at 0°–5°C, the temperature is slowly raised to reflux (36°C). The resulting polymer is extracted by diethyl ether in a Soxhlet apparatus to yield 5.9 gm (dried in a vacuum oven to constant weight). The ether extracts are evaporated to give an oil which upon dissolving in benzene is reprecipitated by *n*-heptane to give 2.5 gm of a soluble polymer. Pyrolysis indicates that the latter contains

TABLE XII

Curing Resin 6[a] and Epikote 828 with Anhydride[b,c]

Epoxy resin	Curing agent	Parts curing agent to 100 parts resin (w/w)	Weight loss on curing[d] (%)	Cured resin				
				Color	Acetone-insol. (%)	Barcol hard. (20°C)	Barcol hard. (105°C)	M.p. (°C)
Epikote 828	Maleic anhydride	41	6.7	Yellow	0.8	38	29	360
Epikote 828	TPSA[e]	111	0.5	Yellow	5.6	25	Rubbery	360
Epikote 828	Phthalic anhydride	61	1.4	Yellow	6.0	45	Rubbery	360
Resin 6	Maleic anhydride	30	2.2	Brown	4.6	53	36	360
Resin 6	TPSA[e]	82	0.4	Dark brown	32.2	Shattered	Rubbery	Decomp., 300
Resin 6	Phthalic anhydride	45	0.3	Brown	17.1	47	35	Decomp., 300

[a] Di-β-naphthol epoxy resin prepared using a 10 molar ratio of epichlorohydrin:dinaphthol and a 2.1 molar ratio of NaOH:dinaphthol at a reaction temperature of 95°C for 1¼ hr to give a resin softening point of 55°C, mol. wt. 413, epoxide groups/100 gm = 0.38, and hydroxyl groups/100 gm = 0.10.

[b] In all cases 0.8 mole of curing agent per 1.0 mole of epoxy group was employed.

[c] Reprinted from V. C. E. Burnop, J. Appl. Polym. Sci. **12**, 699 (1968). Copyright 1968 by the Journal of Applied Polymer Science. Reprinted by permission of the copyright owner.

[d] Curing time 140 hr at 160°C.

[e] TPSA, tetrapropenyl succinic anhydride.

TABLE XIII

Curing agent type	Composition[c] (proportions by weight shown in parentheses)
Aromatic amines	Diglycidyl terephthalate + MDA (2.1:0.5)
	Diglycidyl 2,5-dichloroterephthalate + MDA (2.1:0.3)
	Diglycidyl hexahydroterephthalate + MDA (1.4:0.5)
	Diglycidyl isophthalate + MDA
	Diglycidyl phthalate + MDA (2.5:0.9)
	Triglycidyl trimesoate + MDA (3.6:0.6)
	Triglycidyl trimellitate + MDA (3.6:1.0)
	Tetraglycidyl pyromellitate + MDA (2.4:1.0)
	Diglycidyl diphenoate + MDA (3.0:0.6)
Aliphatic amines	Tetraglycidyl pyromellitate + triethylenetetramine (1.2:0.13 ml)
	Diglycidyl phthalate + triethylenetetramine (TETA) (1.7:0.13 ml)
	Triglycidyl trimellitate + triethylenetetramine (1.8:0.13 ml)
	Triglycidyl trimesoate + triethylenetetramine (1.3:0.13)
	Diglycidyl phthalate + diethylenetriamine (DETA) (1.4:0.15 ml)
	Triglycidyl trimellitate + diethylenetriamine (1.3:0.15 ml)
	Tetraglycidyl pyromellitate + diethylenetriamine (1.2:0.15 ml)
	Diglycidyl terephthalate + DETA (2.0:0.17 ml)
	Triglycidyl trimesoate + DETA (2.4:0.17 ml)
	Diglycidyl phthalate + DETA (2.2:0.17 ml)
	Diglycidyl terephthalate + TETA (2.0:0.28 ml)
	Triglycidyl trimesoate + TETA (2.4:0.14 ml)
	Triglycidyl trimesoate + TETA (2.2:0.10 ml)
Anhydrides	Diglycidyl phthalate + CPDA + BDA (1.4:2.1:0.06 ml)
	Diglycidyl isophthalate + CPDA + BDA (2.8:2.1:0.06 ml)
	Diglycidyl terephthalate + CPDA + BDA (1.7:1.0:0.03 ml)
	Diglycidyl hexahydroterephthalate + CPDA + BDA (0.7:1.1:0.04 ml)
	Diglycidyl 2,5-dichloroterephthalate + CPDA + BDA (1.2:1.0:0.03 ml)
	Triglycidyl trimellitate + CPDA + BDA (1.3:2.1:0.06 ml)
	Tetraglycidyl pyromellitate + CPDA + BDA (1.2:1.0:0.03 ml)
	Triglycidyl trimesoate + CPDA + BDA (1.3:2.1:0.06 ml)
	Diglycidyl terephthalate + pyromellitic dianhydride + BDA (1.7:1.1:0.03 ml)
	Diglycidyl terephthalate + CPDA + BDA (1.7:1.0:0.03 ml)
	Diglycidyl terephthalate + phthalic anhydride + BDA (2:1:0.04 ml)
	Diglycidyl terephthalate + trimellitic anhydride + BDA (1.7:1.3:0.03 ml)

[a] The samples were cured at 100°C for 2 hr and equilibrated 7 days prior to testing.

[b] Reprinted from S. R. Sandler, *J. Appl. Polym. Sci.* **11**, 465 (1967). Copyright 1967 by the Journal of Applied Polymer Science. Reprinted by permission of the copyright owner.

[c] BDA = benzyldimethylamine; MDA = methylenedianiline; CPDA = cyclopentane tetracarboxylic dianhydride.

TABLE XIV
CURING (POLYMERIZATION) OF VINYLALKYL GLYCIDYL ETHERS [99]

Vinylalkyl glycidyl ether	Curing agent or catalyst	Temperature (°C)
2-Vinylthioethyl glycidyl ether	2,2′-Azobisisobutyronitrile	80–100
2-Vinylthioethyl glycidyl ether + styrene	Benzoyl peroxide or 2,2′-azobisisobutyronitrile	80–100
2-Vinylthioethyl glycidyl ether + vinyl acetate	Benzoyl peroxide or 2,2′-azobisisobutyronitrile	80–100
2-Vinylthioethyl glycidyl ether + diethyl maleate	Benzoyl peroxide or 2,2′-azobisisobutyronitrile	80–100
2-Vinyloxyethyl glycidyl ether	Maleic anhydride, benzoyl peroxide	80–100
2-Vinyloxyethyl glycidyl ether + vinyl chloride	Ammonium persulfate	40–56
2-Vinylthioethyl glycidyl ether	Ammonium persulfate	45
n-Butyl acrylate	Sodium hydrosulfite, ferrous sulfate	—

TABLE XV
GEL-TIME DATA FOR PVC–GMA COPOLYMERS[a]

Curing agent	Gel time (min)	Curing conditions	
		Solvent	Temp. (°C)
H_2SO_4	1	THF	25
H_3PO_4	1	THF	25
$H_2NCH_2CH_2NH_2$	10–30	THF	100 (sealed tube)
$H_2N—NH—CH_2CH_2NH—CH_2CH_2NH_2$	10–30	THF	100 (sealed tube)
$HOCH_2CH_2OH$	1200	THF	100 (sealed tube)
$HO\overset{\overset{\displaystyle O}{\|\|}}{C}(CH_2)_4COOH$	10	None	160

[a] Data taken from Technical Bulletin on Nissan Blemmer G 7D-11-1000A. Nippon Oils & Fats Co., Ltd., Yurakucho, Chiyodaku, Tokyo, 1974.

no appreciable oxirane fraction. Additional examples of curing (polymerizations) of vinylalkyl glycidyl ethers are shown in Table XIV.

The glycidyl methacrylate (GMA) copolymers can be cured with conventional epoxy curing agents. Some representative gel-time data is shown in Table XV for a vinyl chloride–GMA copolymer.

6. MISCELLANEOUS METHODS

1. Propylene oxide via oxidation of propylene [100].
2. Molybdenum hexacarbonyl-catalyzed hydroperoxide oxidation of olefins to epoxides [101].
3. Triglycidyl isocyanurate synthesis using dispersed caustic [102].
4. The preparation of *trans*-stilbene oxide from *trans*-stilbene and peracetic acid [103].
5. The preparation of cholesteryl oxides using monoperphthalic acids [104].
6. Epoxy compounds are also made by elimination of HX from halo-hydrins of all types as shown in Eq. (21) [105].

$$(21)$$

7. Silicon [106] and fluoroglycidyl [107] resins have been described.
8. Ester interchange of glycidyl acetate with lower alkyl esters of carboxylic acids may be used to give novel glycidyl esters [108]. The diglycidyl ester of chlorendic acid is obtained by the ester interchange of glycidyl acetate with chlorendic acid [108].
9. Other hydroxy aromatics in addition to bisphenol A, such as phenol [109], cresols [110], and chlorinated compounds useful in flame-retardant applications [111], have been reported to be used to give epoxy compounds.
10. Flame-retardant phenolic polyglycidyl ether resin compositions [112].
11. Epoxidation of allyl-substituted phenols and ethers [113].
12. Curing of epoxidized liquid polybutadiene with ortho titanates [114].
13. Polymerization of vicinal epoxide groups with the aid of organic titanates [115].

14. Titanium esters as catalysts for the reaction of poly(carboxylic acid anhydrides) with polyepoxy compounds [116].

15. Triglycidyl isocyanurate resins [117].

16. Reaction of glycidyl acrylate or methacrylate with bisphenol A (catalyzed by dimethyl-*p*-toluidine) at 60°C [118].

17. Imidazole catalysts in the curing of epoxy resins [119].

18. Curing epoxy resins with phosphoramidates [120].

19. Brominated epoxy resins cured with halogenated anhydrides to give flame-retardant polymers [121].

20. Polyalkylenepolyamino imides as curing agents for epoxy resins [122].

21. Reaction of polyepoxides with finely divided red phosphorus [123].

22. Dibromophenol glycidyl ethers as flame-retardant additives for epoxy resins [124].

23. Polyhalogenated aromatic amine curing agents for epoxy resins that impart flame retardance [125].

24. Reaction product of poly(oxyalkylene amines) and a polyepoxide [126,127].

25. Reaction of fluoroalkyl trimellitic anhydride with epoxy resins to give water-repellent resins [128].

26. Reaction of epibromohydrin with castor oil to give a resin useful for flame-retardant polyurethane formulations [129].

27. Reactions of *O*-benzylbromoperoxycarbonic acid with olefins to give epoxides [130].

REFERENCES

1. S. R. Sandler and W. Karo, "Polymer Syntheses," Vol. 1. Academic Press, New York, 1974.

2. P. Schlack, U.S. Patent 2,131,120 (1938); German Patent 676,117 (1939).

3. P. Castan, U.S. Patent 2,324,483 (1943).

3a. P. Castan, Swiss Patent 211,116 (1940).

4. S. O. Greenlee, U.S. Patent 2,493,486 (1950).

5. F. Whittier and R. J. Lawn, U.S. Patent 2,765,288 (1956).

6. D. Swern, *Chem. Rev.* **45**, 1 (1949).

6a. T. W. Findley, D. Swern, and J. T. Scanlan, *J. Am. Chem. Soc.* **67**, 412 (1945).

7. B. Phillips, "Peracetic Acid and Derivatives." Bulletin P-58-0283. Union Carbide Chem. Co., New York, New York, 1958; Bulletin 4. Food Machinery and Chemical Corp., Becco Chem. Div., 1952 (revised 1957); R. J. Gall and F. P. Greenspan, *Ind. Eng. Chem.* **47**, 147 (1955); Bulletin P61-454. E. I. du Pont de Nemours & Co., Inc., 1954; Bulletin A6282. E. I. du Pont de Nemours & Co., Inc., 1955.

7a. F. P. Greenspan and R. J. Gall, *J. Am. Oil Chem. Assoc.* **33**, 391 (1956).

8. S. R. Sandler and W. Karo, "Organic Functional Group Preparations," Vol. 1, pp. 99–115. Academic Press, New York, 1968.

8a. H. Lee and K. Neville, "Epoxy Resins, Their Applications and Technology." McGraw-Hill, New York, 1957; "Handbook of Epoxy Resins." McGraw-Hill, New York, 1967; *Encycl. Polym. Sci. Technol.* 6, 209 (1967); I. Skeist, "Epoxy Resins." Van Nostrand-Reinhold, Princeton, New Yersey, 1958; M. W. Ranney, "Epoxy and Urethane Adhesives." Noyes Data Corp., New Jersey, 1971; G. R. Somerville and P. D. Jones, *Abstr. 168th Am. Chem. Soc. Meet., Atlantic City, N.J., 1974* Abstract ORPL, 146 in the Organic Coatings and Polymer Chemistry Division (1974).

9. W. F. Fallwell, *Chem. Eng. News* 51 (40), 8 (1973).

10. "Industrial Hygiene Bulletin on Handling Epon Resins and Auxiliary Chemicals," SC 62-33. Shell Chem. Corp., 1962.

11. D. W. Knoll, D. H. Nelson, and P. W. Keheres, *Pap., 134th Am. Chem. Soc. Meet., Chicago, 1958* Division of Paint, Plastics, and Printing Ink Chemistry, Paper No. 5 p. 20 (1958).

11a. B. Dobinson, W. Hoffmann, and B. P. Stark, "The Determination of Epoxide Groups." Pergamon, Oxford, 1969.

12. A. J. Durbetaki, *Anal. Chem.* 28, 2000 (1956).

13. R. R. Jay, *Anal. Chem.* 36, 667 (1964).

14. R. Dijkstra and E. A. M. F. Dahmen, *Anal. Chim. Acta* 31, 38 (1964).

15. A. J. Durbetaki, *Anal. Chem.* 29, 1666 (1957).

16. E. A. Eggers and J. S. Humphrey, Jr., *J Chromatogr.* 55, 33 (1971).

17. J. D. Zech, U.S. Patent 2,538,072 (1951).

18. P. Castan, U.S. Patent 2,444,333 (1948); J. M. Goppel, U.S. Patent 2,810,227 (1957).

19. L. H. Griffin and J. H. Long, U.S. Patent 2,848,435 (1958); H. L. Morson, U.S. Patent 2,921,049 (1960).

20. N. H. Reinking, U.S. Patent 2,943,096 (1960).

21. A. Farnham, U.S. Patent 2,943,095 (1960).

22. N. H. Reinking, B. P. Barth, and F. J. Castner, U.S. Patent 2,951,825 (1960).

23. M. St. Clair, U.S. Patent 2,842,849 (1959).

24. C. G. Schwarzen, U.S. Patent 2,806,016 (1957).

25. J. W. Clark and A. E. Winslow, U.S. Patent 2,897,163 (1959).

26. J. D. Garber, R. E. Jones, and H. C. Reynolds, U.S. Patent 3,025,307 (1962).

27. M. E. Chiddix and R. W. Wynn, U.S. Patent 2,951,829 (1960).

28. G. L. Brode and J. Wynstra, *Pap. 41, 148th Am. Chem. Soc. Meet., Chicago, 1964* p. 17 (1964).

29. S. R. Sandler and F. R. Berg, *J. Appl. Polym. Sci.* 9, 3707 (1965).

30. E. G. G. Weiner and E. Farenhorst, *Recl. Trav. Chim. Pays-Bas* 67, 438 (1948).

31. E. G. G. Weiner and E. Farenhorst, U.S. Patent 2,467,171 (1949).

32. S. R. Sandler and F. R. Berg, *J. Appl. Polym. Sci.* 9, 3708 (1965).

33. R. W. Fourie, K. B. Cofer, and D. M. Sheets, U.S. Patent 3,093,661 (1963).

34. A. Fairbourne, B. P. Gibson, and D. W. Stephens, *J. Chem. Soc.* p. 1965 (1932).

35. "Epichlorohydrin," Tech. Booklet SC 49-35. Shell Chem. Corp., 1949.

36. W. R. Sorenson and T. W. Campbell, "Preparative Methods of Polymer Chemistry," pp. 307–313. Wiley (Interscience), New York, 1961.

37. R. G. Kadesch, *J. Am. Chem. Soc.* 68, 44 (1946).

38. T. H. Rider and A. J. Hill, *J. Am. Chem. Soc.* 52, 1521 (1930).

39. G. Braun, *J. Am. Chem. Soc.* 52, 3185 (1930).

40. S. R. Sandler and F. R. Berg, *J. Chem. Eng. Data* 11, 447 (1966).

41. E. B. Kester and M. E. Lazar, *J. Org. Chem.* 8, 550 (1948).

42. Canadian Industries, British Patent 862,588 (1961).

43. Henkel & Cie, British Patent 735,001 (1955).
44. Henkel & Cie, French Patent 1,086,934 (1955).
45. W. H. M. Nieuwenhuis and P. Bruin (to N. V. De Bataafsche Petroleum Maatschappij), Netherlands Patent 86,753 (1961).
46. G. B. Payne and C. W. Smith, U.S. Patent 2,761,870 (1965).
47. B. Raecke and R. Kohler (to Henkel & Cie), U.S. Patent 3,073,804 (1963).
48. B. Raecke, R. Kohler, and A. Pietsch (to Henkel & Cie), U.S. Patent 2,865,897 (1958).
49. E. C. Shokal and A. S. Mueller (to Shell Development Co.), U.S. Patent 2,895,947 (1959).
50. T. J. Suen and G. Hewlett (to American Cyanamid Co.), U.S. Patent 2,801,232 (1957).
51. I. H. Updegraff (to American Cyanamid Co.), U.S. Patent 2,781,333 (1957).
52. S. R. Sandler, The Borden Chemical Co., Central Research Labs, Philadelphia, Pennsylvania, (unpublished data).
53. G. Maerker, J. Carmichael, and W. S. Port, *J. Appl. Polym. Sci.* **7**, 301 (1963).
54. G. Darzens, *C. R. Hebd. Seances Acad. Sci.* **203**, 1374 (1936).
55. M. S. Newman and B. J. Magerlein, *Org. React.* **5**, 443 (1952).
56. H. E. Zimmerman and H. L. Ahramjian, *J. Am. Chem. Soc.* **82**, 5459 (1960).
57. K. Freudenberg and W. Stoll, *Ann. Chem.* **440**, 41 (1924).
58. E. Bergmann and J. Hervey, *Ber. Dtsch. Chem. Ges. No. 1* **62**, 893 (1929).
59. M. S. Newman and B. J. Magerlein, *Org. React.* **5**, 419 (1952).
60. C. F. H. Allen and J. Van Allan, *Org. Synth., Collect. Vol. 3*, 727 (1955).
61. D. J. Pasto and C. C. Cumbo, *J. Org. Chem.* **30**, 1271 (1965).
62. D. Swern, *Encycl. Polym. Sci. Technol.* **6**, 83 (1967); *Chem. Rev.* **45**, 1 (1949); S. R. Sandler and W. Karo, "Organic Functional Group Preparations," Vol. 1, pp. 110–111. Academic Press, New York, 1968.
63. Hercules Powder Co., British Patent 892,361 (1962).
64. F. P. Greenspan, *High Polym.* **19**, 152–172 (1964); R. W. Rees, U.S. Patent 3,050,507 (1962); J. W. Pearce and J. Kana, *J. Am. Oil Chem. Soc.* **34**, 57 (1957); R. J. Gall and F. P. Greenspan, *ibid.* p. 161; W. Wood and J. Termini, *ibid.* **35**, 331 (1958); F. P. Greenspan and R. E. Light, Jr., Canadian Patent 560,690 (1958).
64a. A. F. Chadwick, D. O. Barlow, A. A. D'Addieco, and J. G. Wallace, *J. Am. Oil Chem. Soc.* **35**, 355 (1958).
65. W. D. Niederhauser and J. E. Koroly, U.S. Patent 2,485,160 (1949); H. Fukutani, M. Tokizawa, H. Okada, and N. Wakabayashi, German Offen. 2,143,071 (1972); *Chem. Abstr.* **77**, P20675S (1972).
66. T. W. Evans and E. C. Shokal, U.S. Patent 2,599,817 (1952).
67. Data taken from "Preparation of O-Type Polymer (III) Solution Copolymerization of Glycidyl Methacrylate," Tech. Bull. 6J12 1000A. Nippon Oil & Fats Co., Ltd., Tokyo, Japan 1974.
68. Data taken from Tech. Bull. 6J12 1000A. Nippon Oils & Fats Co., Ltd., Tokyo, Japan, 1974.
69. I. Uno, Y. Iwakura, M. Makita, and T. Ninomiya, *J. Polym. Sci., Polym. Chem. Ed.* **5**, 2311 (1967).
70. A. I. Goldberg, M. Skoultchi, and J. Fertig, U.S. Patent 3,162,676 (1964).
71. H. Lee and K. Neville, "Handbook of Epoxy Resins," Chapter 5. McGraw-Hill, New York, 1967.
72. "Epon Resins," Technical Brochure SC 71-1. Shell Chem. Corp., Houston, Texas, 1971.

73. L. Schechter and J. Synstra, *Ind. Eng. Chem.* **48**, 94 (1956).
74. R. F. Fischer, *Ind. Eng. Chem.* **52**, 321 (1960) (refer to Scheme I).
75. S. R. Sandler, *J. Polym. Sci.* **5** (A-1), 1481 (1967).
76. M. S. Malinovskii and M. K. Romantserich, *Sb. Statei Obshch. Khim.* **2**, 1366 (1953).
77. P. Bedos, *C. R. Hebd. Seances Acad. Sci.* **183**, 750 (1926).
78. S. O. Greenlee, U.S. Patent 2,521,912 (1950).
79. M. Tachimori and T. Matsushima, *J. Soc. Chem. Ind. Jpn.* **46**, 1270 (1943).
80. J. J. Carnes, U.S. Patent 2,520,381 (1950).
81. I. L. Knonyants and R. N. Sterlin, *Dokl. Akad. Nauk SSSR* **56**, 49 (1947).
82. R. Brunel, *Ann. Chim. Phys.* [8] **6**, 249 (1905).
83. C. E. Adams and B. H. Shoemaker, U.S. Patent 2,372,244 (1945).
84. S. Winstein and R. B. Henderson, *J. Am. Chem. Soc.* **65**, 2196 (1943).
85. M. Mousseron and R. Jacquier, *C. R. Hebd. Seances Acad. Sci.* **229**, 374 (1949).
86. E. M. Bellet and J. E. Casida, *Science* **182**, 1135 (1973); J. E. Casida, *Chem. & Eng. News* **52**, 56 (1974).
87. S. H. Langer, I. N. Elbling, A. B. Finestone, and W. R. Thomas, *J. Appl. Polym. Sci.* **5**, 370 (1961).
88. *Chem. Abstr.* **77** (1972); **78** (1973).
89. *Chem. Abstr., 6th Collect. Index* **51–55** (1957–1961).
90. *Chem. Abstr., 7th Collect Index.* **56–65** (1962–1966).
91. S. R. Sandler and F. R. Berg, U.S. Patent 3,437,671 (1969).
92. J. D. Zech, U.S. Patent 2,581,464 (1952); A. Pannell, U.S. Patent 3,058,921 (1962); R. E. Burge and J. R. Hallstrom, *SPE J.* **20**, 75 (1964).
93. D. A. Rogers, Jr. and D. W. Lewis, U.S. Patent 2,883,395 (1959); R. W. Wegler and G. Frank, U.S. Patent 2,884,406 (1959); A. Karlheinz, R. W. Wegler, and G. Frank, U.S. Patent 2,921,037 (1960).
94. H. L. Bender, A. G. Farnham, and J. W. Guyer, U.S. Patent 2,731,437 (1956).
95. S. O. Greenlee, U.S. Patent 2,698,315 (1954).
96. H. Newey and E. C. Shokal, U.S. Patent 2,575,558 (1951); S. O. Greenlee, U.S. Patent 2,694,694 (1954).
97. A. Damusis, *Pap., 130th Am. Chem. Soc. Meet., Atlantic City, N.J., 1956* Paper 24, Polymer Div.
98. J. F. Harrod, *J. Polym. Sci., Part A* **1**, 385 (1963).
99. G. C. Murdoch and H. J. Schneider, U.A. Patent 3,000,690 (1961).
100. *Chem. & Eng. News* **43**, 40 (1965).
101. M. N. Sheng and J. W. Zajacek, *Abstr. 3rd Middle Atlantic Reg. ACS Meet., Philadelphia, 1968* p. 60H (1968).
102. H. P. Price and G. E. Schroll, *J. Org. Chem.* **32**, 2005 (1967).
103. D. J. Reif and H. O. House, *Org. Synth., Collect. Vol.* **4**, 860 (1963).
104. P. N. Chakravorty and R. H. Levin, *J. Am. Chem. Soc.* **64**, 2317 (1942).
105. R. G. Neville and J. W. Mahoney, *J. Polym. Sci.* **11**, 2029 (1967).
106. D. A. Rogers, Jr. and D. W. Lewis, U.S. Patent 2,883,395 (1959).
107. A. Khlakyan, *Izv. Akad. Nauk SSSR, Ser. Khim.* **1**, 72–75 (1965).
108. H. Newey and D. Holm, U.S. Patent 3,106,536 (1963).
109. D. D. Applegath, *SPE J.* **15**, 75 (1959).
110. T. F. Bradley and H. A. Newey, U.S. Patent 2,716,099 (1955).
111. G. F. D'Alelio, U. S. Patent 2,658,884 (1953).
112. C. G. Schwarzer, U.S. Patent 3,385,908 (1968).
113. S. A. Harrison and D. Aelony, *J. Org. Chem.* **27**, 3311 (1962).
114. C. E. Wheelock and J. E. Wicklatz, U.S. Patent 2,946,756 (1960).

115. R. R. Durst and W. O. Phillips, U.S. Patent 3,312,637 (1967).
116. F. Kugler, O. Ernst, P. Ruf, and W. Seiz, U.S. Patent 3,385,835 (1968).
117. H. V. Freyhold and V. Wehle, U.S. Patent 3,505,262 (1970).
118. R. L. Bowen, U.S. Patent 3,179,623 (1965).
119. A. Farkas and P. F. Strohm, *J. Appl. Polym. Sci.* **12**, 159 (1968).
120. R. R. Hindersiner, U.S. Patent 3,645,971 (1972).
121. J. A. Harrington, U.S. Patent 3,378,434 (1968).
122. E. M. Geiser, U.S. Patent 3,371,099 (1968).
123. T. J. Dijkstra and E. J. W. Vogelzang, U.S. Patent 3,477,982 (1969).
124. K. Jellinek, U. Post, D. Gerlach, and R. Oellig, U.S. Patent 3,775,355 (1973).
125. L. Sobel, L. Parri, and A. Iscard, U.S. Patent 3,707,525 (1972).
126. J. E. Pretka, U.S. Patent 3,175,987 (1965).
127. H. A. Anthes, U.S. Patent 2,982,751 (1961).
128. J. E. Quick, U.S. Patent 3,711, 514 (1973).
129. G. A. Roth, U.S. Patent 3,732,265 (1973).
130. R. M. Coates and J. W. Williams, *J. Org. Chem.* **39**, 3054 (1974).

Chapter 4

SILICONE RESINS
(POLYORGANOSILOXANES OR SILICONES)

I. INTRODUCTION

This chapter describes the topic of silicone resins or polyorganosiloxanes.

$$\left[-\overset{\displaystyle R}{\underset{\displaystyle R}{Si}}-O- \right]_n$$

Other organosilicon monomers will also be mentioned and described where appropriate. The major preparative routes to the organosilicon starting materials involve either the direct route [Eq. (1)] or the indirect route [Eq. (2)]. These starting materials with the proper substituents may be used to give silicone resins as described below.

$$2RCl + Si \xrightarrow{\text{catalyst}} R_2SiCl_2, \text{ etc.} \qquad (1)$$

$$2RM + SiCl_4 \longrightarrow R_2SiCl_2 + 2MCl \qquad (2)$$

$$M = Li, Na, MgX, \text{ etc.}$$

One of the earliest works (1865–1880) involved the reaction of dialkylzinc with silicon tetrachloride [Eq. (3)] [1].

$$Zn + RCl \longrightarrow 2ZnR_2 + SiCl_4 \longrightarrow Si(R)_4 + 2ZnCl_2 \qquad (3)$$

In 1945, Hurd reported the vapor-phase alkylation of silicon halides by the use of zinc and alkyl halides [2].

Much of the rapid development of the polymeric organosiloxanes or silicone resins is credited to Rochow's direct process [3] of preparing dialkyl- or diarylldichlorosilanes as shown in Eqs. (1), (4), and (6).

$$SiO_2 + 2C \longrightarrow Si + 2CO \qquad (4)$$

$$CH_3OH + HCl \longrightarrow CH_3Cl + H_2O \qquad (5)$$

$$2CH_3Cl + Si \xrightarrow{\text{copper catalyst}} (CH_3)_2SiCl_2 \qquad (6)$$

$$\xrightarrow[H_2O]{} [(CH_3)_2SiO]_n + 2HCl$$

In addition, the efforts during World War II resulted in rapid growth in silicone resins which led to the establishment of the Dow Corning Corporation of Midland, Michigan in 1943. General Electric Company established its own silicon department in Waterford, New York, in 1947, followed by Union Carbide and Stauffer. Within the next few years came major developments in Germany, Russia, Great Britain, and many other countries, which are not outlined here for the sake of brevity (see reference [4] for further details).

Some other processes involve the reaction of unsaturated compounds [5–7] ethylene oxide, or carbon monoxide with silicon tetrachloride [Eqs. (7)–(10)].

$$CH{\equiv}CH + SiCl_4 \longrightarrow Cl_3Si{-}CH{=}CHCl \qquad (7)$$

$$CH_2{=}CH_2 + SiCl_4 \longrightarrow Cl_3Si{-}CH_2CH_2Cl \qquad (8)$$

$$\underset{\underset{O}{\diagdown\diagup}}{CH_2{-}CH_2} + SiCl_4 \longrightarrow Cl_3SiOCH_2CH_2Cl \qquad (9)$$

$$CO + SiCl_4 \longrightarrow Cl_3Si\overset{O}{\overset{\|}{C}}{-}Cl \qquad (10)$$

Nucleophilic substitution of the Si—X bond affords a wide variety of derivatives [Eq. (11)] [8].

$$\text{$>$Si—X} + \text{N}^- \longrightarrow \text{$>$Si—N} + \text{X}^- \tag{11}$$

The synthesis of polymeric organosiloxanes (silicone resins) was first reported in a series of classic papers by Professor Kipping and his co-workers at the University College, Nottingham, England [9]. Kipping and his students neglected the organosilicon polymers (characterized by Kipping as oily or gluelike) and were mainly concerned with crystalline or pure liquid organic derivatives of silicon.

Silicones with the empirical structure R_2SiO were at first erroneously considered to be analogous to ketones and were obtained by the hydrolysis of disubstituted silicon chlorides [Eq. (12)] [10].

$$\text{Cl—}\overset{\overset{\displaystyle R}{|}}{\underset{\underset{\displaystyle R}{|}}{Si}}\text{—Cl} \xrightarrow{\text{H}_2\text{O}} \left[\overset{\overset{\displaystyle R}{|}}{\underset{\underset{\displaystyle R}{|}}{Si}}\text{—O—} \right]_n + 2\text{HCl} \tag{12}$$

Kipping's pioneering work may be summarized by the series of reactions shown in Eq. (13).

$$\text{SiCl}_4 + 2\text{ArMgX} \longrightarrow \text{Ar}_2\text{SiCl}_2 \xrightarrow{\text{H}_2\text{O}} \text{Ar}_2\text{Si(OH)}_2 \tag{13}$$

$$\text{Ar}_2\text{Si}{=}\text{O}$$

$$\text{Ar}_2\text{Si(—OSi—)}_n\text{Cl} \quad\quad \left[\overset{\overset{\displaystyle Ar}{|}}{\underset{\underset{\displaystyle Ar}{|}}{Si}}\text{—O—} \right]_n$$

Patnode [11] in 1946 exploited the preparation of dimethylsilicones from dimethyldichlorosilane with or without the cohydrolysis with trimethylchlorosilane to give complex polymers and elastomers. If insufficient water is used then α,ω-dichlorosilanes are obtained [Eqs. (14) and (15)] [12].

$$(\text{CH}_3)_2\text{SiCl}_2 \longrightarrow \text{HO[Si(CH}_3)_2\text{O]}_n\text{Si(CH}_3)_2\text{OH} \tag{14}$$

$$\longrightarrow \text{Cl[Si(CH}_3)_2\text{O]Si(CH}_3)_2\text{Cl}$$

$$(\text{CH}_3)_2\text{SiCl}_2 + (\text{CH}_3)_3\text{SiCl} \longrightarrow \text{CH}_3\text{[(CH}_3)_2\text{SiO]}_n\text{Si(CH}_3)_3 \tag{15}$$

In the case of the methylsilicone resins cross-linking is required in order to attain resinous properties, and hence the R (alkyl):Si ratio is always less than 2. These resins are still called silicone resins since they are derived from the original R_2SiO composition. Available preparative methods can involve:

    1. Hydrolysis of dimethyldichlorosilane and subsequent catalytic oxidation with air to give the desired CH_3:Si ratio [13].

    2. Cocondensation of the hydrolysis products dimethyldichlorosilane with methyltrichlorosilane or silicon tetrachloride [14].

    3. Hydrolysis of the partially methylated silicon tetrachloride (use of CH_3MgCl with $SiCl_4$ to give the desired CH_3:Si ratio) [15].

Depending on the method of preparation and the reactants used the properties will vary considerably.

Methylsilicone rubbers are prepared by curing inorganic-filled dimethylsilicone on a rubber mill to give an insoluble infusible material [16]. Ethylsilicone compound was first reported by Ladenburg [17] and correctly characterized by Kipping [18] as a high molecular weight substance (polymer). Other silicone resins are prepared by similar methods [19]. In contrast, diphenyldichlorosilane on hydrolysis does not yield a polymer but gives diphenylsilanediol, $(C_6H_5)_2Si(OH)_2$, m.p. 148°C [19], or hexaphenylcyclotrisiloxane if acid is present $[(C_6H_5)_2SiO]_3$. The hydrolysis of phenyltrichlorosilane yields a brittle, fusible resin. The cocondensation of the hydrolysis products of phenyltrichlorosilane and diphenyldichlorosilane affords useful phenylsilicone resins. The latter can be chlorinated to improve oxidative resistance.

Alkylsilicones of high R:Si ratios are oily liquids or weak gels when polymerized alone; ratios of 1.5 for a methylsilicone and 1.0 for a butyl- or benzylsilicone are desirable to yield satisfactory resinous and adhesive properties. Since aryl silicones of high R:Si ratio are too brittle it has been found that an alkyl–aryl combination [derived from $R(C_6H_5)SiCl_2$] gives satisfactory results when R:Si is approximately 1.8 [20]. Sometimes similar results are obtained by cocondensation of mixed ethyl- and phenylchlorosilanes [21].

Polyorganosiloxanes find use as adhesives, resins, oils, lubricants, and rubbers.

Silicone oils have the advantage of oxidation stability, low freezing point, and a small temperature coefficient of viscosity [22]. The silicone resins are similarly characterized by chemical inertness, good thermal stability (to 550°C), and resistance to oxidation [15]. Methylsilicone resins are excellent electrical insulators [23]. Methylchlorosilane vapor reacts with many surfaces to render them water repellent [24].

The IUPAC nomenclature rules for polymers and definite silicone compounds are shown below with the common names in brackets [25].

$(C_2H_5)_2Si(OH)_2$ Diethylsilanediol

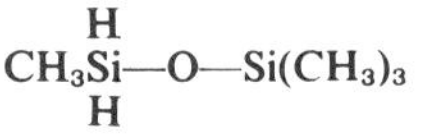

$$CH_3\overset{H}{\underset{H}{Si}}\!-\!O\!-\!Si(CH_3)_3$$ 1,3,3,3-Tetramethyldisiloxane

CCl_3SiH Trichlorosilane [silicochloroform]

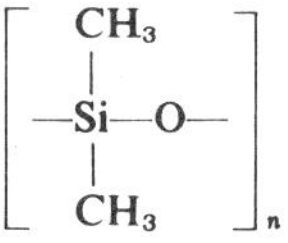

$$\left[\begin{array}{c} CH_3 \\ | \\ -Si\!-\!O- \\ | \\ CH_3 \end{array}\right]_n$$ Poly(dimethylsiloxane)

$$\left[\begin{array}{c} CH_3 \\ | \\ -Si\!-\!O- \\ | \\ C_6H_5 \end{array}\right]$$ Poly(methylphenylsiloxane)

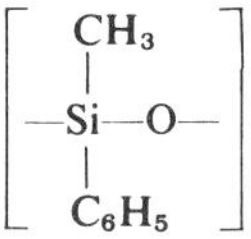

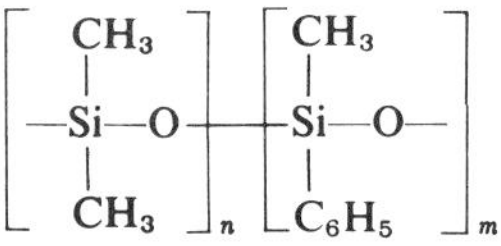

$$\left[\begin{array}{c} CH_3 \\ | \\ -Si\!-\!O \\ | \\ CH_3 \end{array}\right]_n \left[\begin{array}{c} CH_3 \\ | \\ Si\!-\!O- \\ | \\ C_6H_5 \end{array}\right]_m$$ Poly(dimethylsiloxane-ω-methylphenylsiloxane)

$$\begin{array}{ccc} (CH_3)_2Si\!-\!O\!-\!Si(CH_3)_2 \\ |\qquad\quad\; | \\ O\qquad\quad O \\ |\qquad\quad\; | \\ (CH_3)_2Si\!-\!O\!-\!Si(CH_3)_2 \end{array}$$ 1,1,3,3,5,5,7,7-Octamethylcyclotetrasiloxane

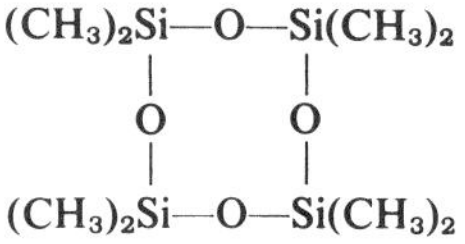

A shorthand abbreviation earlier introduced by Hurd [26] and Patnode and Wilcock [27] is sometimes used to denote some general types of organo-siloxanes and is shown in Table I.

An explanation * of the units used in the abbreviated formulas shown in Table I is given below; note that half of each attached oxygen is included in each unit.

M, the monofunctional unit

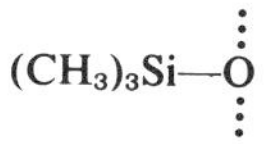

$$(CH_3)_3Si\!-\!\vdots\!O\!\vdots$$

D, the difunctional unit

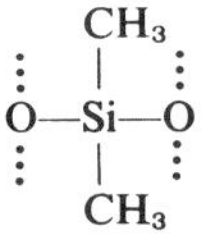

$$\begin{array}{c} CH_3 \\ \vdots\quad | \quad\vdots \\ O\!-\!Si\!-\!O \\ \vdots\quad | \quad\vdots \\ CH_3 \end{array}$$

* This explanation is reprinted from C. B. Hurd, *J. Am. Chem. Soc.* **68**, 364 (1946). Copyright 1946 by the American Chemical Society. Reprinted by permission of the copyright owner.

TABLE I

ABBREVIATIONS OF NAMES OF SILOXANES[a]

Code name	Name of compound
MM	Hexamethyldisiloxane
MDM	Octamethyltrisiloxane
MD_2M	Decamethyltetrasiloxane
MD_3M	Dodecamethylpentasiloxane
MD_4M	Tetradecamethylhexasiloxane
D_4	Octamethylcyclotetrasiloxane
D_5	Decamethylcyclopentasiloxane
D_6	Dodecamethylcyclohexasiloxane
D_7	Tetradecamethylcycloheptasiloxane
MD*M	Hexamethyl-3,3-diphenyltrisiloxane
TM_3	Methyltri(trimethylsiloxy)silane or heptamethyl-3-trimethyl-siloxytrisiloxane

[a] Reprinted in part from C. B. Hurd, *J. Am. Chem. Soc.* **68**, 364 (1946). Copyright 1946 by the American Chemical Society. Reprinted by by permission of the copyright owner.

T, the trifunctional unit

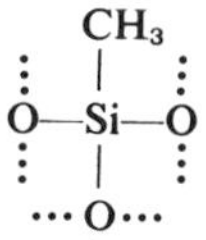

$$\begin{array}{c} CH_3 \\ | \\ O\!-\!Si\!-\!O \\ | \\ \cdots O \cdots \end{array}$$

D*, the difunctional unit including two C_6H_5 radicals

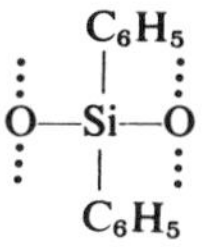

$$\begin{array}{c} C_6H_5 \\ | \\ O\!-\!Si\!-\!O \\ | \\ C_6H_5 \end{array}$$

Thus, decamethyltetrasiloxane

$$\begin{array}{c} \quad\quad CH_3 \quad CH_3 \\ \quad\quad | \quad\quad | \\ (CH_3)_3\!-\!Si\!-\!O\!-\!Si\!-\!O\!-\!Si\!-\!O\!-\!Si(CH_3)_3 \\ \quad\quad | \quad\quad | \\ \quad\quad CH_3 \quad CH_3 \end{array}$$

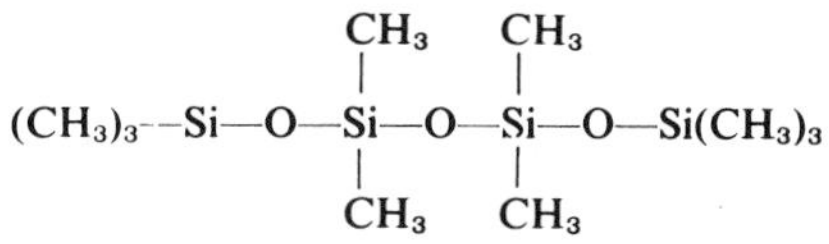

is written MDDM or MD_2M.

Other examples are MD_2M or decamethyltetrasiloxane

$$(CH_3)_3Si\text{—}O\text{—}Si(CH_3)_2\text{—}O\text{—}Si(CH_3)_2\text{—}O\text{—}Si(CH_3)_3$$

and TM_3, or 1,1,1,3,5,5,5-heptamethyl-3-trimethylsiloxytrisiloxane

$$CH_3\text{—}Si[O\text{—}Si(CH_3)_3]_3$$

Several reviews [28–30] and monographs [31] should be consulted for additional details on history and commercial developments.

2. POLYORGANOSILOXANES

The hydrolysis of organochlorosilanes or organoacetoxysilanes by water can be accelerated by small amounts of acids or alkalis. Preferred acid catalysts are those that are easily removed by washing with water such as HCl, HCOOH, $(COOH)_2$, Cl_3CCOOH, and CH_2COOH [32].

$$X\text{—}\underset{\underset{R}{|}}{\overset{\overset{R}{|}}{Si}}\text{—}X + nH_2O \longrightarrow \left[\text{—}\underset{\underset{R}{|}}{\overset{\overset{R}{|}}{Si}}\text{—}O\text{—} \right]_n + 2nHX \qquad (16)$$

For the case of dimethyldichlorosilane the following hydrolysis conditions favor high polymer formation:

1. Salt hydrates [33]
2. Strong hydrochloric acid [27]
3. 50–85% Sulfuric acid [34]
4. Alkaline conditions [27]
5. Alcohol and carboxylic acid [35]

Conditions that form low molecular weight linear or cyclic polysiloxanes are the following:

1. Water [27]
2. Water and ammonia, pH 6.5–8.5 [36]
3. Water and diethyl ether [27]
4. Deficiency of water, dioxane, and ether [37]

The molecular weight distribution of the products has been determined using gel permeation chromatography [38].

Other functional silanes also can be hydrolyzed or reacted by nonhydrolytic techniques to give polyorganosiloxanes. For example, the reactions of chlorosilanes or alkoxy silanes with alcohols, esters, or hydroxyl groups are examples of the nonhydrolytic process [Eq. (17)] [39,39a].

$$\ce{>Si-X + ^- O-Si< \longrightarrow >Si-O-Si<} \tag{17}$$

X = Cl, NH$_2$, OAc, OR, H

In a fashion similar to ring-opening reactions of epoxides [40], 1,2-siloxacycloalkanes are polymerized in the presence of Lewis acids or bases [Eq. (18)] [41]. Zero-order kinetics and an activation energy of about 12 kcal/mole are typical for the 1,2-siloxacycloalkane polymerization.

$$\ce{R2Si-O(CH2)_nCl + 2Li ->[{-2LiCl}] R2Si-(CH2)_n ->[cata-][lyst] [-Si(R)(R)-O(CH2)_nO-]_n} \tag{18}$$

A. Silicone Fluids and Linear Organopolysiloxanes

The silicone fluids are less complex structurally than the cross-linked or elastomeric resins and possess a lower molecular weight. They possess different properties depending on the carbon functional groups.

Highly purified dialkyldichlorosilanes (distillation is required to remove traces of alkyltrichlorosilane) are required in order to keep the viscosity of the polymeric fluids constant. The simplest preparation involves a batch process in which a given amount of dialkyldichlorosilane is fed into 22% aqueous hydrochloric acid with stirring to give the polydialkylsiloxane and 32% aqueous hydrochloric acid [42,42a]. The product is separated from the hydrochloric acid and is then washed with water or with sodium bicarbonate solution. The product is dried by passing through a column of fuller's earth or other absorbent medium. Chain stoppers are used to stabilize the product and are reacted with the hydrolyzate at temperatures of 20°–200°C for several hours in the presence of an acid or basic catalyst. The chain stoppers are typically disiloxanes obtained from trialkylchlorosilane [43,44].

$$\ce{HO-[Si(CH3)(CH3)-O]_n-H + (CH3)3Si-O-Si(CH3)3 ->[H^+ or OH^-][\Delta]}$$

$$\ce{(CH3)3-Si-O-[Si(CH3)(CH3)-O]_n-Si(CH3)3 + H2O} \tag{19}$$

The polymers have a viscosity range from 0.5 cP for hexamethyldisiloxane to several million centipoise for gums used in silicone rubber manufacture. Molecular weight–viscosity relationships have been described earlier [43]. Copolymers have also been described using modified synthetic approaches [45,45a].

Some typical preparations using a variety of starting materials and conditions are shown in Table II. Typical preparations are given in more detail as described below.

Silicon fluid manufacture has been described by Gutoff [42a] (see Figs. 1 and 2). Batch (Fig. 1) and continuous (Figs. 2 and 3) process flow diagrams are also shown. Another continuous silicone oil process was described by Kirk and is shown in Fig. 3.

Kirk [46] described the use of tetrabutylphosphonium hydroxide which brings about rapid rearrangement and polymerization of siloxanes at temperatures up to 110°C. After the polymerization is completed the phosphonium catalyst is decomposed by heating the polymer above 130°C, as shown in Eq. (20).

$$(n\text{-}C_4H_9)_4POH \xrightarrow[\Delta]{130°-170°C} n\text{-}C_4H_{10} + (n\text{-}C_4H_9)_3P{=}O \tag{20}$$

The polymers made by this process are clear, colorless, and stable, and may have viscosities up to 30 million cP. The continuous process shown in Fig. 1 uses this catalytic process.

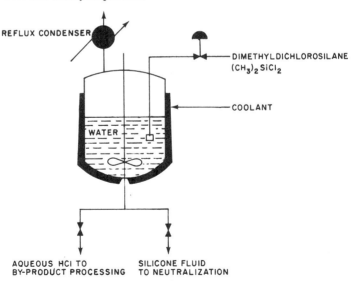

REFLUX CONDENSER

DIMETHYLDICHLOROSILANE
$(CH_3)_2 SiCl_2$

COOLANT

WATER

AQUEOUS HCl TO
BY-PRODUCT PROCESSING

SILICONE FLUID
TO NEUTRALIZATION

FIG. 1. Batch process for silicone fluid manufacture. [Reprinted from R. Gutoff, *J. Ind. Eng. Chem.* **49**, 1807 (1957). Copyright 1957 by the American Chemical Society. Reprinted by permission of the copyright owner.]

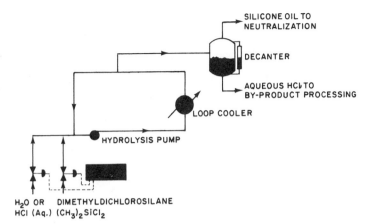

FIG. 2. Continuous process for silicone fluid manufacture. [Reprinted from R. Gutoff, *J. Ind. Eng. Chem.* **49**, 1807 (1957). Copyright 1957 by the American Chemical Society. Reprinted by permission of the copyright owner.]

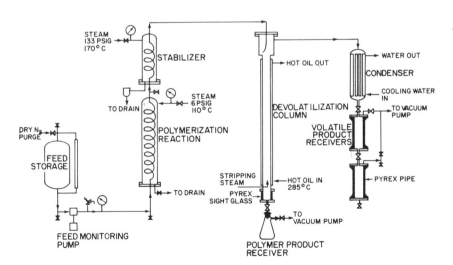

FIG. 3. A coil-type heat exchanger was used in the continuous process for silicone oils. [Reprinted from N. Kirk, *J. Ind. Eng. Chem.* **51**, 515 (1959). Copyright 1959 by the American Chemical Society. Reprinted by permission of the copyright owner.]

TABLE II

CONDENSATION POLYMERIZATION OF ALKYLHALOSILANES AND RELATED SILANES TO ALKYLPOLYSILOXANES (SILICONES)

Silane derivatives	Concentration	Water (gm)	Solvent	Catalyst	Temp. (°C)	Time (hr)	% Yield of alkyl-polysiloxane	B.p. or m.p. (°C; mm Hg)	Other characteristics	Ref
$(CH_3)_2SiCl_2$	40 ml	120	—	—	15–20	—	Trimer, 0.5 Tetramer, 92.0 Pentamer, 6.7 Hexamer, 1.6 Polymeric residue, 49.2	134; 20 175; 20 210; 20 245; 20	— — — —	a a a a b
$(CH_3)_3Si-O-Si-(CH_3)_3[(CH_3)_2SiO]_4$	1.0 moles	—	—	10 ml H_2SO_4 (conc.)	25	4	—	153–229; 20	—	a
$(CH_3)_2SiCl_2$	26.2 moles	2650	—	—	10–15	$\frac{1}{2}$–1	—	—	—	c
$CH_3Si-SiCl_3$	3.8 moles	2650	—	—	10–15	$\frac{1}{2}$–1	—	—	—	c
$(CH_3)_2Si(OC_2H_5)_2$	9.41 moles	254	—	75 gm NaOH	40–65	$1\frac{1}{2}$	98	—	—	d
$(CH_3)_3SiOC_2H_5$	9.41 moles	50	—	75 gm NaOH	Reflux 100°	2	98	—	—	d
$(CH_3)_2Si(OC_2H_5)_2$	1.95 moles	36	—	0.2 gm NaOH	30–60	1–2	83	—	Viscosity = 81 cP at 25°C	d
$(CH_3)_3SiOC_2H_5$	9.95 moles	36	—	0.2 gm NaOH	60–100	1–2	83	—	Viscosity = 81 cP at 25°C	d
OH OH $(CH_3)_2Si-O-Si(CH_3)_2$	0.10 moles	—	100 ml C_6H_6	1.0 gm PTSA	80	$3\frac{1}{2}$	100	—	—	e
$(CH_3)_3SiCl$	6.4 moles	—	1200 gm toluene	—	25	1–2	—	—	Viscous translucent liquid	f
CH_3SiCl_3	6.4 moles	—	1200 gm butanol	—	25	1–2	—	—	Viscous translucent liquid	f
$(CH_3)_2SiCl_2$	11.0 moles	—	1200 gm toluene	—	25	1–2	—	—	Viscous translucent liquid	f
$SiCl_4$	2.2 moles	—	1200 gm butanol	—	25	1–2	—	—	Viscous translucent liquid	f
p-Phenylenebis(dimethylsilanol)	50 gm	—	35 ml C_6H_6	0.1 gm n-Hexyl-ammonium 2-ethyl-hexoate	80	6	100	—	Intrinsic viscosity = 2.0 cP (toluene, 25°C)	g

$C_6H_5Si(OC_2H_5)_3$	0.5 moles	1.5 moles MIBK, 50 gm Air or aq. HCl	0.51 mole $(C_2H_5)_4$NOH	25	192	14	—	Intrinsic viscosity dl/gm = 0.14 Av. mol. wt. = 1310	h
$(C_6H_5)(C_2H_5)$—Si—$(OC_2H_5)_2$	—	—	100 ppm KOH, 0.5% DMSO	190	24	—	—	—	i
$[(CH_3)_4SiO]_4$	—	—	—	95	—	—	—	—	j
CH₃ | $C_{10}H_{21}$—Si—Cl | CH₃	1.0 moles	402 gm toluene	60	—	—	56	—	—	m
CH₃—Si—Cl₂ | H	3.25 moles	402 gm toluene	60	—	—	56	—	—	m
$[(CH_3)_2Si$—$O]_4$	1000 gm	—	0.14 gm KOH-isopropanol complex	165	½	100	—	Intrinsic viscosity 1.5% (mol. wt. = 804,600)	k
	—	—	—	150	3½	100	—	Intrinsic viscosity 1.5% (mol. wt. = 804,600)	k
CH₃—SiCl₃	2990 gm	98 gm acetone	—	55	2	95	m.p. 70	—	l
	720	4048 gm toluene	—	55	2	95	m.p. 70	—	l

[a] W. Patnode and D. F. Wilcock, *J. Am. Chem. Soc.* **68**, 358 (1946).

[b] The residue can be pyrolyzed at 350°–450°C to give 82% of the tetramer.

[c] D. F. Wilcock, *J. Am. Chem. Soc.* **69**, 477 (1947).

[d] M. J. Hunter, E. L. Warrick, J. F. Hyde, and C. C. Currie, *J. Am. Chem. Soc.* **68**, 2289 (1946).

[e] G. R. Lucas and R. W. Martin, *J. Am. Chem. Soc.* **74**, 5225 (1952).

[f] D. W. Scott, *J. Am. Chem. Soc.* **68**, 356 (1946).

[g] R. L. Merker and M. J. Scott, *J. Polym. Sci., Part A* **2**, 15 (1964).

[h] J. F. Brown, Jr., L. H. Vogt, Jr., and P. I. Prescott, *J. Am. Chem. Soc.* **86**, 1120 (1964).

[i] J. F. Hyde and R. C. DeLong, *J. Am. Chem. Soc.* **63**, 1194 (1941).

[j] G. D. Cooper, *J. Polym. Sci., Part A-1* **4**, 603 (1966).

[k] E. L. Warrick, U.S. Patent 2,634,252 (1953).

[l] Société Industrielle des Silicones, Netherlands Patent Appl. 6,609,046 (1966).

[m] A. L. Culpepper, U.S. Patent 3,532,730 (1970).

2-1. Preparation of a Copolymer of Organosiloxane (Silicone) by the Condensation of Dichlorosilanes [47]

$$
\begin{array}{c}
\underset{\underset{CH_3}{|}}{\overset{\overset{C_2H_4C_6H_5}{|}}{Cl-Si-Cl}} \;+\; \underset{\underset{CH_3}{|}}{\overset{\overset{CH_3}{|}}{Cl-Si-Cl}} \;\xrightarrow[H_2O]{(CH_3)_3-SiCl}
\end{array}
$$

$$
(CH_3)_3SiO\left[\underset{\underset{CH_3}{|}}{\overset{\overset{C_2H_4-C_6H_5}{|}}{Si-O}}\right]_x\left[\underset{\underset{CH_3}{|}}{\overset{\overset{CH_3}{|}}{Si-O}}\right]_y Si(CH_3)_3
$$

(Phenylethyl)methyldichlorosilane (220 gm, 1.0 mole), 258 gm (2.0 moles) of dimethyldichlorosilane, and 32.2 gm (0.30 mole) of trimethylchlorosilane are mixed and added over a period of 75 min to 85 gm (4.7 moles) of water with vigorous stirring. During hydrolysis the temperature is kept below 0°C. The water is later removed by stripping at 150°C over a period of 225 min. The oily product is cooled to 110°C and 3.4 gm (0.05 mole) of 87% potassium hydroxide is added. The oil is cooled and then acidified with hydrochloric acid. Excess sodium bicarbonate is added to neutralize the hydrochloric acid. The product oil was separated by filtration from the solids. The resulting oil is sparged for 2 hr at 200°C with a nitrogen gas stream at a flow rate of 5 liters/min. The final oil product has a viscosity of 408 centistokes at 25°C and a 2:1 ratio of dimethylsiloxy groups to (phenylethyl)methylsiloxy groups.

Other preparations where the ratio of groups is varied are also given by Paten [47].

2-2. Polydimethylsiloxane by the Polymerization of Octamethylcyclotetrasiloxane [48]

$$
[(CH_3)_2Si-O-]_4 \;\xrightarrow{\text{base}}\; [-(CH_3)_2Si-O-]_n \tag{21}
$$

To a resin flask equipped with a thermometer, stirrer, and reflux condenser is added 1000 gm of octamethylcyclotetrasiloxane. The siloxane is heated to 165°C and then 0.14 gm of a potassium hydroxide–isopropanol complex (neutral equivalent = 193.5) is added to give a Si:K ratio of 4470:1. In 25 min the stirrer begins to stall and the polymer is now heated for $3\frac{1}{2}$ hr at 150°C to complete polymerization. The resulting polymer has an intrinsic viscosity of 1.57 dl/gm corresponding to a molecular weight of 804,600.

2-3. DMSO–Base-Catalyzed Bulk Polymerization of (D₄) Octamethylcyclotetrasiloxane [49]

$$
\begin{array}{c}
\underset{\substack{|\\O\\|}}{\overset{\substack{CH_3\\|}}{CH_3-Si}}-O-\underset{\substack{|\\O\\|}}{\overset{\substack{CH_3\\|}}{Si}}-CH_3 \\
CH_3-\underset{\substack{|\\CH_3}}{Si}-O-\underset{\substack{|\\CH_3}}{Si}-CH_3
\end{array}
\xrightarrow[\text{DMSO}]{\text{KOH}}
\left[\underset{\substack{|\\CH_3}}{\overset{\substack{CH_3\\|}}{-Si}}-O-\right]_n
\tag{22}
$$

To a 100 ml round-bottomed flask equipped with a mechanical stirrer and situated in a constant temperature bath is added a 50-ml sample of (D₄) octamethylcyclotetrasiloxane. A 10 ml/min stream of dry nitrogen is directed over the surface of the liquid while 100 ppm potassium hydroxide (1% potassium hydroxide is in D₄) and then 0.1 wt.% dimethyl sulfoxide are added through a hypodermic syringe to each. The mixture is stirred at 95° ± 1°C and after 25 min the viscosity increases. After an additional 65 min a stiff gum forms that stops the mechanical stirrer. The reaction is stopped by adding iodine. The polymer is dissolved in benzene, stirred with silver powder, filtered through a bed of Celite, and the solvent is removed by vacuum stripping. The intrinsic viscosity is measured in toluene, and the number-average molecular weight is calculated using the Barry relationship [Eq. (23)] [50].

$$
[\eta] = 2 \times 10^{-4} M_n^{0.66}
\tag{23}
$$

The products usually have number-average molecular weights (M_n) of the polymer from 4×10^5 to 1.1×10^6.

The condensation of disiloxanediols is usually not the common method of preparing polyorganosiloxanes, and the following preparation is given only to illustrate the technique.

2-4. Condensation Polymerization of Tetramethyldisiloxane-1,3-diol [51]

$$
\underset{\substack{|\\CH_3}}{\overset{\substack{CH_3\\|}}{HO-Si}}-O-\underset{\substack{|\\CH_3}}{\overset{\substack{CH_3\\|}}{Si}}-OH + H^+
\xrightarrow{C_6H_5}
\left[\underset{\substack{|\\CH_3}}{\overset{\substack{CH_3\\|}}{-O-Si}}-O-\underset{\substack{|\\CH_3}}{\overset{\substack{CH_3\\|}}{Si}}-\right]_n + H_2O
\tag{24}
$$

To a flask equipped with a Birdwell–Stirling water trap and condenser are added 16.6 gm (0.1 mole) of tetramethyldisiloxane-1,3-diol, 100 ml of dry benzene, and 100 gm of p-toluenesulfonic acid as catalyst. The mixture is refluxed for about 0.5 hr to give the theoretical amount of water (1.8 ml) and polymer (properties not reported).

Cyclic methylpolysiloxanes and hexamethyloldisiloxane can be equilibrated under catalytic (acid) conditions to give linear methylpolysiloxanes as shown in Eq. (25).

$$[(CH_3)_2SiO]_n + [(CH_3)_3Si]_2O \xrightarrow{\text{H}_2\text{SO}_4} (CH_3)_3SiO[(CH_3)_2SiO]_nSi(CH_3)_3 \quad (25)$$

The reaction is carried out using no external heat source and just involves mixing the ingredients as described below in Preparation 2-5. The preparation below is only illustrative of the technique, which can be applied to higher molecular weight cyclic products rather than to the tetramer shown.

2-5. *Linear Methylpolysiloxanes by Equilibration Rearrangement of Cyclic Methylpolysiloxane* [27]

To a resin flask equipped with a mechanical stirrer are added 162.3 gm (1.0 mole) of hexamethyldisiloxane, 74.1 gm (0.25 mole) of $[(CH_3)_2SiO]_4$, and 10 ml of 95.5% sulfuric acid. The mixture is vigorously stirred for 4 hr at room temperature. Then 25 ml of water is added and the mixture vigorously stirred and allowed to stand to give a separation of layers. The upper layer is separated from the lower aqueous (34 ml) layer and washed twice with 25-ml portions of water, dried overnight over anhydrous potassium carbonate, filtered, and fractionally distilled (280 ml of crude product). The original tetramer is not present in the isolated product as shown in Fig. 4 and Table III. It would be interesting to repeat this analysis today using gas and liquid chromatography techniques.

TABLE III
LINEAR METHYLPOLYSILOXANES[a]

Compound	B.p. °C (760 mm)	(20 mm)	M.p. (°C)	d^{20} (gm/ml)	n_D^{20}
$CH_3[(CH_3)_2SiO]_2Si(CH_3)_3$	153	—	~ -80	0.8200	1.3848
$CH_3[(CH_3)_2SiO]_3Si(CH_3)_3$	194	—	~ -70	0.8536	1.3895
$CH_3[(CH_3)_2SiO]_4Si(CH_3)_3$	229	—	~ -80	0.8755	1.3925
$CH_3[(CH_3)_2SiO]_5Si(CH_3)_3$	—	142	< -100	0.8910	1.3948
$CH_3Si[(CH_3)_3SiO]_2$	190	—	~ -80	0.8497	1.3880

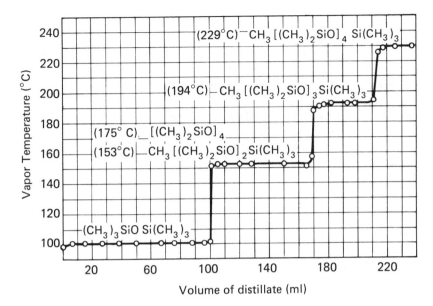

FIG. 4. Fractional distillation of methylpolysiloxanes. [Reprinted from W. Patnode and D. F. Wilcock, *J. Am. Chem. Soc.* **68**, 358 (1946). Copyright 1946 by the American Chemical Society. Reprinted by permission of the copyright owner.]

2-6. Polydimethylsiloxane with Terminal Trimethylsiloxy End Groups [52]

$$
\underset{\substack{\big| \\ CH_3}}{\overset{\substack{CH_3 \\ \big|}}{C_2H_5-O-Si-OC_2H_5}} + C_2H_5-O-Si(CH_3)_3 \xrightarrow{\text{H}_2\text{O}} (CH_3)_3SiO\left[\overset{\substack{CH_3 \\ \big|}}{\underset{\substack{\big| \\ CH_3}}{Si-O}}\right]_n Si(CH_3)_3
$$

$$+ \ 2C_2H_5OH \quad (26)$$

To a 1-liter three-necked flask equipped with a mechanical stirrer, dropping funnel, thermometer, and reflux condenser are added 288.6 gm (1.95 moles) of diethoxydimethylsilane and 5.9 gm (0.05 mole) of ethoxytrimethylsilane. Then 36 gm (2.0 moles) of water containing 0.2 gm of sodium hydroxide is added dropwise over a 30-min period while the mixture is vigorously agitated. The temperature rises spontaneously from 25°C to 60°C in approximately 1 hr. The reflux condenser is exchanged for a downward condenser and the alcohol distilled off with the aid of heat until the residue temperature rises to 100°C (169 gm of alcohol is obtained out of a calculated 180 gm). To the oily residue is added 150 ml of 20% hydrochloric acid and the mixture is refluxed for 4 hr. The oil and acid layers are separated and the oil layer is washed with water until free of acid. The fluid is heated to 250°C at 1 mm Hg pressure under distillation conditions to give a clear, colorless fluid residue of 123.4 gm (83%) with a viscosity of 81 centistokes at 25°C. The molecular weights of the products varied with viscosity as shown in Table IV.

Aromatic siloxanes react somewhat differently to give cyclic structures in some cases rather than polymers. For example, reaction of dipotassium diphenylsilanolate with diphenyldiacetoxysilane gave a mixture of crystalline products, hexaphenylcyclotrisiloxane, and octaphenylcyclotetrasiloxane [39a]. Hydrolysis of diphenyldiacetoxysilane (0.1 N sodium hydroxide) gave a 92% yield of pure diphenylsilanediol [39a].

B. Silicone Elastomers (Rubbers)

Silicone elastomers are basically high molecular weight linear polydialkyl-siloxanes that have been cross-linked or cured by either free radical catalysts (peroxides, etc.) or reactive multifunctional molecules [see Eqs. (27) and (28)].

$$
\begin{bmatrix} & \underset{|}{\overset{R}{\text{R}}} & \underset{|}{\overset{R}{\text{R}}} & \\ -O-\underset{|}{Si}-O-\underset{|}{Si}-O- \\ & CH_3 & CH_3 \end{bmatrix}_n \xrightarrow[120°C]{\substack{1.5-2\% \\ benzoyl \\ peroxide}} \quad \begin{array}{cc} R & R \\ | & | \\ -O-Si-O-Si-O- \\ | & | \\ CH_2 & CH_3 \\ | & | \\ CH_2 & CH_3 \\ | & | \\ -O-Si-O-Si- \\ | & | \\ R & R \end{array} \qquad (27)
$$

$$
\begin{bmatrix} CH_3 \\ | \\ -Si-O- \\ | \\ CH=CH_2 \end{bmatrix}_n \xrightarrow[120°C]{\substack{1\% \\ benzoyl \\ peroxide}} \begin{bmatrix} CH_3 \\ | \\ -Si-O- \\ | \\ CH-CH_2R \end{bmatrix}_m \longrightarrow \begin{array}{c} CH_3 \\ | \\ -Si-O- \\ | \\ CH-CH_2R \\ | \\ CH-CH_2R \\ | \\ -Si-O- \\ | \\ CH_3 \end{array} \qquad (28)
$$

R = C_6H_5COO or C_6H_5

In one example curing is usually accomplished by heating a polydimethyl-siloxane with a filler (ZnO, Ti_2O, Fe_2O_3, S, O_2, etc.) and benzoyl peroxide in a mold in the absence of air to about 150°C [53–55]. The polymer may also be postcured at 250°C for at least 4 hr in a circulating air oven to improve its compression set and abrasion resistance. The silicone elastomers or rubbers

TABLE IV
PHYSICAL PROPERTIES OF LINEAR HIGH POLYMERS[a]

Viscosity (centistokes, 25°C)	ASTM[b] pour point	Density (25°C)	Refr. index (25°C)	Approx.[c] mol. wt.	Exp. coeff. $K \times 1000$ per °C (25°-100°C)
10	−67	0.937	1.399	1200	1.035
20	−60	0.947	1.400	1900	1.025
50	−55	0.952	1.402	3700	1.000
100	−55	0.965	1.4030	6700	0.969
200	−53	0.968	1.4031	11300	0.968
350	−50	0.969	1.4032	15800	0.966
500	−50	0.969	1.4033	19000	0.965
1000	−50	0.970	1.4035	26400	0.963

[a] Reprinted from M. J. Hunter, E. L. Warrick, J. F. Hyde, and C. C. Currie, *J. Am. Chem. Soc.* **68**, 2284 (1946). Copyright 1946 by the American Chemical Society. Reprinted by permission of the copyright owner.

[b] American Society for Testing and Materials.

[c] "Viscometric Investigation of Dimethylsiloxane Polymers," presented before the High Polymer Division of the American Physical Society, New York, N.Y., January 26, 1946.

are useful at high and low temperature ranges where the conventional rubbers lose their properties.

Curing reactions also include the reaction of Si—H with Si–vinyl polymers [56].

Bouncing putty is based on a polydimethylsiloxane polymer modified with boric acid, additives, fillers, and plasticizers to give a material that on shock behaves like an elastic material but flows like a viscous liquid on slow application of pressure.

Room temperature vulcanizing silicone rubbers (RTV) are also available commercially and are used as adhesives, caulks, and coatings [57]. The RTV systems usually improve the reaction of silanols with ethyl silicate catalyzed by tin compounds (dibutyltin dilaurate) to eliminate ethyl alcohol with the formation of the resin [Eq. (29)]. The time required for the final cure is about an hour at 150°C or 24 hr at room temperature.

The base polymers are mainly high molecular weight linear polydimethylsiloxanes prepared by the base- or acid-catalyzed polymerization of cyclic dimethylsiloxanes $[(CH_3)_2SiO—]_n$ (see Section 2,A). The copolymerization of methylvinylcyclosiloxanes with dimethylsiloxanes affords polymers with methylvinylsiloxy groups that can be rapidly cross-linked by peroxides such as 2,4-dichlorobenzoyl peroxide or other systems as described in Preparation 2-7.

2-7. *Preparation of a Silicone Rubber* [58]

To a polymerization vial is added 20 gm of **I**, 5 gm of **II**, and 2.5 gm of **III**. A trace of chloroplatinic acid solution is added, and the mixture is evacuated to remove bubbles and cured to give a clear solid rubber.

2-8. *Preparation of a Fluorinated Organosiloxane Rubber* [59]

One-hundred parts by weight of a polysiloxane gum is milled with 30 parts by weight of a fumed silica and 2 parts by weight of benzoyl peroxide until a uniform mix is obtained. The resulting product is cured for 10 min at 125°C, 24 hr at 150°C and 24 hr at 200°C to give an elastomeric rubber showing good swell resistance to ASTM Oil No. 3.

NOTE: Similar techniques are used to convert dimethyl- or higher alkyl-siloxanes to rubbers [60].

C. Silicone Resins

Silicone resins differ from the fluids or elastomers in that they are highly cross-linked and contain a high proportion of silicon atoms with one or no organic substituent groups. A typical starting reaction solution may contain CH_3SiCl_3, $(CH_3)_2SiCl_2$, C_6H_5—$SiCl_3$, $(C_6H_5)_2SiCl_2$, and sometimes also $SiCl_4$ dissolved in toluene. The use of various starting materials affects the properties of the final resin in the following ways: (1) trifunctional siloxy groups give harder, less organic-soluble polymers; (2) difunctional siloxy groups give softer, more organic-soluble polymers; and (3) monophenyl-siloxanes give more organic-soluble and thermally stable polymers than the methylsiloxanes.

The toluene solution of the chlorosilanes is hydrolyzed, and the aqueous hydrochloric acid is separated, washed, and then heated in the presence of a mild condensation catalyst to adjust the resin to the proper viscosity and cure time. Fillers can be added prior to the removal of the solvent and isolation of the final resin. A resin can also be prepared from phenyltrichlorosilane alone to give a tough, infusible resin that can be cast from solvents to give a clear film [61,61a].

The hydrolysis of the chlorosilane blend is usually carried out in a diluent to moderate the hydrolysis reaction. Too much tri- or tetrachlorosilane or too little solvent may cause the reaction mixture to gel or the product to become prematurely insoluble. Low molecular weight alcohols are used to control the degree of condensation by forming alkoxysilanes rather than silanols. The alkoxysilanes are later converted to siloxanes at a controlled rate during processing. The hydrolysis reaction is quite complicated and still requires further study [61–63].

The final cure step of the silicone resin solution involves either removal of solvent or addition of a metal soap (tin acetate) or amine catalyst prior to heat curing with subsequent solvent evaporation.

2-9. *Silicone Resins from Phenyltrichlorosilane* [61,64]

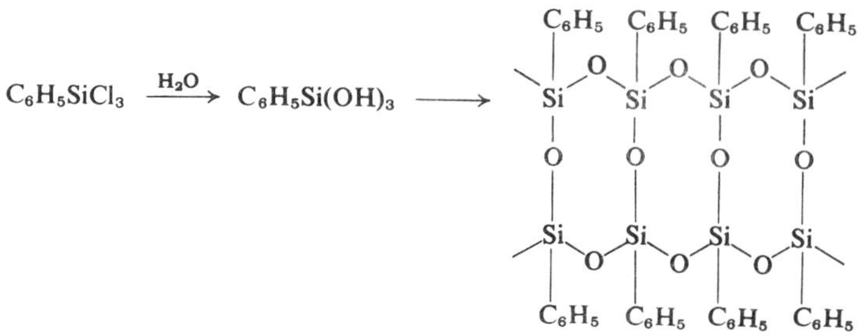

$$C_6H_5SiCl_3 \xrightarrow{H_2O} C_6H_5Si(OH)_3 \longrightarrow$$

(32)

A mixture of the hydrolyzate of phenyltrichlorosilane (300 gm), 0.1% potassium hydroxide, and an equal weight of toluene is refluxed for 16 hr while water is removed in a Dean & Stark trap. The reaction mixture is cooled, filtered, and the product precipitated into ligroin or methanol to give approximately 99.9% condensation of the silanol analyzing for $(C_6H_5—SiO_{3/2})_x$. The polymer obtained had the following properties: $[\eta] = 0.12$ dl/gm in benzene, $M_n = 14,000$, mol. wt. $= 26,000$. Similar polymers were prepared from phenyltriethoxysilane, base catalyzed by hydrolysis [65].

More recently Brown [62] found that phenylsilanetriol (from the phenyltrichlorosilane hydrolysis reaction) underwent polycondensation in a selective manner. In aqueous solution the major reaction pathway is through *cis,cis,cis*-cyclotetrasiloxanetetrol and the epimeric tetracyclo[7.7.1.1^{3,31}.1^{5,11}]octasiloxanediols and then to intermolecular condensation polymers of these. Side reactions give 8- to 12-unit cagelike polycyclics in polar solvent and 20- to 30-unit siloxane polycyclics in less polar solvents. The structure of the high polymers formed upon complete condensation resembles a string of beads (the beads being the polycyclic blocks), rather than the double chain (ladder type) obtained on equilibration or the irregular structure of randomly connected units as previously assumed.

Condensations in less polar solvents gave molecular distributions as shown in Table V.

2-10. *Condensation of n-Amyltriethoxysilane* [65]

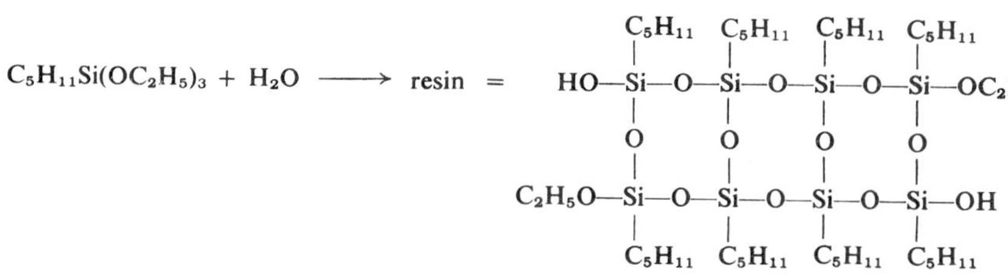

$$C_5H_{11}Si(OC_2H_5)_3 + H_2O \longrightarrow resin =$$

(33)

TABLE V
APPROXIMATE MOLECULAR DISTRIBUTIONS IN
PHENYLTRICHLOROSILANE HYDROLYSATES[a]

Solvent	Conc. ratio[b]	Temp. (°C)	% in degree of polymerization range		
			8–12	16–30	> 30
Aq. acetone	4	25	90	10	—
Ether	53	35	50	50	—
Toluene	67	23	10	60	30
Toluene	116	50	8	37	55

[a] Reprinted from J. F. Brown, Jr., *J. Am. Chem. Soc.* **87**, 4317 (1965). Copyright 1965 by the American Chemical Society. Reprinted by permission of the copyright owner.
[b] Grams of PhSiCl$_3$ per 100 ml of solvent.

To a flask are added 117 gm (0.5 mole) of *n*-amyltriethoxysilane, 400 ml of isobutyl methyl ketone, and 27 ml (1.5 moles) of water. Then 0.10 ml of 35.4% hydrochloric acid is added and the solution is refluxed for 3 hr. The water is removed by azeotropic distillation to give 6.5 gm (10%), b.p. 180°–200°C (< 5 mm Hg, n_D^{20} 1.4503, mol. wt. 596) and 52.5 gm (80%) of a clear, viscous liquid (mol. wt. 988).

2-11. Silicone Resins by Cyclohydrolysis of Dimethyldichlorosilane and Methyltrichlorosilane [66]

$$(CH_3)_2SiCl_2 + CH_3SiCl_3 \xrightarrow{H_2O}$$

$$\begin{array}{c} \text{CH}_3 \quad \text{CH}_3 \quad \text{CH}_3 \\ | \qquad | \qquad | \\ -\text{Si}-\text{O}-\text{Si}-\text{O}-\text{Si}-\text{O}- \\ | \qquad \qquad | \\ \text{CH}_3 \qquad \quad \text{CH}_3 \\ \qquad \qquad | \\ \qquad \qquad \text{O} \\ \qquad \qquad | \quad \text{CH}_3 \quad \text{CH}_3 \\ \qquad \qquad | \qquad | \qquad | \\ -\text{O}-\text{Si}-\text{O}-\text{Si}-\text{O}-\text{Si}-\text{O}- \\ \qquad | \qquad | \qquad | \\ \qquad \text{CH}_3 \quad \text{CH}_3 \quad \text{O} \\ \qquad \qquad \qquad \qquad | \end{array}_n \qquad (34)$$

A mixture of 4.5 gm (0.035 mole) of dimethyldichlorosilane and 1.95 gm (0.013 mole) of methyltrichlorosilane in 50 ml of ether is hydrolyzed by pouring into 100 gm of cracked ice. The ether is evaporated and the residue heated in air to give a glossy, infusible, and insoluble solid with a C:Si ratio of 1.3. The resin gives silicon when heated at 300°–400°C.

3. MISCELLANEOUS SILICONE RESIN PREPARATIONS AND COMPOSITIONS

1. Preparation of *N*-silyl-substituted polyurea [67].
2. Preparation of silane polymers based on vinylsilanes [68].
3. Preparation of silazane polymers [69].
4. Siloxane polyether copolymer surfactants [70].
5. Siloxanes with SiH bonds [71].
6. Poly(silarylenesiloxanes) [72].
7. Condensation of toluene diisocyanate with 4,4,6,6-tetramethyl-5-oxa-4,6-disila-1,9-nonenediol [73].
8. Silicone release agent [74].
9. Random poly(*m*-carborarylenesiloxane) copolymers [75].
10. Fluorinated organosilicon rubber [75a].
11. Polymeric cyclosilazanes [76].
12. Siloxane emulsions [77].
13. Polyalkylene polysiloxane block copolymer [78].
14. Organosilicon sulfides [79].
15. Block copolymers [80].
16. Preparation of polysiloxane resins without solvents [81].
17. Poly(dimethylsilmethylene) [82].
18. Flame-retardant polycarbonates containing diorganopolysiloxanes to prevent dripping [83].
19. Urea-functional organosilicon compounds [84].
20. Heat-curable aminoorganosilicon compositions and the cured elastomers obtained from them [85].
21. Nonionic organopolysiloxanes [86].
22. Preparation of cyclic organopolysiloxanes of the formula [87]:

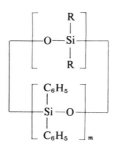

23. Condensation of organohydrogenpolysiloxane with α-olefins [88].

REFERENCES

1. C. Friedel and J. M. Crafts, *Liebigs Ann. Chem.* **136**, 203 (1865); C. Friedel and A. Landenburg, *ibid.* **159**, 259 (1871); **203**, 251 (1880).
2. D. T. Hurd, *J. Am. Chem. Soc.* **67**, 1545 and 1813 (1945).
3. E. G. Rochow, *J. Am. Chem. Soc.* **67**, 963 (1945); B. A. Bluestein, U.S. Patent 2,887,502 (1957).
4. R. Meals, *Kirk-Othmer, Encycl. Chem. Technol., 2nd Ed.* Vol. 18, p. 221 (1969); W. Noll, "Chemistry and Technology of Silicones." Academic Press, New York, 1968.
5. I. I. Shtetter, Russian Patent 44,934 (1935).
6. W. I. Patnode and R. O. Sauer, U.S. Patents 2,381,137 and 2,381,138 (1945).
7. R. O. Sauer, U.S. Patent 2,381,139 (1945).
8. L. Birkofer, A. Ritter, and H. Dickopp, *Chem. Ber.* **96**, 1473 (1963).
9. F. S. Kipping and L. L. Lloyd, *J. Chem. Soc.* **79**, Paper No. 1, 449 (1901); F. S. Kipping and J. T. Abrams, *J. Chem. Soc.* p. 81, Paper No. 51 (1944); E. G. Rochow, "An Introduction to the Chemistry of the Silicones," Chapter 4. Wiley, New York, 1946.
10. F. S. Kipping, *Proc. Chem. Soc., London* **28**, 243 (1912); *J. Chem. Soc.* **101**, 2106 (1912).
11. W. Patnode, *J. Am. Chem. Soc.* **68**, 358 and 360 (1946).
12. D. F. Wilcock, *J. Am. Chem. Soc.* **68**, 691 (1946); W. Patnode and D. F. Wilcock, *ibid.* p. 362.
13. J. F. Hyde and R. C. DeLong, *J. Am. Chem. Soc.* **63**, 1194 (1941).
14. E. G. Rochow and W. F. Gilliam, *J. Am. Chem. Soc.* **63**, 798 (1941).
15. E. G. Rochow, U.S. Patent 2,258,218 (1941).
16. D. W. Scott, *J. Am. Chem. Soc.* **68**, 2294 (1946).
17. A. Ladenburg, *Liebigs Ann. Chem.* **164**, 311 (1872); C. Friedel and J. M. Crafts, *Ann. Chim. Phys.* [4] **9**, 5 (1866).
18. R. Robinson and F. S. Kipping, *J. Chem. Soc.* **93**, 439 (1908); G. Martin and F. S. Kipping, *ibid.* **95**, 302 (1909).
19. C. A. Burkhard, *J. Am. Chem. Soc.* **67**, 2173 (1945); E. G. Rochow, U.S. Patent 2,258,219 (1941).
20. E. G. Rochow, U.S. Patent 2,258,222 (1942).
21. J. F. Hyde and R. C. DeLong, *J. Am. Chem. Soc.* **63**, 1194 (1941); J. F. Hyde, U.S. Patent 2,371,050 (1945).
22. D. F. Wilcock, *Mech. Eng.* **66**, 739 (1944).
23. J. H. Davis and D. E. W. Rees, *Proc. Inst. Electr. Eng.* **112**, 1607 (1965).
24. W. I. Patnode, U.S. Patent 2,306,222 (1943).
25. W. Noll, "Chemistry and Technology of Silicones," pp. 9–16. Academic Press, New York, 1968; C. Eaborn, "Organosilicon Compounds," pp. 2–9. Academic Press, New York, 1960.
26. C. B. Hurd, *J. Am. Chem. Soc.* **68**, 364 (1946).
27. W. I. Patnode and D. F. Wilcock, *J. Am. Chem. Soc.* **68**, 358 (1946).
28. E. G. Rochow, "An Introduction to the Chemistry of the Silicones." Wiley, New York, 1946.
29. H. K. Lichtenwalner and M. N. Sprung, *Encycl. Polym. Sci. Technol.* **12**, 464 (1970).
30. R. Meals, *Kirk-Othmer, Encycl. Chem. Technol., 2nd Ed.* Vol. 18, p. 221 (1969).
31. W. Noll, "Chemistry and Technology of Silicones." Academic Press, New York, 1968.

32. R. R. McGregor and E. L. Warrick, German Patent 881,404 (1942).
33. J. G. E. Wright U.S. Patent 2,452,416 (1944).
34. Rhône-Poulenc, French Patent 950,582 (1947).
35. P. Simons and W. Noll, German Patent 895,650 (1951).
36. G. R. Lucas and R. W. Martin, *J. Am. Chem. Soc.* **74**, 5225 (1952).
37. W. I. Patnode, U.S. Patent 2,381,366 (1942).
38. F. Rodriguez, R. A. Kulakowski, and O. K. Clark, *J. Ind. Eng. Chem.* **5**, 121 (1966).
39. H. J. Fletcher and G. L. Constant, Canadian Patent 560,981 (1955); R. H. Leitheiser, U.S. Patent 3,122,579 (1960); L. H. Sommer, L. Q. Green, and F. C. Whitmore, *J. Am. Chem. Soc.* **71**, 3253 (1949); S. Nitzsche and M. Wick, German Patent 930,481 (1953); N. F. Orlov and B. N. Dolgov, *Proc. Acad. Sci. USSR* **125**, 266 (1959); B. Smith, *Sven. Kem. Tidskr.* **65**, 101 (1953); L. Ceyzeriat and P. Dumont, German Patent Appl. 1,121,329 (1957); T. N. Ganina, A. A. Zhadanov, and S. A. Pavlov, *Dokl. Akad. Nauk SSSR* **102**, 85 (1955); J. B. Rust and C. A. Mackenzie, U.S. Patent 2,506,616 (1945).
39a. Y. Nakaido and T. Takiguchi, *J. Org. Chem.* **26**, 4144 (1961).
40. S. R. Sandler and W. Karo, "Polymer Syntheses," Vol. 1, Chapter 6. Academic Press, New York, 1974.
41. W. H. Knoth and R. V. Lindsey, *J. Am. Chem. Soc.* **80**, 4106 (1958); G. Koerner and G. Rosomy, *Makromol. Chem.* **102**, 146 (1967).
42. J. F. Brown, Jr. and G. M. Slusarczak, *J. Am. Chem. Soc.* **87**, 931 (1965).
42a. R. Gutoff, *Ind. Eng. Chem.* **49**, 1807 (1957).
43. D. F. Wilcock, *J. Am. Chem. Soc.* **69**, 477 (1947).
44. N. Kirk, *Ind. Eng. Chem.* **51**, 515 (1959).
45. E. D. Brown, U.S. Patent 3,179,619 (1965).
45a. K. E. Palmanteer and E. D. Brown, U.S. Patent 3,050,492 (1962).
46. N. Kirk, *J. Ind. Eng. Chem.* **51**, 515 (1959).
47. A. S. Paten, U.S. Patent 3,221,040 (1965).
48. E. L. Warrick, U.S. Patent 2,634,252 (1953).
49. G. D. Cooper, *J. Polym. Sci., Part A-1* **4**, 603 (1966).
50. A. J. Barry, *J. Appl. Phys.* **17**, 1020 (1946).
51. G. R. Lucas and R. W. Martin, *J. Am. Chem. Soc.* **74**, 5225 (1952).
52. M. J. Hunter, E. L. Warrick, J. F. Hyde, and C. C. Currie, *J. Am. Chem. Soc.* **68**, 2284 (1946).
53. J. G. E. Wright and C. S. Oliver, U.S. Patent 2,448,565 (1948).
54. J. Marsden, U.S. Patent 2,445,794 (1945).
55. S. Nitzsche and M. Wick, U.S. Patent 3,127,363 (1964); L. B. Bruner, U.S. Patent 3,035,016 (1962); L. Ceyzeriat, U.S. Patent 3,133,891 (1964).
56. M. E. Nelson, U.S. Patents 3,020,260 (1962); 3,249,581 (1966).
57. C. A. Berridge, U.S. Patent 2,843,555 (1958); R. A. Pike, U.S. Patent 3,046,294 (1962); "RTV Silicone Rubber," Tech. Brochure No. S-3C. General Electric Co., Waterford, New York, 1964; "Silicone Rubber Adhesive/Sealants for Industrial Applications," Tech. Brochure No. S-2D. General Electric Co., Waterford, New York, 1967.
58. M. E. Nelson, U.S. Patent 3,249,581 (1966).
59. E. D. Brown, U.S. Patent 3,179,619 (1965).
60. W. J. O'Malley, *Adhes. Age* p. 17 (1975).
61. J. F. Brown, Jr., L. H. Vogt, Jr., A. Katchman, J. W. Eustance, K. M. Kiser, and K. W. Krantz, *J. Am. Chem. Soc.* **82**, 6194 (1960).
61a. R. N. Meals and F. M. Lewis, "Silicones." Van Nostrand-Reinhold, Princeton, New Jersey, 1959; F. J. Modic, *Adhes. Age* p. 5 (1962).

62. J. F. Brown, Jr., *J. Am. Chem. Soc.* **87**, 4317 (1965).
63. M. M. Sprung and F. O. Guenther, U.S. Patent 3,017,385 (1962).
64. J. F. Brown, Jr., L. H. Vogt, Jr., and P. I. Prescott, *J. Am. Chem. Soc.* **86**, 1120 (1964).
65. M. M. Sprung and F. O. Guenther, *J. Polym. Sci.* **28**, 17 (1958).
66. E. G. Rochow and W. F. Gilliam, *J. Am. Chem. Soc.* **63**, 798 (1941).
67. J. F. Klebe, *J. Polym. Sci., Part B* **2**, 1079 (1964).
68. J. W. Curry, *J. Am. Chem. Soc.* **78**, 1686 (1956); *J. Org. Chem.* **26**, 1308 (1961).
69. R. E. Burks, Jr., *Encycl. Polym. Sci. Technol.* **12**, 569 (1970).
70. B. Kanner, W. G. Reid, and I. H. Petersen, *Ind. Eng. Chem., Prod. Res. Dev.* **6**, 88 (1967).
71. F. O. Stark and G. E. Vogel, U.S. Patent 3,249,585 (1966).
72. Wacker-Chemie G.m.b.H., French Patent 1,470,795 (1967).
73. W. H. Knoth, Jr. and R. V. Lindsey, Jr., *J. Am. Chem. Soc.* **80**, 4107 (1958).
74. E. R. Martin, U.S. Patent 3,883,628 (1975).
75. S. Papetti and H. A. Schroeder, U.S. Patent 3,463,801 (1969).
76. W. Fink, U.S. Patent 3,431,222 (1969).
77. J. Cekada, Jr. and D. R. Weyenberg, U.S. Patent 3,532,729 (1970).
78. G. Rosomy and J. Wassermeyer, U.S. Patent 3,532,732 (1970).
79. K. M. Lee, U.S. Patent 3,532,733 (1970).
80. R. C. Antomen, U.S. Patent 3,294,718 (1966).
81. S. Miroslav, J. Cermak, and J. Miksa, Czech Patent 151,697 (1973).
82. W. A. Krimer, *J. Polym. Sci., Part A-1* **4**, 444 (1966).
83. General Electric Co., French Patent 2,246,602 (1975).
84. R. H. Krahnke, U.S. Patent 3,772,351 (1973).
85. R. L. Schank and T. C. Williams, U.S. Patent 3,449,289 (1969).
86. D. B. Morgan and E. J. Vickers, U.S. Patent 3,278,485 (1966).
87. P. I. Prescott and T. G. Selin, U.S. Patent 3,317,578 (1967).
88. E. D. Brown, Jr., U.S. Patent 3,418,353 (1968).

ALKYD RESINS

I. INTRODUCTION

Alkyd resins are prepared from polyhydric alcohols (3 or more hydroxyls), polybasic acids, and monobasic fatty acids (saturated and unsaturated) [1]. They are soluble in hydrocarbon solvents and are used in the coating industry. Polyesters prepared from dihydric alcohols are not included in the area of alkyd resins in this chapter and have been described in Vol. I of this series [2].

Alkyd-type resins were first described in 1847 by Berzelius [3], who obtained a brittle resinous polymer by the reaction of tartaric acid and glycerol. In 1853 Berthelot prepared the glycerol ester of camphoric acid [4]. In 1856 von Bemmelen [5] studied resins obtained by heating glycerol with succinic acid, citric acid, and a mixture of succinic and benzoic acids. Other early investigators were Dubus [6], Lourenco [7], and Furaro and Danesi [8]. In 1901 Smith [9] described a solid transparent glycerol phthalate resin that was insoluble in water but soluble in hot glycerol. Removal of the glycerol *in vacuo* afforded a resinous mass containing glycerol and phthalic anhydride in a molecular ratio of approximately 2:3. Further heating caused the resinous material to yield a puffy brittle mass. General Electric Laboratories carried out a series of intensive investigations on the glycerol–phthalic anhydride reactions leading to the production of useful resins of the heat-convertible type as summarized in the work of Callahan [10], Friberg [11], Arsem [12], Dawson [13], and Howell [14]. The latter workers found that replacing some of the phthalic anhydride with monobasic acids leads to more flexible resins with better solubility properties. Kienle [15,15a] in 1929 and 1933 prepared glycerol–phthalate resins containing some unsaturated fatty acids; these came into use as protective coatings that were cured by air drying or heating at low baking temperatures. The commercial production [16] of phthalic anhydride by the catalytic oxidation of naphthalene accelerated progress in commercially manufactured alkyd resins. Other developments that helped in the progress of alkyd resin technology were the introduction of new solvents; lauric acid; soya and tall oil fatty acids; maleic, fumaric, and isophthalic acids; and polyhydric alcohols such as pentaerythritol, dipentaerythritol, and trimethylolethane. The structure of a typical phthalic anhydride alkyd resin may be represented as

Gel permeation chromatography of alkyd resins has been reported and the molecular weight distribution of products has been studied [17]. Molecular weights (determined by the Rast method) are also found to depend on the cooling schedule and data have been recently reported in this area [18].

Fundamental studies on the glycerol–phthalic anhydride reaction were first made by Kienle [15] who found that the gel point occurs at about 78% esterification.

An explanation of the cross-linking behavior in systems with more than two functional groups reacting at esterification was first formulated by Carothers. In simplified form his equation states that as the molecular weight becomes infinite (at the gel point), $p = 2/f$, where p is the extent of the reaction and f is the average degree of functionality or the average number of functional groups per molecule, considering only the use of stoichiometric equivalents of interacting functional groups. Excess functional groups such as hydroxyl are not truly reactive since there are no carboxyl groups which they can react with. In systems with a functionality of 3 or more, gelation can be avoided by limiting the extent of the reaction or using proportions that deviate from stoichiometric proportions. This method is commonly used with monofunctional reagents (fatty acids) to reduce the overall functionality. For example, the Carothers equation when applied to stoichiometric amounts of glycerol and phthalic anhydride is

$$2 \text{ moles glycerol} + 3 \text{ moles phthalic anhydride}$$

$$\underbrace{2 \times 3\text{OH}}_{6} \quad + \quad \underbrace{3 \times 2\text{COOH}}_{6} \quad = 12 \qquad (1)$$

$$f = \frac{12 \text{ functional groups}}{5 \text{ moles}} = 2.4 \quad \text{(average number of functional groups per molecule reacting)}$$

$$p = \frac{2}{f} = \frac{2}{2.4} = 83\% \quad \text{(extent of reaction where gelation should occur; 78\% found by Kienle [15])} \qquad (2)$$

When $f = 2$, $p = 100\%$, and the equation predicts that no gelation occurs when bifunctional compounds are used. Therefore, alkyd chemists tend to use formulations in which f = approximately 2.0 in order to get alkyd resins which will not gel during manufacture. In other words, the molar equivalents of hydroxyl should equal the molar equivalents of carboxyl groups. In actual practice an excess of hydroxyl equivalents is used to give specific properties, especially in baking coatings. For example, a typical alkyd with 20% excess of hydroxyl and a functionality of 2 is prepared from phthalic anhydride, 1.0 mole (2 equivalents); glycerol, 1.0 mole (3 equivalents); and soya fatty acids (soya oil), 0.5 mole (1.5 equivalents).

Flory [19] and Jonason [20] have also derived equations to describe the gelation point of alkyd resins.

Measurements of melt viscosity and intrinsic viscosity are useful in process control work [21].

Some of the typically used raw materials are shown in Table I.

The manufacture of alkyd resins usually involves alcoholysis of the oils with a polyhydric alcohol followed by a batch process in which the first step is the esterification to a specified end point and then solution into a solvent thinning tank. The esterification may be done in the absence (fusion process) or presence (solvent process) of solvent but in both cases an inert gas (nitrogen or carbon dioxide) must be bubbled through the reaction mixture to facilitate removal of water of esterification and to minimize oxidation.

TABLE I
Representative Alkyd Resin Starting Materials

Raw material	M.p. (°F)	B.p. (°F)	Mol. wt.	Sapon. no.	Equiv. wt.
Polyhydric compounds					
Glycerin	—	554	92.1	—	31
Ethylene glycol	—	386	62.1	—	31
Diethylene glycol	—	473	106.1	—	53
Propylene glycol	—	374	76.1	—	38
Trimethylolethane	395	—	120.1	—	40
Trimethylolpropane	136	—	134.2	—	44.7
Pentaerythritol	504	—	136.2	—	45.5
Acids–Anhydrides [a]					
Phthalic anhydride	270	544	148.1	—	74
Phthalic acid	375–410	—	166.1	—	83
Isophthalic acid	650	—	166.1	—	83
Terephthalic acid	Sublimes 570	—	166.1	—	83
Trimellitic anhydride	334	—	192.1	—	64
Trimellitic acid	419–423	—	210.1	—	70
Maleic anhydride	266	—	116.1	—	49
Fumaric acid	548	—	116.1	—	49
Adipic acid	306	—	146.1	—	73
Oils					
Linseed	—	—	—	188–196	280
Soya	—	—	—	189–195	280
Dehydrated castor	—	—	—	188–194	293
Safflower	—	—	—	188–194	293
Tall	—	—	—	170–180	285

[a] Operations with all acids or anhydrides should be carried out in a well-ventilated hood using proper personal protection (gloves, apron, chemical goggles).

The fusion process has the advantage that there is less of a fire hazard and that there is a lower equipment investment involved. However, the solvent process requires less time in clean-up of the kettle, entails lower raw material losses, and gives a lighter colored product.

The equipment used is either stainless steel or stainless steel clad and is corrosion resistant. An example of a typical equipment set-up is shown in Fig. 1, where the sparge is either nitrogen or carbon dioxide.

Other manufacturing equipment has been described by Martens [22] and Hovey [23]. Typical laboratory and manufacturing equipment is described in a review by Monsanto [24,25a]. Continuous methods of producing alkyds have not gained widespread usage [25].

Quality control in the manufacture of alkyd resins involves appropriate rapid tests to determine the acid number, viscosity of solutions of known concentrations in mineral spirits or xylene (Gardner–Holdt viscosity tubes used for comparison), color (Gardner 1933 Color Standards), and cure time. A target range 2–5 units above the acid number units of gelation is desirable. Most manufactured alkyd resins have an acid value (number of mg of potassium hydroxide required to neutralize 1 gm of resin) of less than 15.

Once laboratory results have established the relationship of acid number and solution viscosity to reaction time, both are plotted on graph paper and are used in manufacturing operations (see Fig. 7). Nomograms for alkyd manufacture have also been published [25a].

The first alkyd resins sold commercially were by the General Electric Company under the trade name "Glyptal." The production of alkyds has risen from 100 million lb in 1940 to 545 million lb in 1957 [26] to 734 million lb in 1973 [27]. In 1973 the phthalic anhydride-type alkyd resins amounted to 691 million lb and the other polybasic acid types amounted to 43 million lb of resin used. In addition, styrene alkyd polyesters amounted to 32 million lb.

The unmodified glyceryl phthalate alkyd resin has only limited solubility unless it is reacted during its preparative stage with oils or fatty acids. The oil used is either a drying, semidrying, or nondrying oil. In addition to glycerol, pentaerythritol is the most important of the lesser-used polyhydric alcohols. The resulting resins are more durable than the glyceryl alkyds and are faster setting. Other polybasic acids such as sebacic and adipic are used to give soft resins that are useful as products that are polymeric plasticizers for harder materials [28].

Today alkyds are among the most important and widely used resins in the coating field. They are outstanding in terms of their economy, versatility of formulation and application, and durability. Large quantities of oil-modified glyceryl phthalate alkyds find use in interior and exterior enamels, paints, automotive finishes, industrial finishes, marine coatings, and appliance enamels [28]. Often the oil-modified alkyds are coreacted with either styrene or silicones to give resins with modified properties. The alkyd resins can also be combined with other resins such phenolic resins, amino resins, epoxies, acrylics, and polyurethanes to give special coating materials. Alkyds are also used for molding materials that require good arc and arc track resistance, insulation resistance, good flammability ratings (UL94-VO), and excellent mechanical, thermal dimensional, and electrical properties [29]. Miscellaneous uses of alkyds have been summarized by Martens [30].

Some of the typical reactions that take place during an alkyd resin preparation and curing are (1) esterification, (2) alcoholysis, (3) ester interchange or transesterification, (4) etherification, (5) free radical addition reactions, (6) Diels–Alder reactions, (7) decarboxylations, and (8) polymerization via free radical initiation during curing (air-drying alkyds).

The addition of other resins such as urethane, epoxy, or phenolic resins to the final resin can lead to further condensation reactions with hydroxyl, carboxyl, and olefinic reaction sites.

A typical formulation for an alkyd paint is shown below.

White gloss enamel [31]	Parts
Composition	
Acrylic-modified TMA Resin [74% solids (NVM)][a]	270
Cymel XM-1116[b]	86
	(add before water)
	is added)
Water	300
Triethylamine	18
Rutile titanium dioxide[c]	257
Combine and pebble mill for 24 hr. Then add 150 parts water to give 1081 parts enamel.	
Properties	
The initial viscosity in a #4 Ford Cup (sec)	29
Initial pH	7.7
Pigment:binder ratio	0.9:1.0
Melamine:resin ratio	3:7
Wt.% NVM	50.2
The bake cycle is 20 min at 350°F to give a durable, weather-resistant coating.	

[a] TMA, trimellitic anhydride; NVM, nonvolatile material.
[b] Available from American Cyanamid Co.
[c] TiPure R-900, available from E. I. du Pont de Nemours & Co.

Some important reviews [32,33] and monographs [22,24,30] on alkyd resin production provide additional details.

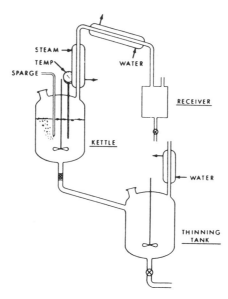

Fɪɢ. 1. Alkyd resin processing equipment. [Reprinted from "Amoco TMA as Primers for Alkyd-Melamine Enamels and Acrylic Lacquers," Tec. Bull. TMA 25a. Amoco Chemical Corp., Chicago, Illinois, 1974.]

2. UNMODIFIED ALKYDS

A. Alkyds Based on Phthalic Anhydride or Phthalic Acid

The earlier alkyd resins (glyceryl phthalate) were all of the heat-convertible variety; i.e., on heating the resin was converted to a nonfusible type [15]. More recently an unmodified alkyd refers to a resin prepared from a dibasic acid, a polyol, and an oil or fatty acid. In some cases a monofunctional acid such as benzoic acid has been added to reduce functionality [22]. Oxygen- or air-convertible alkyd resins are prepared by replacing part of the polybasic acid with the proper amount of an oxidizable unsubstituted fatty acid or acids such as linolenic, oleostearic, or others that are commercially available. With these two types of systems, alkyds are used to prepare coatings that cure upon baking or air drying.

Alkyds are very versatile and can be blended with many other resins [34–34b]. However, alkyds are not highly chemical resistant and are affected by water [35]. They have fair resistance to petroleum solvents and oils but are removed (dissolved) either by polar solvents (ketones, esters) or chlorinated

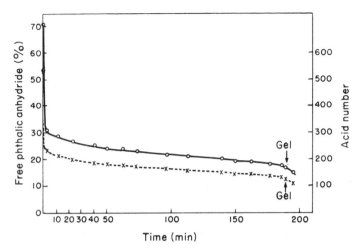

FIG. 2. Decrease of free acidity with time ($T = 195°C$; solid line, free phthalic anhydride; broken line, acid number). [Reprinted from R. H. Kienle and A. G. Hovey, *J. Am. Chem. Soc.* **51**, 509 (1929). Copyright 1929 by the American Chemical Society. Reprinted by permission of the copyright owner.]

compounds. Alkyds are thermally stable to about 300°F but become brittle at this temperature.

Kienle and Hovey [36] studied the reaction of phthalic anhydride (3.0 moles) with glycerol (2.0 moles) and found that by titration of aliquot samples the free acidity is reduced to approximately one-half within one minute after the start of the reaction, as shown in Fig. 2.

Kienle and co-workers also found that phthalic acid gives similar results to phthalic anhydride and reacts at a similar rate once 50% esterification has been attained [37].

This is true since both reactions give a common intermediate product as shown in Eq. (3).

Infrared spectra of phthalate acid and anhydride are shown in Fig. 3.

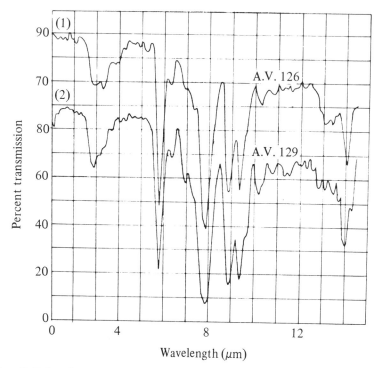

FIG. 3. Infrared spectra: (1) glyceryl phthalate (acid), A.V. = 129; (2) glyceryl phthalate (anhydride), A.V. = 126. [Reprinted from R. H. Kienle, P. A. Van Der Menlen, and F. E. Petke, *J. Am. Chem. Soc.* **61**, 2268 (1939). Copyright 1939 by the American Chemical Society. Reprinted by permission of the copyright owner.]

2-1. Preparation of Poly(glyceryl phthalate) Alkyd Resin [36]

To a resin flask equipped with a thermometer, mechanical stirrer, condenser, Dean & Stark trap, and nitrogen bubbler are added 46.0 gm (0.50 moles) of glycerol and 111 gm (0.75 moles) of phthalic anhydride. The reaction mixture is heated to 195°C for $1\frac{1}{2}$ hr while a slow stream of nitrogen is bubbled through it. The acid value is about 170 at this point and the product is still soluble in acetic acid, acetone, and other solvents. Further heating at 195°C for another $1\frac{1}{2}$ hr causes the resin to set to an insoluble gel (acid no. = 119–136). The reaction data are shown in Table II along with percent esterification data in Table III and Fig. 4. The Dean & Stark trap in the above preparation contained 7.0 gm of a liquid consisting of water and some white crystals. Titration with $N/25$ potassium hydroxide using phenolphthalein as an indicator showed 0.085 gm of phthalic anhydride (0.076%).

Since phthalic anhydride is insoluble in most oils (linseed, safflower, etc.) but soluble in monoglycerides, the oils are first alcoholized with the polyol

TABLE II

ISOTHERMS OF FREE ACIDITY CHANGE
WITH TIME AT 195°C[a]

Time (sec)	Time (min)	Free anhy-dride (%)	Acid number
0	0	70.67	535
215	3.58	31.05	235
720	12.16	28.99	218
1320	23.0	27.02	212
2440	40.67	25.25	191
3420	50.0	24.51	185
3980	63.0	23.96	181
4380	73.0	23.42	177
5790	96.5	22.23	169
6780	113.0	21.47	163
8340	139.0	21.02	159
9000	150.0	19.93	151
9720	162.0	19.84	149
10,610	177.0	19.18	145
11,150	186.0	18.31	139
11,340	189.0	17.94	136
11,700	195.0	15.81	119

[a] Reprinted from R. H. Kienle and A. G. Hovey, *J. Am. Chem. Soc.* **51**, 509 (1929). Copyright 1929 by the American Chemical Society. Reprinted by permission of the copyright owner .

(glycerin or others) in the presence of a catalyst to form monoglycerides. The use of catalysts (litharge, calcium hydroxide, etc.) at 525°–550°F gives satisfactory results provided a condenser is used to prevent loss by volatilization of the ingredients. The reaction is complete when 4 volumes of methanol are soluble in 1 volume of the monoglyceride product at the boiling point of methanol. Samples are taken every 5 min and checked. The reaction usually takes about 20 min. The catalyst is usually 0.1–0.15% of the glycerin calculated on the basis of the metal content. The calcium catalyst is preferred from the standpoint of efficiency and resin color. After the alcoholysis step the polycarboxylic acid (phthalic, etc.) is added with or without solvent to complete the alkyd preparation. The reaction is usually carried out at 450°F and an inert gas is used throughout the entire process to facilitate water removal and to prevent color formation.

Safflower oil, because of its outstanding quick drying and excellent initial color with minimum after-yellowing, has come into wider use than has linseed oil. An analysis of the properties of the various oils is given in Table IV.

A typical oil-modified alkyd resin synthesis using linseed oil is given in Preparation 2-2.

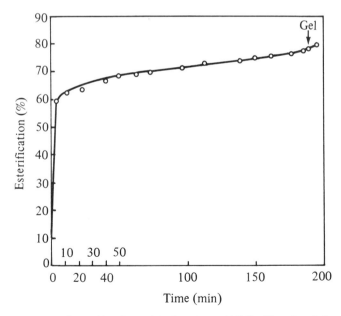

FIG. 4. Increase of esterification with time ($T = 125°C$). [Reprinted from N. H. Kienle and A. G. Hovey, *J. Am. Chem. Soc.* **51**, 509 (1929). Copyright 1929 by the American Chemical Society. Reprinted by permission of the copyright owner.]

TABLE III

VARIATION OF PROPERTIES OF PRODUCTS WITH TIME AT 195°C[a]

Time (sec)	Time (min)	Flow pt. (°C)	Refr. index	Acid number	Sapon. number	Ester value	Ester (%)
0	0	—	—	(535)	—	0	0.0
215	3.58	64	—	235	584	349	59.8
720	12.16	69	—	218	582	364	62.5
1380	23.0	76	1.56	212	583	371	63.6
2440	40.67	78	—	191	570	379	66.5
3420	50.0	81	—	185	589	404	68.5
3980	63.0	82.5	—	181	591	410	69.4
4380	73.0	83	—	177	588	411	69.9
5740	96.5	86	—	169	589	420	71.4
6780	113.0	91	—	163	605	442	73.0
8340	139.0	96	—	159	611	452	73.8
9000	150.0	96	—	151	604	453	75.0
9720	162.0	105	—	149	605	456	75.5
10,610	177.0	106	—	145	608	463	76.1
11,150	186.0	118	1.58	139	614	475	77.3
11,340	189.0	Gel	—	136	585	449	77.8
11,700	195.0	Gel		119	574	455	79.3

[a] Reprinted from R. H. Kienle and A. G. Hovey, *J. Am. Chem. Soc.* **51**, 509 (1929). Copyright 1929 by the American Chemical Society. Reprinted by permission of the copyright owner.

TABLE IV

COMPONENT FATTY ACIDS IN OILS USED IN COATING COMPOSITIONS[a]

Fatty acid	Percent acid in various oils			
	Linseed[d]	Safflower[e]	Soybean[f]	Dehydrated castor[f]
Linolenic	52.0	0.0	9:0	—
Linoleic	16.0	78.0	51.0	65.0
Conjugated[b]	—	—	—	22.0
Olein	22.0	13.0	25.0	7.5
Saturates[c]	10.0	9.0	15.0	0.6
Hydroxy acids	—	—	—	5.0

[a] Reprinted from A. E. Rheineck and L. O. Cummings, *J. Am. Chem. Soc.* **43**, 409 (1966). Copyright 1966 by the American Oil Chemists Society. Reprinted by permission of the copyright owner.

[b] 9,11-Octadecadienoic acid.

[c] Comprises myristic, palmitic, stearic, and archidic acids and small amounts of other monoenoic acids.

[d] Liquid chromatography chart published by Archer-Daniels-Midland, Minneapolis, Minn. 1961. Mol. wt. 878.

[e] W. Ibrahim, J. Iverson, and D. Firestone, *J. Assoc. Off. Agric. Chem.* **47**, 776 (1964).

[f] R. L. Terrill, *J. Am. Oil Chem. Soc.* **27**, 471 (1950).

2-2. Preparation of Linseed Oil-Modified Poly(glyceryl phthalate) Alkyd Resin by the Fusion Process [38]

$$\tag{5}$$

To a resin kettle equipped as in Preparation 2-1 and containing a carbon dioxide sparge are added 336 gm of linseed oil (0.383 mole, 1.15 equiv.) and 80 gm (0.87 mole, 2.6 equiv.) of glycerin, and the mixture is heated to 225°F. Then 0.15 gm of calcium hydroxide catalyst is added and the alcoholysis completed at 450°F in about 30–40 min. (One volume of resin is soluble in 4 parts of boiling methanol.) The remainder of the glycerin (28 gm = 0.305 mole, 0.915 equiv.) is added and after 10 min 250 gm (1.69 moles, 3.39 equiv.) of phthalic anhydride is added. The temperature of the reaction mixture is raised to 450°F and the batch held at 450°F until the reaction is complete (no condenser used), i.e., until an acid value of 5–7 is obtained and a 50% mineral spirits sample has a Gardner viscosity of W (Gardner Standard Tubes used). The total processing time is approximately 5–6 hr.

A typical safflower oil alkyd involves the following formulation [39].

Safflower oil, 797 gm	⎫
Litharge, 0.4 gm	⎬ Alcoholysis
Pentaerythritol, 225 gm	⎭
Phthalic anhydride, 346 gm	

After condensation to a resin the mixture is made up to 70% solids in mineral spirits. The solution dries in 30 hr to a tack-free film.

A short oil alkyd (35–45% oil content and >35% phthalic anhydride) prepared from soya oil by a solvent cook procedure is given below (Preparation 2-3). A short alkyd is characterized by having fast setting and fair air drying but good baking, adhesion, color, color retention, and gloss. These alkyds also have fair flexibility ratings.

2-3. Preparation of a Soya Oil-Modified Poly(glyceryl phthalate) Alkyd Resin by the Solvent Process [40]

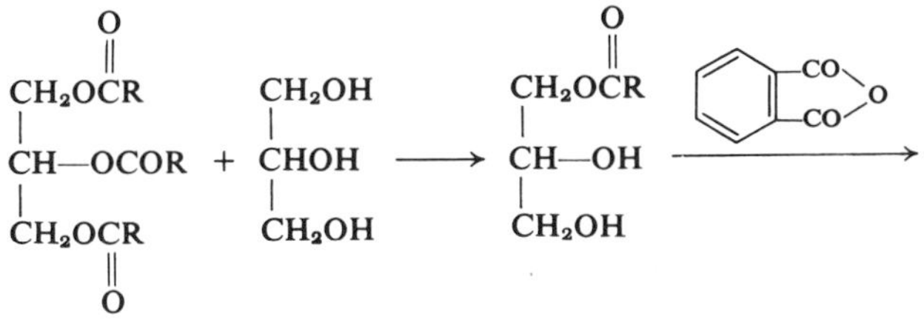

(Soya oil; see Table IV)

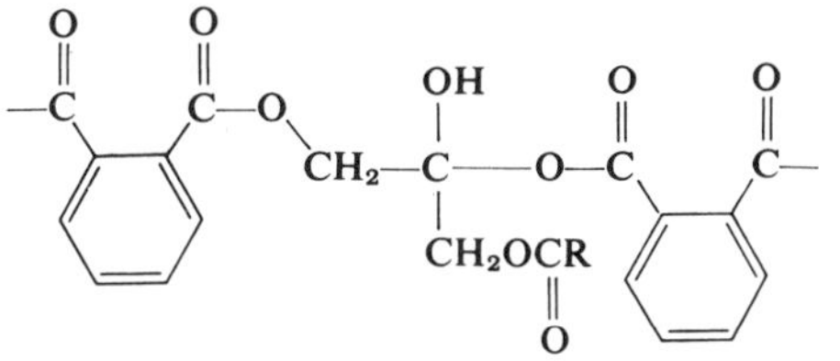

(6)

To a resin kettle equipped as in Preparation 2-1 are added 160 gm (0.183 mole) of soya oil, 110 gm (1.21 moles) of glycerol, and 0.25 gm of litharge catalyst. The reaction mixture is heated to 233°C while carbon dioxide is bubbled through the reaction and the temperature is held at 233°C until the mixture gives a clear methanol test (alcoholysis is complete if 1 volume of the alcoholysis product is soluble in 4 volumes of methanol). The mixture is cooled to 138°C and 180 gm (1.21 moles) of phthalic anhydride is added along with 18 gm of xylene. The mixture is heated to reflux for approximately 6 hr until 22.0 gm (1.21 moles) of water is collected in the Dean & Stark trap. Then 440 gm of xylene is added to give a 50% solids solution with an acid value of 4.0. This resin has 37.5% excess hydroxyls and a functionality of 1.95. It will form a hard film on air drying in 4–5 hr. The use of 0.03–0.07% cobalt naphthenate is recommended as an aid to speed drying.

A medium oil alkyd (46–55% oil content and 30–35% phthalic anhydride in final resin) is one that has good water impermeability, fair baking and color, usual application, and is soluble in aliphatic solvents such as mineral spirits and naphthas. The modifying oil commonly used here is linseed or soya.

A typical fusion process formulation for a medium oil alkyd resin is as follows [41].

	Parts	Equivalents
Pentaerythritol	100	2.92
		(27% excess OH)
Tall oil fatty acid	200	0.69
Phthalic anhydride	120	1.61
		(resin functionality
		2.08)

The mixture is heated to 550°F and after 1 hr of heating nitrogen is bubbled through until an acid value of 10 is obtained (approximately 6 hr). The mixture is diluted to 50% solids with mineral spirits.

A long oil (56–70% oil content; 20–30% phthalic anhydride in final resin) alkyd resin is one that has good flexibility, solubility, and brush application. This alkyd resin requires aliphatic solvents such as mineral spirits and is slow drying.

A typical fusion process formulation of a long oil alkyd is given below [42].

	Parts	Equivalents
Glycerin	80	2.57
Linseed oil	330	1.22

Both glycerin 80 gm (2.57 moles) and linseed oil 360 gm (1.22 moles) are heated to 260°C and then 160 gm (1.94 equiv.) of phthalic anhydride is added. The mixture is cooked until an acid value of 10 is reached and then the resin is cooled and diluted to 70% solids with mineral spirits.

As mentioned earlier benzoic acid (or *p-tert*-butylbenzoic acid) is sometimes added to alkyd resin formulations to control the molecular weight by acting as a molecular chain stopper [43,43a]. A typical formulation of such an alkyd is shown in Preparation 2-4.

2-4. *Preparation of an Alkyd Resin with Benzoic Acid to Control Molecular Weight* [43]

In a resin kettle equipped as previously described are heated 94.5 gm (0.65 mole) of alkali-refined soya oil, 23.5 gm (0.26 mole) of glycerin (98%), and 0.05 gm of calcium hydroxide for $\frac{1}{2}$ hr at 450°F with good stirring and using a carbon dioxide blanket. Alcoholysis is checked by mixing 1 part of the reaction mixture with 4 parts of anhydrous boiling methanol. A clear solution of the boiling methanol indicates proper alcoholysis. At this point

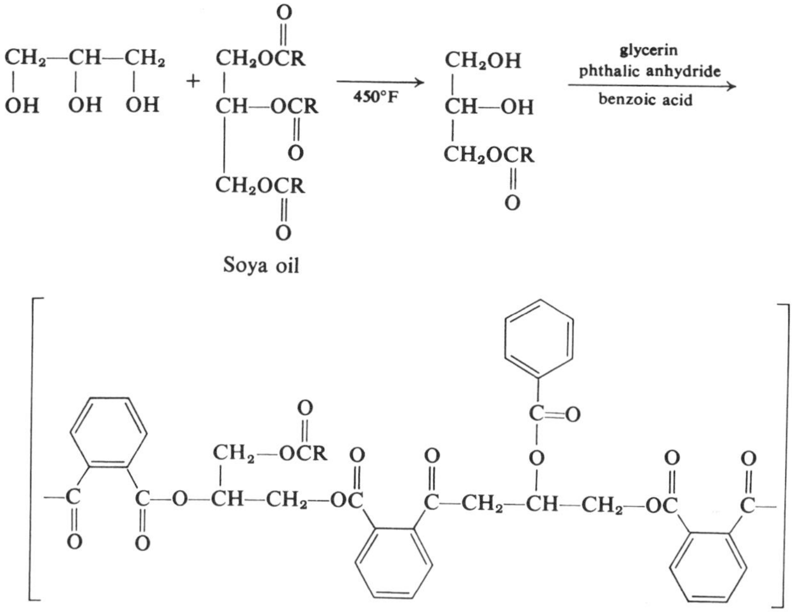

$$
\begin{array}{c}
\underset{\substack{| \quad | \quad | \\ \text{OH} \quad \text{OH} \quad \text{OH}}}{\text{CH}_2-\text{CH}-\text{CH}_2} \;+\; \underset{\substack{| \\ \text{CH}-\text{OCR} \\ \underset{\text{O}}{\overset{\|}{}} \\ \text{CH}_2\text{OCR} \\ \underset{\text{O}}{\overset{\|}{}}}}{\overset{\overset{\text{O}}{\|}}{\text{CH}_2\text{OCR}}} \;\xrightarrow{\;450°F\;}\; \underset{\substack{| \\ \text{CH}-\text{OH} \\ | \\ \text{CH}_2\text{OCR} \\ \underset{\text{O}}{\overset{\|}{}}}}{\text{CH}_2\text{OH}} \;\xrightarrow[\text{benzoic acid}]{\substack{\text{glycerin} \\ \text{phthalic anhydride}}}
\end{array}
$$

Soya oil

(7)

are added 39.0 gm (0.42 mole) of glycerin (98%), 113.0 gm (0.76 mole) of phthalic anhydride, and 16.7 gm (0.14 mole) of benzoic acid (95%) [glycerin: phthalic anhydride 2.4:3 (20% excess glycerin)]. While carbon dioxide is bubbled through, the reaction mixture is heated slowly from 290°F to 440°F over a 1-hr period, and heating at 440°F is continued until the acid number of 13.6 is reached. The product is diluted to 50% solids with xylene to give a resin that air dries in 215 min (with the aid of 0.5 gm lead or 0.05 gm cobalt naphthenate-type catalyst) to a dry, hard film. Baking for $\frac{1}{2}$ hr at 250°F speeds the curing time to 33 min.

Trimethylolpropane $[\text{C}_2\text{H}_5\text{C}(\text{CH}_2\text{OH})_3]$ (m.p. 58.8°C) can be substituted for glycerin in many alkyd resin preparations to give improved heat and light stability to the final products. Some typical formulations (1–4) are shown below and are prepared by the solvent cook procedure to an acid number of 14–17 [44].

	Formulations			
Parts by weight	(1)	(2)	(3)	(4)
Soybean oil	39.1	29.9	—	—
Glycerin	20.9	—	—	—
Trimethylolpropane	—	30.2	33.6	31.86
Phthalic anhydride	140	140	39.7	35.6
Tall oil fatty acids	—	—	26.7	32.5

TABLE V

PREPARATION OF ALKYD RESINS BY THE REACTION OF GLYCEROL WITH VARIOUS ACIDS[a]

Glyceryl polymeride (alkyd resin)	Moles resin	Reaction temp. (°C)	Gelation time (min)	Water evolved (cm³)	Acid value at gelation (obs.)	Acid value (calc.)		Percent esterification	Average mol. wt.
						Dimer	Tetramer		
Phthalate (acid)	$\frac{1}{8}$	190	160	10.80	120	139	117	78.6	910–1120
Succinate	$\frac{1}{8}$	190	64	9.62	178–189	183	154	74.5–76.0	1075–1085
Adipate	$\frac{1}{8}$	190	83	9.64	152–159	155	130	74.0–75.5	1260
Sebacate	$\frac{1}{8}$	190	107	9.78	115–118	119	99	77.7	1450
Maleate	$\frac{1}{4}$	160	57	5.49	230	187	157	68.5	1050[b]

[a] Reprinted in part from R. H. Kienle and F. E. Petke, *J. Am. Chem. Soc.* **63**, 481 (1944). Copyright by the American Chemical Society. Reprinted by permission of the copyright owner.

[b] R. H. Kienle and F. E. Petke, *J. Am. Chem. Soc.* **62**, 1053 (1940).

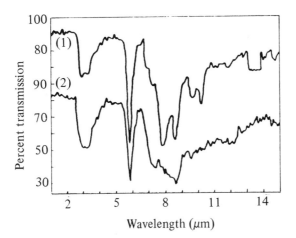

FIG. 5. Infrared spectra: (1) glyceryl maleate, A.V. = 231; (2) glyceryl succinate, A.V. = 182. [Reprinted from R. H. Kienle and F. E. Petke, *J. Am. Chem. Soc.* **62**, 1053 (1940). Copyright 1940 by the American Chemical Society. Reprinted by permission of the copyright owner.]

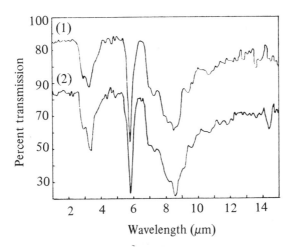

FIG. 6. Infrared spectra: (1) glyceryl adipate, A.V. = 159; (2) glyceryl sebacate, A.V. = 118. [Reprinted from R. H. Kienle and F. E. Petke, *J. Am. Chem. Soc.* **63**, 481 (1941). Copyright 1941 by the American Chemical Society. Reprinted by permission of the copyright owner.]

The molecular weight distribution in alkyd resins prepared using castor oil or hydrogenated castor oil has been reported [45].

B. Alkyds Based on Other Acids

Kienle and Petke were among the first to study the effect on the ratio and properties of alkyd resin formation of acids other than phthalic acid or anhydride. In all cases where either aliphatic or aromatic acids were used the greatest time involved the first half of the reaction. The acid values near gelation were between the calculated values for the dimer and tetramer. The percent esterification attained at gelation was about the same for the different dibasic acids (approximately the same amount of water evolved). A summary of some other acids in comparison to phthalic anhydride is shown in Table V.

The infrared spectra of the alkyd resins from the aliphatic acids are similar as seen in Figs. 5 and 6.

2-5. *Preparation of Glyceryl Adipate* [46]

$$
\begin{array}{c}
\text{CH}_2\text{—CH—CH}_2 \ + \ \text{HOC—(CH}_2)_4\text{—C—OH} \longrightarrow \\
\underset{\text{OH}}{|} \quad \underset{\text{OH}}{|} \quad \underset{\text{OH}}{|}
\end{array}
$$

$$
\left[
\begin{array}{c}
\text{CH}_2\text{—O—} \\
| \\
\text{CH—OH} \\
\text{CH}_2\text{O—(CH}_2)_4\text{—C—OCH}_2 \\
\text{CH—OC—(CH}_2)_4\text{—C—} \\
\text{—C—(CH}_2)_4\text{—C—OCH}_2
\end{array}
\right] \quad (8)
$$

In a resin flask equipped as earlier described and equipped with a nitrogen bubbler are heated 23.0 gm (0.25 mole) of glycerol and 54.8 gm (0.375 mole) of adipic acid. The experimental data for this alkyd resin are found in Table V and Fig. 7. The infrared spectrum of the product is shown in Figs. 5 and 6.

The alkyd resins prepared from the aliphatic polyacids [47] (sebacic, azelaic, and adipic) were used as caulking compounds during World War II, particularly for aircraft. These products also have valuable features as non-migrating plasticizers for film formers such as cellulose derivatives [48].

The addition of oils gives resins that may be useful for coated fabric

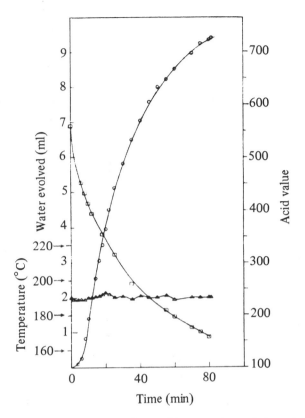

FIG. 7. Reaction data for glyceryl adipate with proportions of glycerol 23.0 gm and adipic acid 54.8 gm; △, temperature; □, acid value; ○, water evolved. [Reprinted from R. H. Kienle and F. E. Petke, *J. Am. Chem. Soc.* **63**, 481 (1941). Copyright 1941 by the American Chemical Society. Reprinted by permission of the copyright owner.]

applications and in lacquers for rubber articles such as cable coatings and automotive finishes [47].

In recent years the availability of several aromatic polycarboxylic acids has been exploited in specialty alkyd resin preparations [49,49a]. For example, trimellitic acid-based alkyds are used in alkyd–melamine enamels and acrylic lacquers [49]. Isophthalic acid can be used alone or in conjunction with tri-mellitic anhydride or phthalic anhydride [50]. Pyromellitic acid and pyro-mellitic anhydride have also been suggested as partial replacements for phthalic anhydride in long or short oil alkyds to give more water-resistant (also caustic- and gasoline-resistant) coatings [51].

Some typical formulations are described in Table VI.

TABLE VI

PREPARATION OF ALKYD RESINS BASED ON AROMATIC POLYCARBOXYLIC ACIDS

Parts by weight							
Polycarboxylic acid or anhydride	Polyol	Oil	Catalyst	Temp. (°F)	Time (hr)	Acid no.	Ref.
Trimellitic anhydride (307)–adipic acid (78)	Neopentyl glycol (388)	Tall oil fatty acids (304)	—	360–380	5–6	50–55	a
Trimellitic anhydride (336)–adipic acid (85)	Neopentyl glycol (424)	Coconut oil fatty acids (242)	—	360	8	50–55	b
Isophthalic acid (274)–trimellitic anhydride (70)–acrylic prepolymer (50% NVM)[e] (432)	Trimethylolpropane (273)	Safflower oil (160)–castor oil (79)	Litharge (0.2)	400–460	8	51–52	b
Isophthalic acid (243)	Glycerin (112)	Soya oil (420)	Litharge (0.168)	450	6	11	c
Trimellitic anhydride (156)	Glycerin (80)	Linseed oil (756)	—	400	6–8	5	d

[a] " Amoco TMA in Primers for Alkyd-Melamine Enamels and Acrylic Lacquers," Tech. Bull. TMA 25a. Amoco Chem. Corp., Chicago, Illinois, 1974.

[b] "Water-Soluble Resins for Industrial Topcoats Using Amoco TMA DO 870," Tech. Bull. TMA-20b. Amoco Chem. Corp., Chicago, Illinois, 1974.

[c] C. R. Martens, "Alkyd Resins," p. 139. Van Nostrand-Reinhold, Princeton, New Jersey, 1961.

[d] J. R. Eisner, R. S. Taylor, and B. A. Bolton, *Paint Varn. Prod.* **49**, 54 (1959).

[e] NVM, nonvolatile material.

2-6. Preparation of a Trimellitic Anhydride-Based Alkyd Resin No. 3712NT [49]

COOH

$+ (CH_2)_4(COOH)_2 + HOCH_2{-}\underset{\underset{CH_3}{|}}{\overset{\overset{CH_3}{|}}{C}}{-}CH_2OH + $ tall oil fatty acids $\longrightarrow$

$$R = \text{neopentyl} \tag{9}$$

$C\!\sim$ derived from tall oil fatty acids

CAUTION: All operations with trimellitic anhydride (or acid) must be carried out in a well-ventilated hood using proper personal protection such as gloves, laboratory apron–coat, and chemical goggles. Exposure to trimellitic anhydride may cause eye or skin irritation or lung irritation (pulmonary edema).

A resin kettle is charged with 307 gm (1.6 moles) of trimellitic anhydride (TMA), 388 gm (2.85 moles) of neopentyl glycerol, 78 gm (0.535 mole) of adipic acid, and 304 gm (1.07 equiv.) of tall oil fatty acids [97.6% fatty acid, 1.2% rosin acid (acid number 197)] and agitation and inert gas sparging is begun. The temperature is raised to 370–380°F over a 1–2-hr period to control the evolution of water. The temperature is held at 370°–380°F to obtain an acid number of 55–60. (The resin is insoluble in a conventional benzene-ethanol mixture; therefore, acid numbers are determined in acetone. Two to three hours are required to obtain this acid number after reaching 370°–380°F. The resin is cooled to 360°–370°F and held at that temperature until the acid number is 50–55 (see Fig. 8). The total processing time is 5–6 hr, and 1000 gm of resin is obtained. The resin is diluted with a mixture of 85 parts water to 15 parts *tert*-butyl alcohol and solubilized with *N,N*-dimethylethanolamine to give a pH 7–8 and a 40% solids (NVM) resin of Gardner Color 2–4 and Gardner–Holdt viscosity Z-2.

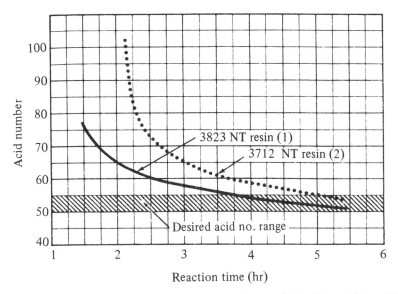

FIG. 8. Reaction time by acid number. (1) Heated to 380°F in 1½ hr; held at 380°F for 2 hr, lowered to 360°F, and held for properties. (2) Heated to 370°F in 1 hr and held for properties. Resin 3823 NT is similar to 3712 NT but contains TMA (244 gm), neopentyl glycol (352 gm), adipic acid (123 gm), tall oil fatty acids (362 gm). It is processed in a manner similar to resin 3712 NT. [Reprinted from "Amoco TMA in Primers for Alkyd–Melamine Enamels and Acrylic Lacquers," Tech. Bull. TMA 25a. Copyright by the Amoco Chemical Corporation, Chicago, Illinois, 1974. Reprinted by permission of the copyright owner.]

3. MODIFIED ALKYDS

Alkyd resins have been reported to be modified by a variety of starting materials (rosin acid–maleic anhydride adducts) and other resins with reactive or exchangeable groups. Some typical resins used to modify alkyds are [52] phenolic resins (OH groups), polyamide resins (NH_2 groups), epoxy resins [34a,53] (epoxy and OH groups), polyurethane resins (NCO groups), amine resins (melamine–formaldehyde or urea–formaldehyde) [34], and vinyl resins containing reactive groups.

$$(HOCH_2CH_2)_2NCR' \ + \ R(COOH)_2 \ \xrightarrow[140°C]{xylene}$$

$$H \left[OCH_2CH_2N-CH_2CH_2OCRCO \right]_n OH \quad (10)$$
$$\underset{COR'}{|}$$

TABLE VII
AIR-DRIED FILM PROPERTIES OF POLYESTER AMIDES [a,b]

Acid	Sward rocker hardness 3 days	10 days	25 days	Dry to touch (hr)	Tack free (hr)	Resistance to 5% NaOH (min)	Resistance to xylene (hr)
Terephthalic	5	20	46	1	5	51	220
Phthalic	5	16	32	2.7	21	13	220
Fumaric	20	28	48	3.5	8	4	220
Maleic	8	28	34	32	48	4	32
Itaconic	8	22	38	3	21	4	32
Dimer T[c]	—	—	—	90	500	—	—
Dimer C[c]	—	—	—	168	500	—	—
Azelaic	—	—	—	250	250	9	220
Brassylic	—	—	—	72	500	3	220
Soy alkyd[d]	8	14	18	2	22	6	0

[a] Contains 0.5% Pb + 0.01% Co naphthenates used as driers.

[b] Reprinted from L. E. Gast, W. J. Schneider, and J. C. Cowan, *J. Am. Oil Chem. Soc.* **43**, 418 (1966). Copyright 1966 by the Journal of the American Chemists Society. Reprinted by permission of the copyright owner.

[c] T = prepared by thermal process; C = prepared by a catalytic process.

[d] Sixty-five percent oil-length alkyd.

Recently, polyester amides were prepared by heating linseed diethanolamide in refluxing xylene with dibasic acids or anhydrides to give modified alkyd resins [54,55].

Some typical acids used and the air-dried film properties obtained are shown in Table VII.

3-1. *Preparation of Polyester Amide by the Reaction of N,N-Bis(2-hydroxyethyl)linseed Amide and Dibasic Acids* [54]

Method A: Preparation of N,N-Bis(2-hydroxyethyl)linseed Amide. To a round-bottomed flask fitted with a stirrer, thermometer, nitrogen inlet tube, and dropping funnel is added 210 gm (0.2 mole) of distilled diethanolamine (b.p.$_{25}$ 170°C, n_D^{28} 1.4740). The contents are warmed to 100°–105°C and 0.13 gm (0.0024 mole) of sodium methoxide catalyst is added. Linseed methyl ester (29.5 gm, 0.1 mole) is then added dropwise over a 15-min period. After the addition the reaction is continued under reduced pressure (20 mm Hg) to facilitate removal of methanol. The reaction mixture is cooled, dissolved in ether, washed with 15% aqueous sodium chloride, and dried over sodium sulfate. The ether solution is filtered and the ether removed to afford 34 gm (93%) of a yellow-orange oil.

TABLE VIII
PROPERTIES OF POLYESTER AMIDES OF HELA AND
DIBASIC ACIDS[a]

| | Gardner | | | |
| | Viscosity | Color | Iodine value | Acid no. |
Acid				
Terephthalic	≫Z10	17	117	13.3
Phthalic	>Z10	15	114	6.6
Fumaric	Z5–6	12	135	1.0
Maleic	≫Z10	13	122	21.3
Itaconic	Z7–8	15	125	1.4
Dimer T[b]	V–W[c]	10	116	8.0
Dimer C[b]	Z7	14	148	19.0
Azelaic	Z6	15	107	17.0
Brassylic	Z8–9	15	95	12.1

[a] Reprinted from L. E. Gast, W. J. Schneider, and J. C. Cowan, *J. Am. Oil Chem. Soc.* **43**, 418 (1966). Copyright 1966 by the Journal of the American Oil Chemists Society. Reprinted by permission of the copyright owner.
[b] T = prepared by thermal process; C = prepared by catalytic process.
[c] Fifty percent solids in toluene.

Method B: Alkyd Resin Preparation. To a round-bottomed flask fitted with a mechanical stirrer, condenser, Dean & Stark trap, and nitrogen inlet/outlet tubes are added 0.05 mole of N,N-bis(2-hydroxyethyl)linseed amide (HELA), 50 ml of xylene, and 0.05 mole of a given dibasic acid. The reaction mixture is refluxed (140°–150°C) while nitrogen is bubbled slowly through it until the theoretical amount of water is collected. The xylene is removed under reduced pressure in a rotating evaporator to give polyester amides with properties as described in Table VIII.

It should be noted that in preparing epoxy alkyd resins the functionality of a typical epoxy resin (Epon 1001) is 6 due to 2 OH groups and 2 epoxy groups (4 functional sites). A typical recipe is as follows [56].

	Moles used in alkyd resin
Reaction product of Epon 1001 (Shell Chemical Co.) (1.0 mole) and fatty acid (3.0 moles)	0.25
Glycerin	0.75
Phthalic anhydride	1.00
Fatty acid	1.00
Functionality/mole $= \frac{6}{3} = 2.0$	

4. ALKYDS MODIFIED WITH VINYL MONOMERS

The unsaturated sites in alkyds are capable of undergoing further reactions such as Diels–Alder (maleic acid or anhydride [22,57], cyclopentadiene [58]) and polymerization [59] in the presence of catalysts or other monomers, styrene [60], vinyl toluene [61,61a], methyl methacrylate [62], acrylic [62], and acrylonitrile [63].

The availability of styrene at low cost in large commercial quantities led to its use in applications where quick drying, chemical resistance and good weathering alkyds were required [64]. Mixtures of several monomers also have been occasionally used provided copolymerization between monomers is possible [65].

In preparing styrenated alkyds there are four possible methods [66]:

1. Prestyrenation of reactive oil, followed by monoglyceride formation with a polyol and subsequent reaction with a dibasic acid
2. Prestyrenation of a fatty acid followed by alkyd formation
3. Prestyrenation of unsaturated monoglyceride before alkyd formation
4. Poststyrenation of an oil-modified alkyd

The last method [67] is used in the presence of solvent and the alkyd is prepared in the normal fashion but the reaction is stopped short to allow further esterification to take place during the styrenation procedure.

Since styrene-modified alkyds are the most important commercially an example of a typical preparation is described below (Preparation 4-1).

Rohm & Haas Co. has suggested the use of methacrylates in alkyd resin modification [68].

4-1. Preparation of a Styrene-Modified Linseed–Trimethylolethane Alkyd [61a]

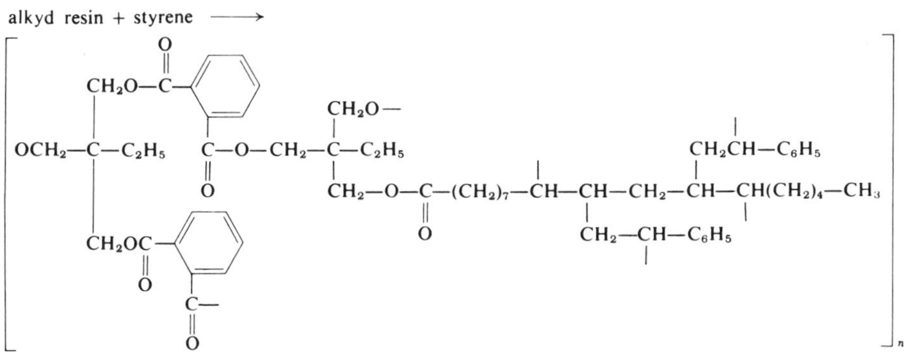

alkyd resin + styrene ⟶

(11)

Method A: To a 1-liter resin flask equipped with stirrer, thermometer, Dean & Stark trap, and condenser are added 130.0 gm (0.88 mole) of phthalic anhydride, 260.5 gm (0.925 mole) of linseed oil (fatty acids), 114.5 gm (0.956 mole) of trimethylolethane, and 30 ml of xylene. The reaction mixture is heated to 230°C over a 1-hr period and kept there until the water of esterification is removed and an acid value of 13 ± 3 is reached. The reaction mixture is cooled to 130°C and diluted to 70% solids with xylene.

Method B: To 200 gm of the above alkyd resin (70% xylene solution) in a 500 ml three-necked round-bottomed flask equipped with a thermometer, stirrer, and reflux condenser is added at reflux over a 2-hr period 140 gm (1.34 moles) of styrene containing 3.5 gm of di-*tert*-butyl peroxide. The reflux is continued for a total of 6 hr and then the reaction cooled to 130°C and diluted to 60% solids with xylene.

In the latter preparation the alkyd resin is prepared by the simultaneous esterification of linseed fatty acids and phthalic anhydride with trimethylolethane.

In another procedure the alkyd resin is prepared in the usual manner by alcoholysis of the fatty acid with the polyol followed by phthalic anhydride condensation. Then the alkyd, diluted in xylene to 66% solids, is reacted with styrene monomer using di-*tert*-butyl peroxide catalyst [24].

5. ALKYDS MODIFIED
WITH OTHER THERMOSETTING RESINS

Nonoxidizing (nondrying) alkyds can be cured to very durable finishes by baking with amino resins such as urea [69] or melamine–formaldehyde [70]. Phenolic-modified alkyds are used for coatings requiring alkali resistance. In addition alkyd resins can be modified with epoxies and polyurethanes or other thermosetting resins to give coatings of improved performance [71–73]. For example, urethane oils are prepared by reacting diglycerides with toluene diisocyanate. The resulting oils are used in alkyd resin preparations or as additives for coatings.

A typical example [74] involves the preparation of an alkyd resin from phthalic anhydride (712 gm), trimethylolpropane (399 gm), isodecyl alcohol (277 gm), and xylene (22 gm) and heating (190°–280°C) to remove water azeotropically and give a resin of acid value 23. The product is diluted with a 4:1 xylene:butyl alcohol mixture to 75% solids. A lacquer is prepared by mixing 16 gm of the above alkyd resin solution with 3 gm of a methylated melamine–formaldehyde condensate along with 7.5 gm of titanium oxide and

0.225 gm of *p*-toluenesulfonic acid monohydrate. Sufficient xylene:butanol (4:1) is added to give a 55% solids solution. A film of this solution is dried for 30 min at 120°C. The film is hard and stable to 10% sodium hydroxide.

6. MISCELLANEOUS ALKYD PREPARATIONS

1. Imide-modified alkyds [75].
2. Alkyd resins based on glycerol allyl ether [76,77].
3. Tetrachlorophthalic anhydride alkyd resins with fire-retardant properties [78].
4. Maleic anhydride–cyclopentadiene adduct (3,6-endomethylene 1,2,3,6-tetrahydrophthalic anhydride) in alkyd resin preparations [79].
5. Polyethylene glycol-containing alkyd resins [80].
6. Reaction of alkyd resin terminal hydroxyl groups with ketene [81] and acetic anhydride [82].
7. Oilless alkyds [83].
8. Aminoplast resin for alkyd resin [84].
9. Ultraviolet light-hardenable pigmented alkyd resin lacquer preparations [85].
10. Zinc chromate-containing alkyd resin [86].
11. Triphenyl phosphite-containing alkyd resin of improved color and drying properties [87].
12. Transesterification catalysts in alkyd resin preparation [88].
13. Water-soluble alkyd resins [89].
14. Mill base alkyd resin formulations [90].
15. Water-dispersible alkyd resins [64].

REFERENCES

1. R. H. Kienle and A. G. Hovey, *J. Am. Chem. Soc.* **51**, 509 (1929).
2. S. R. Sandler and W. Karo, "Polymer Syntheses," Vol. 1, pp. 55–72. Academic Press, New York, 1974.
3. J. Berzelius, *Rapp. Annu. Inst. Geol. Hong.* **26**, 1 (1847).
4. M. M. Berthelot, *C. R. Hebd. Seances Acad. Sci.* **37**, 398 (1853).
5. J. von Bemmelen, *J. Prakt. Chem.* **69**, 84 and 93 (1856).
6. H. Dubus, *Philos. Mag.* [4] **16**, 438 (1858); *Jahrb. Fortschr. Chem.* p. 431 (1856).
7. A. V. Lourenco, *Ann. Chim. Phys.* [3] **7**, 67 and 313 (1863).
8. A. Furaro and F. Danesi, *Gazz. Chim. Ital.* **10**, 56 (180); *Jahrb. Forstchr. Chem.* p. 799 (1880).

9. W. Smith, *J. Soc. Chem. Ind., London* **20**, 1073 (1901).
10. M. J. Callahan, U.S. Patents 1,108,329; 1,108,330; 1,108,332; 1,091,627; 1,091,628; and 1,091,732 (1914).
11. L. H. Friberg, U.S. Patent 1,119,592 (1914).
12. W. C. Arsem, U.S. Patents 1,098,776 and 1,097,777 (1914).
13. E. S. Dawson, U.S. Patent 1,141,944 (1915).
14. K. B. Howell, U.S. Patent 1,098,728 (1914).
15. R. H. Kienle and C. S. Ferguson, *Ind. Eng. Chem.* **21**, 349 (1929).
15a. R. H. Kienle, U.S. Patent 1,893,873 (1933).
16. H. D. Gibbs, *J. Ind. Eng. Chem.* **11**, 1031 (1919).
17. G. Christensen and P. Fink-Jensen, *J. Chromatogr. Sci.* **12**, 59 (1974).
18. T. Nagata, *J. Appl. Polym. Sci.* **13**, 2277 and 2601 (1969).
19. P. J. Flory, *J. Am. Chem. Soc.* **63**, 3083 (1941).
20. M. Jonason, *J. Appl. Polym. Sci.* **4**, 120 (1960).
21. E. G. Bobalek, E. R. Moore, S. S. Levy, and C. C. Lee, *J. Appl. Polym. Sci.* **8**, 625 (1964).
22. C. R. Martens, "Alkyd Resins," pp. 51–59. Van Nostrand-Reinhold, Princeton, New Jersey, 1961.
23. A. G. Hovey, *Ind. Eng. Chem.* **41**, 730 (1949).
24. "The Chemistry and Processing of Alkyd Resins." Monsanto Co., 1952 (reprinted 1972).
25. E. C. Haines, U.S. Patent 2,396,698 (1944); E. Beres, *Przem. Chem.* **50**, 273 (1971); *Chem. Abstr.* **75**, 49921p (1971).
25a. R. Seaborne, *Paint Technol.* **19**, 6 (1955).
26. *U.S. Tariff Comm., Rep.* [2] **203**, 40 (1957).
27. *U.S. Tariff Comm., Prelim. 1973 Data* p. 3 (1974).
28. A. G. Roberts, *Build. Sci. Ser., Natl. Bur. Stand. (U.S.)* **7**, 43 (1968).
29. W. A. Dannels, *Mod. Plast.* **51**, (10A), 15–16 (1974).
30. C. R. Martens, "Alkyd Resins," pp. 132–151. Van Nostrand-Reinhold, Princeton, New Jersey, 1961.
31. Data obtained from "Water-Soluble Resins for Industrial Topcoats Using Amoco TMA (Trimellitic Anhydride)," Tech. Bull. TMA-20b. Amoco Chem. Corp., Chicago, Illinois, 1974.
32. R. G. Mraz and R. P. Silver, *Encycl. Polym. Sci. Technol.* **1**, 663 (1964), *Kirk-Othmer, Encycl. Chem. Technol., 2nd Ed.* Vol. 1, p. 851 (1963).
33. T. C. Patton, "Alkyd Resin Technology." Wiley (Interscience), New York, 1962.
34. J. N. Butler, *Paint Varn. Prod.* **49**, 47 (1959).
34a. T. Mika, *Paint Varn. Prod.* **49**, 51 (1959).
34b. J. S. Long, *Paint Varn. Prod.* **49**, 60 (1959).
35. E. G. Bobalek, *Paint Varn. Prod.* **49**, 31 (1959).
36. R. H. Kienle and A. G. Hovey, *J. Am. Chem. Soc.* **51**, 509 (1929).
37. R. H. Kienle, P. A. Van Der Menlen, and F. E. Petke, *J. Am. Chem. Soc.* **61**, 2268 (1939).
38. E. S. Dawson, U.S. Patent 1,888,849 (1932); Monsanto Co., "The Chemistry and Processing of Alkyd Resins," p. 38, 1952.
39. A. E. Rheineck and L. O. Cummings, *J. Am. Chem. Soc.* **43**, 409 (1966).
40. C. R. Martens, "Alkyd Resins," pp. 71–73. Van Nostrand-Reinhold, Princeton, New Jersey, 1961.
41. C. R. Martens, "Alkyd Resins," pp. 74–75. Van Nostrand-Reinhold, Princeton, New Jersey, 1961.

42. C. R. Martens, "Alkyd Resins," pp. 77–78. Van Nostrand-Reinhold, Princeton, New Jersey, 1961.
43. R. L. Heinrich, D. A. Berry, R. L. Christian, and E. R. Meuller, *Am. Chem. Soc., Div. Paint, Plast. Chem., Pap.* **19**, 241 (1959).
43a. J. T. Geoghegan and W. E. Bambrick, *J. Paint Technol.* **44**, 84 (1972); C. R. Martens, "Alkyd Resins," pp. 80–82. Van Nostrand-Reinhold, Princeton, New Jersey, 1961.
44. "Technical Brochure on Trimethylolpropane." Celanese Chem. Co., (1971).
45. D. H. Solomon and J. J. Hopwood, *J. Appl. Polym. Sci.* **10**, 993 (1966).
46. R. H. Kienle and F. E. Petke, *J. Am. Chem. Soc.* **63**, 481 (1941).
47. Staff Report, *Paint Varn. Prod.* **51**, 35 (1961).
48. E. M. Beavers, *Ind. Eng. Chem.* **41**, 738 (1949).
49. "Amoco TMA in Primers for Alkyd-Melamine Enamels and Acrylic Lacquers," Tech. Bull. TMA 25a. Amoco Chem. Corp., Chicago, Illinois, 1974.
49a. J. R. Eisner, R. S. Taylor, and B. A. Bolton, *Paint Varn. Prod.* **49**, 54 (1959).
50. "Water-Soluble Industrial Metal Primers and Maintenance Enamels Based on Amoco TMA," Tech. Bull. TMA-29b. Amoco Chem. Corp., Chicago, Illinois, 1974.
51. "Pyromellitic Dianhydride and Pyromellitic Acid *PMDA/PMA* A-15191," Tech. Bull. E. I. du Pont de Nemours & Co., Inc.
52. C. R. Martens, "Alkyd Resins," pp. 85–104. Van Nostrand-Reinhold, Princeton, New Jersey, 1961.
53. G. R. Somerville and O. S. Herr, *Ind. Eng. Chem.* **49**, 1080 (1957).
54. L. E. Gast, W. J. Schneider, and J. C. Cowan, *J. Am. Oil Chem. Soc.* **43**, 419 (1966).
55. W. J. Schneider, L. E. Gast, V. E. Johns, and J. C. Cowan, *J. Paint Technol.* **44**, 58 (1972).
56. C. R. Martens, "Alkyd Resins," p. 99. Van Nostrand-Reinhold, Princeton, New Jersey, 1961.
57. E. Foster, U.S. Patent 2,305,224 (1943); F. B. Root, U.S. Patent 2,559,465 and 2,559,466 (1951).
58. E. L. Kropa, U.S. Patent 2,409,633 (1946).
59. H. J. Lanson, *Paint Varn. Prod.* **49**, 25 (1959).
60. L. E. Wakeford, D. H. Hewitt, and F. Armitage, *J. Oil Colour Chem. Assoc.* **29**, 324 (1946); H. Thielker, *Seifen, Oele, Fette, Wachse* **93**, 281 (1967).
61. E. G. Bobalek, U.S. Patent 2,470,787 (1949); C. F. Prickett, *Am. Chem. Soc., Div. Org. Coat. Plast. Chem., Pap.* **14** (2), 151 (1954).
61a. W. M. Kraft, *Am. Chem. Soc., Div. Org. Coat. Plast. Chem., Pap.* **14** (2), 195 (1954).
62. I. G. Farbenindustrie, British Patent 369,915 (1930).
63. J. C. Petropoulos, L. E. Cadwell, and W. F. Hart, *Am. Chem. Soc., Div. Org. Coat. Plast. Chem., Pap.* **14** (2), 183 (1954); J. C. Petropoulos, *Ind. Eng. Chem.* **49**, 379 (1957).
64. M. Kronstein and H. A. Taylor, *Paint Varn. Prod.* **51**, 47 (1961).
65. F. Benner, *Offic. Dig., Fed. Paint Varn. Prod. Clubs* **31**, 1143 (1959).
66. F. S. Leutner, E. L. Brazet, and E. G. Bobalek, *Am. Chem. Soc., Div. Paint, Plast. Print. Ink Chem., Pap.* **14** (2), 156 (1954).
67. L. E. Wakeford and D. Helmsley, U.S. Patent 2,392,710 (1946).
68. "The Methacrylation of Alkyd Resins." Rohm & Haas Co., Philadelphia, Pennsylvania, 1958.
69. Shell International Research, Belgian Patent 653,112 (1965); *Chem. Abstr.* **64**, 8492 (1966).
70. H. P. Wohnsiedler, *Am. Chem. Soc., Div. Org. Coat. Plast. Chem., Pap.* **20** (2), 53 (1960).

71. A. G. Roberts, *Build. Sci. Ser.*, *Natl. Bur. Stand.* (*U.S.*) **7**, 43–45 (1968).

72. C. R. Martens, "Alkyd Resins," pp. 118–131. Van Nostrand-Reinhold, Princeton, New Jersey, 1961.

73. A. E. Rheineck, *J. Am. Oil Chem. Soc.* **36**, 574 (1965).

74. Imperial Chemical Industries, Ltd., Netherlands Patent Appl. 6,503,380 (1965); *Chem. Abstr.* **64**, 8493 (1966).

75. H. J. Wright and R. N. Dupuis, *Ind. Eng. Chem.* **38**, 1303 (1946).

76. H. Dannenberg, T. F. Brandley, and T. W. Evans, *Ind. Eng. Chem.* **41**, 1709 (1949).

77. T. W. Evans and D. E. Adelson, U.S. Patent 2,399,214 (1946).

78. D. J. Mehta and H. F. Payne, *Paint Varn. Prod.* **30**, 6 (1950).

79. H. Asai and T. Okuda, Japan Kokai 74/53,279 (1974); *Chem. Abstr.* **82**, 32615p (1975).

80. R. M. Christenson and D. P. Hart, U.S. Patent 2,853,459 (1958); D. L. Light, U.S. Patent 2,471,346 (1949); R. S. Robinson, U.S. Patent 2,586,092 (1952).

81. E. I. du Pont de Nemours & Co., British Patent 419,373 (1934).

82. C. S. Fuller, U.S. Patent 2,275,260 (1942).

83. T. M. Powanda, U.S. Patent 3,829,530 (1974).

84. B. Kosteve, German Offen. 2,261,654 (1974).

85. H. Brose, H. DieFrich and H. Pelshenki, German Offen. 1,966,796 (1969).

86. J. Boxall and J. A. von Fraunhofen, *Paint Manuf.* **42**, 13–16 (1973).

87. Societa Italiana Resine S.p.A., British Patent 1,240,230 (1971); *Chem. Abstr.* **75**, 130485 (1971).

88. H. Krasemacher, German Patent 1,520,747 (1969).

89. G. Bras and R. Daviaud, French Patent 1,560,418 (1969); *Chem. Abstr.* **71**, 126116b (1969).

90. M. D. Garret, *Paint Varn. Prod.* **60**, 37 (1970).

POLYACETALS AND POLY(VINYL ACETALS)

I. INTRODUCTION

Polyacetals may be prepared by the reaction of aldehydes with either polyols [including poly(vinyl alcohol) which affords poly(vinyl acetals)] or polythiols as shown in Eq. (1). Polyacetals are similarly prepared except that ketones are used, as shown in Eq. (2).

$$n\text{RCH}{=}\text{O} + n\text{HX}{-}\text{R}'{-}\text{XH} \xrightarrow{-n\text{H}_2\text{O}} \left[{-}\text{R}'{-}\text{X}{-}\overset{\displaystyle R}{\underset{\displaystyle |}{\text{CH}}}{-}\text{X}{-} \right]_n \tag{1}$$

$$n\text{R}_2\text{C}{=}\text{O} + n\text{HXR}'{-}\text{XH} \xrightarrow{-n\text{H}_2\text{O}} \left[{-}\text{R}'{-}\text{X}{-}\overset{\displaystyle R}{\underset{\displaystyle R}{\underset{\displaystyle |}{\overset{\displaystyle |}{\text{C}}}}}{-}\text{X}{-} \right]_n \tag{2}$$

X = O or S

The methods used to synthesize the acetal and ketal functional groups have already been considered in a previous volume by the authors [1]. In addition the polymerization of aldehydes to give polyethers as shown in Eq. (3) has been described in Vol. 1 of this series [2].

$$n\text{RCH}{=}\text{O} \longrightarrow \left[{-}\overset{\displaystyle R}{\underset{\displaystyle |}{\text{CH}}}{-}\text{O}{-} \right]_n \tag{3}$$

Carothers was one of the first to exploit the polyacetal-forming reaction in his attempt to prepare polyformals by the reaction of formaldehyde [3] with glycols. He found that diols below 1,5-pentanediol led to cyclic formals whereas the principal product from 1,6-hexanediol and higher molecular weight diols was a polyformal [3] [Eq. (4)]. Similar products were also formed by the acetal interchange reaction when the appropriate diol was used [4].

$$ n\text{ROCH}_2\text{OR} + n\text{HOR}'\text{OH} \xrightarrow[\substack{\text{acid} \\ \text{catalyst}}]{-2n\text{ROH}} [-\text{R}'-\text{OCH}_2-\text{O}-]_n \qquad (4) $$

Read [5] in 1912 reported obtaining an apparently high molecular weight solid by the strongly acid-catalyzed reaction of pentaerythritol and glyoxal. In 1951 Orth [6] reacted terephthaldehyde or 1,4-cyclohexanedione with 2,6-dioxaspiro[3,3]heptane to give polyspiroacetal resins [Eq. (5)]. In 1962 Cohen and Lavin [7–7b] prepared similar polyspiroacetal resins from pentaerythritol and dialdehydes.

$$ (5) $$

Acrolein has been reacted with polyols such as pentaerythritol in the presence of an acid catalyst to give monoallylidenepentaerythritol and diallylidenepentaerythritol [8]. Reaction of the diallylidenepentaerythritol with diols [Eq. (6)] or pentaerythritol itself yields polyacetal resins [9]. Isolation of the diallylidenepentaerythritol permits its use for reaction with specific diols [10,10a].

$$ (6) $$

Another common method for the synthesis of polyacetals involves trans-acetalation reactions of diols with acetylenic compounds [11] or vinyl ethers [12]. The reaction of bis(chloromethyl) ether with diols in the presence of pyridine to give polyacetals is not recommended since bis(chloromethyl) ether has been reported to be carcinogenic [13].

Polyacetals or polyketals can be made by most of the methods shown in Schemes 1 and 2 starting with diols, polyols, or polythiols in place of the alcohols [1]. The use of poly(vinyl alcohol) gives either poly(vinyl acetals) or poly(vinyl ketals) and is described in Section 7 in this chapter.

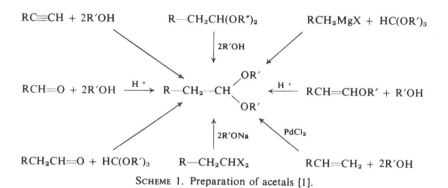

SCHEME 1. Preparation of acetals [1].

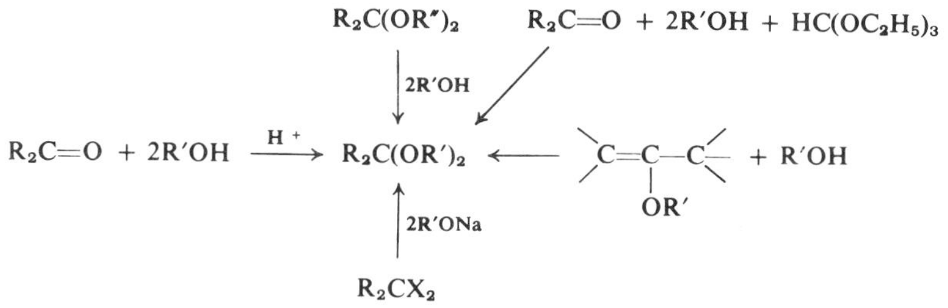

SCHEME 2. Preparation of ketals [1].

Polyacetals have not been exploited to their fullest potential as other polymeric systems have and much research still remains to be done [14]. Some typical uses are as prepolymers for polyurethane elastomers [15,15a], in dimensional stabilization of cellulosic fabrics [16], paper [17], and segmented cellulose films [17]; as synthetic lubricants [18]; as coatings for wire [19]; as wash-and-wear finishes [20], in solvent-resistant coatings [21]; and in castings, laminates, coatings [10a,22], and varnishes [23].

Polyacetals have earlier been briefly reviewed in several sources which should be consulted for further details [4].

2. POLYACETALS FROM FORMALDEHYDE

Polyacetals derived from diols or polyols using formaldehyde or its derivatives are known as polyformals. Cyclic formals are formed when a five-, six-, or seven-membered ring can be formed [24].

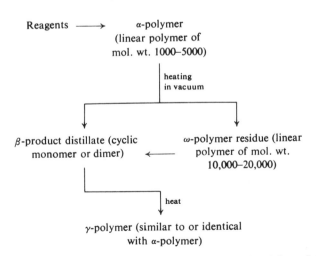

Hill and Carothers found that the acetal interchange reaction is carried out easily and that the reaction is reversible [3].

Heating a glycol with dibutyl formal in the presence of an acidic catalyst at about 150°C results in an acetal interchange reaction evidenced by the distillation of butanol as shown in Eq. (7). Trimethylene or tetramethylene glycol are converted to mobile, easily distillable liquids which are mostly cyclic monomers. When the glycol is pentamethylene, or a higher one, the residue remaining upon distillation of the alcohol is a viscous liquid. The various relationships of forming α-, β-, γ-, and ω-polymers for acetals (also for other macrocyclic esters and anhydrides) are shown in Scheme 3.

Reagents $\longrightarrow$ α-polymer
(linear polymer of
mol. wt. 1000–5000)

heating
in vacuum

β-product distillate (cyclic ω-polymer residue (linear
monomer or dimer) $\longleftarrow$ polymer of mol. wt.
10,000–20,000)

heat

γ-polymer (similar to or identical
with α-polymer)

SCHEME 3. Preparation of various polyacetal forms. [Reprinted from J. W. Hill and W. H. Carothus, *J. Am. Chem. Soc.* **57**, 927 (1935). Copyright 1935 by the American Chemical Society. Reprinted by permission of the copyright owner.]

Heating the α-polymer results in either ω-polymer or β-products which are cyclic monomer or dimer products. In some cases all are formed to some extent depending on catalysis and temperature of reaction. Some representative data on formals and polyformals as prepared by Hill and Carothers [3] are shown in Table I.

2-1. Preparation of Decamethylene Formal [3]*

$$CH_2(O-Bu)_n + HO(CH_2)_{10}OH \longrightarrow [O-(CH_2)_{10}-O-CH_2-]_n + 2nBuOH \quad (8)$$

Dibutyl formal (0.2 mole) with a 5% excess of decamethylene glycol and 0.1 gm of ferric chloride gives a fairly rapid reaction at 165°C (bath). The temperature is raised to 200°C during 2 hr and heating continued for $1\frac{1}{2}$ hr in a good vacuum. The distillate is 98% of the theoretical yield calculated as butyl alcohol; the yield of residual α-polymer is 103%. When cold the latter is a light brown, rather hard wax. When dissolved in hot ethyl acetate (150 ml for 17.5 gm), it separates in the form of a microcrystalline powder which is soluble in chloroform, benzene, carbon tetrachloride, and xylene, and insoluble in alcohol, ether, petroleum, hydrocarbons, and acetone.

(a) β-Product. Eight grams of the crude α-polymer is heated in a 250-ml suction flask provided with a test tube through which water can be circulated to act as an internal condenser (see Note a); the pressure is about 1 mm Hg. After 48 hr 2 gm of distillate has collected. It is a pasty mixture of liquid and crystals having a pleasant camphoraceous odor. The solid portion after crystallization from alcohol is odorless and melts at 93°–94°C. This is the cyclic dimer.

(b) ω-Polymer. The residue from the above is a hard, very tough, opaque, leatherlike mass. It melts (became transparent) at 58°–63°C, but at this temperature it is too stiff to flow and shows considerable resistance to deformation. At slightly higher temperatures, it can be drawn out into thin strips or filaments which can be stretched and cold drawn (see Note b). The product then shows fiber orientation and also exhibits parallel extinction between crossed Nicols. The cold-drawn material is exceedingly strong, tough, and pliable.

(c) γ-Polymer. The cyclic dimer (0.5 gm) with a trace of camphorsulfonic acid heated at 150°C soon becomes very viscous and the characteristic odor of the cyclic monomer appears. After an hour, the melt when cooled sets to a hard wax, which is easily electrified when powdered. Purified out of benzene,

---

* Procedure 2-1 is reprinted from J. W. Hill and W. H. Carothers, *J. Am. Chem. Soc.* **57**, 925 (1935). Copyright 1935 by the American Chemical Society. Reprinted by permission of the copyright owner.

TABLE I

DATA ON FORMALS[a]

| Name of formal | M.p. (°C) | Formula | Analytical data | | | | | |
| | | | Calc. | | | Found | | |
			C	H	Mol. wt.	C	H	Mol. wt. in freezing benzene
Tetramethylene, monomer	112–117 (b.p.)	$C_5H_{10}O_2$	58.8	9.9	102	59.6	10.0	103
Decamethylene, α-polymer	56–57	$(C_{11}H_{22}O_2)_x$	70.9	11.9	$(186)_x$	68.9	11.7	2190
Decamethylene, dimer	93–94	$(C_{11}H_{22}O_2)_2$	70.9	11.9	372	70.8	12.1	368
Pentamethylene, dimer	55–56	$(C_6H_{12}O_2)_2$	62.1	10.4	232	61.8	10.5	262[b]
Pentamethylene, monomer	40–44 (b.p., 11 mm Hg)	$(C_6H_{12}O_2)_2$	62.1	10.4	—	62.0	10.8	—
Hexamethylene, dimer	71–72	$(C_7H_{12}O_2)_2$	64.6	10.8	260	64.0	10.6	257
Nonamethylene, dimer	68–69	$(C_{10}H_{20}O_2)_2$	69.8	11.6	344	69.8	11.6	334[b]
Tetradecamethylene, α-polymer	68–69	$(C_{15}H_{30}O_2)_x$	74.4	12.4	$(242)_x$	73.1	12.1	2480
Tetradecamethylene, dimer	103.5–104	$(C_{15}H_{30}O_2)_2$	74.4	12.4	484	74.4	12.5	503
Triethylene glycol, monomer	18–20	$C_7H_{14}O_4$	51.9	8.6	162	51.6	8.4	161

[a] Reprinted from J. W. Hill and W. H. Carothers, *J. Am. Chem. Soc.* **57**, 925 (1935). Copyright 1935 by the American Chemical Society. Reprinted by permission of the copyright owner.
[b] In boiling benzene.

it separates as a microcrystalline powder, m.p. 58°–59°C; molecular weight observed in freezing benzene, 2580.

NOTES: (a) See J. W. Hill and W. H. Carothers, *J. Am. Chem. Soc.* **55**, 5035 (1935). (b) See W. H. Carothers and J. W. Hill, *J. Am. Chem. Soc.* **55**, 1580 (1932).

Some typical examples of polyformals prepared by the direct reaction of formaldehyde or paraformaldehyde with polyols are shown in Table II.

2-2. Preparation of the Polyformal of 1,5-Pentanediol [25]

$$\text{HOCH}_2\text{CH}_2\text{CH}_2\text{CH}_2\text{CH}_2\text{OH} + \text{CH}_2\!=\!\text{O} \longrightarrow \text{H}\,\text{+OCH}_2\text{CH}_2\text{CH}_2\text{CH}_2\text{CH}_2\text{OCH}_2\,\text{+}_n\text{OH}$$

(9)

To a three-necked round-bottomed flask equipped with a mechanical stirrer, condenser, and vacuum outlet are added 208 gm (2.0 moles) of 1,5-pentanediol, 180 gm (2.2 moles) of 37% formaldehyde (formalin) and 1 gm of sulfuric acid. The mixture is heated to 95°C and vacuum is applied. Heating is continued for 3 hr at 100°C to afford a wax, m.p. 25°–45°C, hydroxyl number 69.5 and mol. wt. 1610.

In place of sulfuric acid, *p*-toluenesulfonic acid or sodium bisulfate may be used as catalysts. Polyacetals of other diols, such as diethylene glycol, dipropylene glycol, 2-methyl-1,5-pentanediol, polyethylene glycol, 5-(2-hydroxyethoxy)-4-hydroxymethyl-2,4-dimethyl-1-pentanol, and liquid poly-(oxytetramethylene) glycol (from tetrahydrofuran; mol. wt. 340), may be prepared by this procedure.

2-3. Preparation of the Polyformal of Diethylene Glycol [26]

$$\text{HOCH}_2\text{CH}_2\text{OCH}_2\text{CH}_2\text{OH} + \text{CH}_2\!=\!\text{O} \longrightarrow \text{H}\,\text{+OCH}_2\text{CH}_2\text{O}\!-\!\text{CH}_2\text{CH}_2\text{OCH}_2\,\text{+}_n\text{OH}$$

(10)

To a three-necked round-bottomed flask equipped with a mechanical stirrer, condenser, and Dean & Stark trap are added 106 gm (1.0 mole) of diethylene glycol, 33 gm (1.0 mole) of 91% paraformaldehyde, 0.1 ml of concentrated sulfuric acid, and 20 gm of toluene. The mixture is heated under reflux; 18 gm (4.0 moles) of water is collected. The toluene is removed under reduced pressure (20 mm Hg) and heating is continued to 150°C. The resulting product is neutralized with dilute sodium hydroxide to pH 7. The polymer is soluble in water and toluene and has a molecular weight of 480. The product has an OH equivalent of 220 and n_D^{30} of 1.462. The product is suggested as useful as a textile finishing agent for paper, cotton [27], and leather.

TABLE II

PREPARATION OF POLYFORMALS

Polyol (moles)	CH$_2$=O (moles)	Solvent (ml)	Catalyst (gm)	Temp. (°C)	Time (hr)	M.p. (°C)	Mol. wt.	Ref.
1,5-Pentanediol 2.0	37% Formalin 2.2	—	H$_2$SO$_4$ 1.0	95–100	3	25–45	1610	a
Diethyleneglycol 1.0	91% Paraformaldehyde 1.0	Toluene 2.0	H$_2$SO$_4$ 0.1	118°	2	—	480	b
				150° (20 mm)	2	—	—	b
	5.0	6.0	—	PTSA 5.0	60–80 (10–20 mm)	12	—	1670	c
					24	—	4220	c
HO(CH$_2$)$_n$OH (cyclic diol) 1.0	Trioxane (CH$_2$=O) 3–4	—	5 N H$_2$SO$_4$ 200–600	—	—	—	3000–8000	d
	3–4	—	5 N H$_2$SO$_4$ 200–600	—	—	—	—	e
1,6-Hexanediol 1.0	CH$_2$=O 3–4	—	5 N H$_2$SO$_4$ 200–600	—	—	26–28	958	e
2-Butyne-1,4-diol 0.5	Paraformaldehyde 0.5	Toluene 440	PTSA 0.3–0.75 FeCl$_3$ 0.5	118	1–2	55–60	620–2300	f
				75–115 (1–15 mm Hg)	3–8	55–60	1000–2500	f

[a] Hudson Foam Plastics Corp., British Patent 850,178 (1960).
[b] B. H. Kress, U.S. Patent 2,786,081 (1957).
[c] H. von Brachel, H. Esser, and E. Muller, German Patent 1,066,734 (1959).
[d] Imperial Chemical Industries, Ltd., Belgian Patent 625,781 (1963).
[e] J. Lichtenberger and J. Hincky, Bull. Soc. Chim. Fr. p. 854 (1961).
[f] Thiokol Chem. Corp., British Patent 886,982 (1962); D. D. Perry, G. Golub, R. D. Dwyer, and P. F. Schaeffer, Adv. Chem. Ser. 54, 118 (1965).

2-4. Preparation of the Polyformal of Tetrabromobisphenol A Diethanol [28,29]

$$\text{HOCH}_2\text{CH}_2\text{O}-\underset{\underset{\text{Br}}{\big|}}{\overset{\overset{\text{Br}}{\big|}}{\bigcirc}}-\underset{\overset{\text{CH}_3}{\big|}}{\overset{\overset{\text{CH}_3}{\big|}}{\text{C}}}-\underset{\underset{\text{Br}}{\big|}}{\overset{\overset{\text{Br}}{\big|}}{\bigcirc}}-\text{OCH}_2\text{CH}_2\text{OH} + \text{CH}_2{=}\text{O} \xrightarrow{-\text{H}_2\text{O}}$$

$$\left[-\text{OCH}_2\text{CH}_2\text{O}-\underset{\underset{\text{Br}}{\big|}}{\overset{\overset{\text{Br}}{\big|}}{\bigcirc}}-\underset{\overset{\text{CH}_3}{\big|}}{\overset{\overset{\text{CH}_3}{\big|}}{\text{C}}}-\underset{\underset{\text{Br}}{\big|}}{\overset{\overset{\text{Br}}{\big|}}{\bigcirc}}-\text{OCH}_2\text{CH}_2\text{OCH}_2-\right]_n \tag{11}$$

To a 1-liter, three-necked round-bottomed flask equipped with a mechanical stirrer, Dean & Stark trap, and condenser are added 400 gm (0.63 mole) of tetrabromobisphenol A diethanol and 460 ml of benzene. The reaction mixture is heated to reflux for 2 hr to form a clear solution. The temperature is lowered to 65°C and 20.6 gm (0.65 mole) of 95% paraformaldehyde and 2.0 gm of p-toluenesulfonic acid are added. The mixture is stirred for $1\frac{1}{2}$ hr at 65°C and then the temperature is raised to reflux to remove 11.4 ml (0.63 mole) of water azeotropically in about $3\frac{1}{2}$ hr. The mixture is cooled to room temperature and 0.5 ml of 15 M ammonium hydroxide is added with vigorous agitation in order to neutralize the catalyst. The polymer is isolated by slowly adding the viscous benzene solution to 8 liters of methanol. The resultant white powder is washed with 500 ml of methanol and dried at 50°C in a vacuum oven. The polymer has an inherent viscosity of 0.12 dl/gm and a softening point of 80°C.

3. POLYACETALS FROM SUBSTITUTED ALDEHYDES

The use of an aldehyde higher than formaldehyde, such as acetaldehyde or benzaldehyde, affords polyacetals containing a side group, as shown in Eq. (12).

$$\text{HO(CH}_2)_n\text{OH} + \text{RCH}{=}\text{O} \longrightarrow \left[-(\text{CH}_2)_n\text{O}-\underset{\underset{\text{R}}{\big|}}{\text{CH}}-\text{O}-\right]_n + \text{H}_2\text{O} \tag{12}$$

R = alkyl, aryl, and substituted derivatives [$\text{CH}_2{=}\text{CH}$ (acrolein), etc.]

Depending on the polyol used, linear, cyclic, or spiro polyacetals are produced. Mixtures of diols may be used to yield random copolymeric polyacetals.

The condensation reactions are usually carried out in the melt or at the boiling temperature of a solvent in the presence of an acid catalyst such as *p*-toluenesulfonic acid (PTSA). The solvents usually used are toluene or benzene, so that the water of condensation can be azeotropically removed.

Acetal interchange is also used with the higher aldehydes and for the self-condensation of hydroxyacetals to give polyacetals.

The use of dialdehydes and polyhydroxy compounds yields cyclic acetals, as shown in Eq. (13).

$$
\begin{array}{c}
\text{HOCH}_2 \\
\quad\quad \text{CH-R-CH} \\
\text{HOCH}_2
\end{array}
\begin{array}{c}
\text{CH}_2\text{OH} \\
\\
\text{CH}_2\text{OH}
\end{array}
\ + \ O{=}CH{-}R'{-}CH{=}O \ \xrightarrow{-2H_2O}
$$

$$
\left[-CH \begin{array}{c} CH_2{-}O \\ \\ CH_2{-}O \end{array} CH{-}R'{-}CH \begin{array}{c} O{-}CH_2 \\ \\ O{-}CH_2 \end{array} CH{-}R{-} \right] \quad (13)
$$

R = CH_2, alkylene, arylene
R' = aliphatic and aromatic groups

A. Linear Polyacetals

Linear poly(alkylene acetals) are formed by the reaction of an aldehyde with a diol [Eq. (14)].

$$
HOR{-}OH \ + \ R'CH{=}O \ \xrightarrow{H^+} \ \left[-R{-}OCH{-}O \right]_n ROH \quad (14)
$$
$$
\qquad\qquad\qquad\qquad\qquad\quad R'
$$

Schonfeld [30] has described these polymers as being flexible and having a low second-order transition point. The ether oxygen is said to help to increase bond rotation and thus overall chain flexibility, resulting in low-melting polymers. Schonfeld [30] kept the diol constant (1,5-pentanediol) and varied the aldehyde. Earlier Schonfeld [15] reported that 1,5-pentanediol is the lowest molecular weight α,ω-diol which polymerized rather than cyclized when reacted with paraformaldehyde. The polyacetals were prepared by the earlier described procedure [15a] using *p*-toluenesulfonic acid and toluene to aid the reaction and removal of water by azeotropically distilling the water. The polyacetals derived from *n*-propionaldehyde, *n*-butyraldehyde, and *n*-valeraldehyde decomposed at about 125°C at 2 mm Hg pressure. The polyacetal derived from paraformaldehyde was stable under these conditions.

The molecular weights of the polymers were determined by end-group analysis for hydroxyl. Since the polymers are linear and terminate at each end

TABLE III
PREPARATION OF POLYACETALS [30]

Aldehyde	Hydroxyl number	$\overline{M}_n$	Viscosity [Cp (Brookfield) at 25°C]
n-Valeraldehyde	90.5	1245	630
n-Butyraldehyde	77	1460	470
n-Propionaldehyde	137	825	715
Paraldehyde	121	930	880

with a hydroxyl group, the number-average molecular weight is calculated from Eq. (15), where the hydroxyl number is defined as the number of

$$\overline{M}_n = \frac{2 \times 56.1 \times 100}{\text{hydroxyl number}} \tag{15}$$

milligrams of potassium hydroxide equivalent to the hydroxyl content of one gram of the polyol. Some typical results are shown in Table III.

In a similar manner, thiodiethylene glycol is reacted with benzaldehyde to give a polyacetal of number-average molecular weight of 1610, as described below.

3-1. Preparation of a Polyacetal by the Reaction of Benzaldehyde with Bis(2-hydroxyethyl) Sulfide [15a]

$$C_6H_5CH=O + HOCH_2CH_2S-CH_2CH_2OH \xrightarrow{-H_2O}$$

$$HOCH_2CH_2\left[S-CH_2CH_2-O-\underset{\underset{C_6H_5}{|}}{CHO}-CH_2CH_2\right]_n SCH_2CH_2OH$$

$$\tag{16}$$

To a resin flask equipped with a mechanical stirrer, thermometer, condenser, Dean & Stark trap, and nitrogen inlet/outlet are added 53.0 gm (0.5 mole) of benzaldehyde, 61.0 gm (0.5 mole) of bis(2-hydroxyethyl) sulfide, 300 ml of toluene, and 0.25 gm of p-toluenesulfonic acid. The mixture is heated to reflux and the water (9.0 ml) of condensation is removed azeotropically over a period of 70 hr. The first 4.5 ml of water (0.25 mole) is obtained in 4 hr and the remaining 0.25 mole is obtained after 3 days of reaction. The polymer is isolated by distillation of the toluene at atmospheric pressure, and the product is then heated to 180°C at 1.0 mm Hg for 5 hr. A liquid, nonviscous polymer is isolated with a 69.4 average hydroxyl number, corresponding to a number-average molecular weight of 1610. The infrared spectrum showed absorption at 9.0 μm for ether linkage and a peak at 2.9 μm for the presence of the hydroxyl group.

The latter polyacetal is also useful for the reaction with 2,4-toluene-diisocyanate to give an isocyanate-terminated prepolymer (structure **III**) [15a].

$$\left(CH_3 - \underset{NCO}{\overset{}{\bigcirc}} - NHCOO - CH_2CH_2 \right)_2 \left[-S - CH_2CH_2O - \underset{\overset{|}{C_6H_5}}{CH} - O - CH_2CH_2 - \right]_n S$$

III

Other linear polyacetals are those derived from acrolein [31], and these products are also useful as starting materials for polyurethane resins [32].

Linear polyacetals are also produced by the acid (sulfuric acid or Friedel–Crafts)-catalyzed addition of diols to divinyl ethers as shown in Eq. (17) at temperatures below $-10°C$ [33].

$$HO - R' - OH + CH_2{=}CHO - R - OCH{=}CH_2 \longrightarrow$$

$$\left[-\underset{\overset{|}{CH_3}}{CH} - O - R - O - \underset{\overset{|}{CH_3}}{CH} - OR' - O- \right]_n \quad (17)$$

Depending on the mole ratio of reactants used the polyacetals have either both terminal vinyl groups, both terminal hydroxyl groups, or one vinyl ether and one hydroxyl group. The reaction can be run at $-10°$ to $20°C$ using acid salts of strong acids, e.g., alkali metal bisulfate or α-haloethers. Organic acids with dissociation constants of 10^3 to 10^1 are used if a reaction temperature of $20°{-}180°C$ is desired.

CAUTION: Bischloromethyl ether has been reported to have carcinogenic properties.

Polyacetals are stable in neutral or alkaline media but are hydrolyzed under acid conditions.

B. Cyclic Polyacetals

The reaction of dicarboxyl compounds and a tetral afford cyclic polyacetals as described earlier. Spiroacetals result when pentaerythritol is used as the polyol [Eq. (18)] [34,35].

$$\begin{matrix} HOCH_2 \\ \\ HOCH_2 \end{matrix} \underset{}{\overset{}{C}} \begin{matrix} CH_2OH \\ \\ CH_2OH \end{matrix} + O{=}CH - R - CH{=}O \xrightarrow{-2H_2O}$$

$$\left[-CH \underset{O-CH_2}{\overset{O-CH_2}{\diagdown}} C \underset{CH_2-O}{\overset{CH_2-O}{\diagup}} CH - R - \right]_n \quad (18)$$

Glutaraldehyde tetraethyl acetal can also be reacted with monopenta-erythritol to give a linear polyspiroacetal [36] as shown in Eq. (19).

$$
\begin{array}{c}
\text{HO—CH}_2 \\
\hspace{1.2cm} \diagdown \\
\hspace{1.4cm} \text{C} \\
\hspace{1.2cm} \diagup \hspace{0.3cm} \diagdown \\
\text{HO—CH}_2 \hspace{1.1cm} \text{CH}_2\text{OH}
\end{array}
\begin{array}{c}
\text{CH}_2\text{OH}
\end{array}
+ (\text{C}_2\text{H}_5\text{O})_2\text{CH—CH}_2\text{CH}_2\text{CH}_2\text{—CH}(\text{OC}_2\text{H}_5)_2 \longrightarrow
$$

$$
\left[\begin{array}{c} \text{—CH} \begin{array}{c}\text{O—CH}_2 \hspace{0.8cm} \text{CH}_2\text{—O} \\ \diagup \hspace{1cm} \text{C} \hspace{1cm} \diagdown \\ \text{O—CH}_2 \hspace{0.8cm} \text{CH}_2\text{—O} \end{array} \text{CH—CH}_2\text{CH}_2\text{CH}_2\text{—} \end{array} \right]_n \hspace{1cm} (19)
$$

Other dialdehyde tetraalkyl acetals, such as adipaldehyde tetrapropyl acetal and succinaldehyde tetramethyl acetal, can be similarly reacted with penta-erythritol to give polyspiroacetals [7,36].

Dihydroxyacetals undergo self-condensation to give a low molecular weight polyacetal [37], as illustrated in Eq. (20) for 5-hydroxy-4-hydroxymethyl-pentanol diethyl acetals.

$$
\begin{array}{c}
\text{CH}_2\text{OH} \\
| \\
\text{HOCH}_2\text{CH—CH}_2\text{CH}_2\text{CH}(\text{OC}_2\text{H}_5)_2
\end{array}
\xrightarrow{\text{H}^+}
\left[\begin{array}{c}\text{—CH} \begin{array}{c}\text{O—CH}_2 \\ \diagup \hspace{1cm} \diagdown \\ \text{O—CH}_2 \end{array} \text{CH—CH}_2\text{CH}_2\text{—}\end{array} \right]_n
$$

$$
+ \; 2n\text{C}_2\text{H}_5\text{OH} \hspace{1cm} (20)
$$

As mentioned earlier, Read [5], Orth [6], and later Kropa and Thomas [38] described some early preparations of polyspiroacetals. More recently Abbott and co-workers [35], Kress [39], and others [36,40] have also described polyspiroacetals prepared under various conditions. Cohen and Lavin re-peated some of Orth's work and extended the polyspiroacetal preparations even further as described in Table IV, Refs. (a) and (b).

The polyspiranes give materials useful for surface coatings as well as for electrical wire enamels [35]. Glutaraldehyde–pentaerythritol polyspiroacetals have been combined with dimethylolethyleneurea in wash-and-wear finishes for cotton [39].

3-2. Preparation of the Polyacetal by the Reaction of Pentaerythritol with a Mixture of 1,3- and 1,4-Cyclohexane Dialdehydes [35]

$$
\begin{array}{c}
\text{CH}_2\text{OH} \\
| \\
\text{HOCH}_2\text{—C—CH}_2\text{OH} \\
| \\
\text{CH}_2\text{OH}
\end{array}
+ \; \text{O}{=}\text{CH—}\bigcirc\text{—CH}{=}\text{O} \longrightarrow
$$

$$
\left[\begin{array}{c} \text{—CH} \begin{array}{c}\text{O—CH}_2 \hspace{0.8cm} \text{CH}_2\text{—O} \\ \diagup \hspace{1cm} \text{C} \hspace{1cm} \diagdown \\ \text{O—CH}_2 \hspace{0.8cm} \text{CH}_2\text{—O}\end{array} \text{CH—}\bigcirc\text{—} \end{array} \right]_n \hspace{1cm} (21)
$$

TABLE IV

PREPARATION OF POLYSPIROACETALS

Dialdehyde	Polyols	Method of preparation	Yield (%)	M.p. (softening point, °C; Fisher–Johns)	DP	Mol. wt.	Ref.
Glyoxal	Pentaerythritol	Transacetylation	64	>310	1.5	252	a
Malonaldehyde	Pentaerythritol	Transacetylation	70	>300	21.6	3750	a
Glutaraldehyde	Pentaerythritol	Aqueous condensation	92	260–270	37.5	4850	a
3-Methylglutaraldehyde	Pentaerythritol	Aqueous condensation	98	205–208	8.3	8350	a
Terephthaldehyde	Pentaerythritol	Aqueous condensation	55	>300	38.5	2000	a
Glutaraldehyde	Dipentaerythritol: pentaerythritol 24:76	Aqueous condensation	93	181–186	—	1200	b
1,1,6-Tetraethoxyhexane	Pentaerythritol	Direct esterification	—	199–203	—	—	c
1,1,5,5-Tetraethoxypentane	Pentaerythritol	Direct esterification	—	180–183	—	—	c
Glyoxal	Pentaerythritol	Aqueous condensation	—		—	—	d
Glutaraldehyde tetraethylacetal	Pentaerythritol	Transesterification	50	245	—	—	e
Adipaldehyde tetrapropylacetal	Pentaerythritol	Transesterification	50	245	—	—	e
Succinaldehyde tetramethylacetal	Pentaerythritol	Transesterification	50	235	—	—	e
1,3- and 1,4-Cyclohexane dialdehyde	Pentaerythritol	Direct esterification	—	—	—	—	f

[a] S. M. Cohen and E. Lavin, *J. Appl. Polym. Sci.* **6**, 503 (1962); S. M. Cohen and E. Lavin, U.S. Patent 2,963,464 (1960).
[b] S. M. Cohen, C. F. Hunt, R. E. Kass, and A. H. Markhart, *J. Appl. Polym. Sci.* **6**, 508 (1962); A. H. Markhart, C. F. Hunt, and E. Lavin, German Patent 1,210,964 (1966).
[c] Soc. of European Research Assoc., S.A., Belgian Patent 568,181 (1958).
[d] E. L. Kropa and W. M. Thomas, U.S. Patent 2,643,236 (1953).
[e] D. B. Capps, U.S. Patent 2,889,290 (1959).
[f] L. S. Abbott, D. Faulkner, and E. E. Hollis, U.S. Patent 2,739,972 (1956).

To a flask equipped with mechanical stirrer, condenser, and vacuum connection are added 6.8 gm (0.05 mole) of pentaerythritol and 7.0 gm (0.05 mole) of a mixture of 1,3- and 1,4-cyclohexane dialdehydes. After refluxing for 1 hr, the mass is heated under a nitrogen stream for 1 hr at 110°–180°C. The mass becomes very viscous and then solid and finally the temperature is raised to 240°C. The temperature is maintained at 240°–250°C for 1 hr and the pressure reduced to 12 mm Hg. The temperature is now raised to 260°C over a 1½-hr period and the mixture is finally cooled to give a brown hard resin of an extremely high softening point (> 250°C).

3-3. *Polymerization of 5-Hydroxy-4-hydroxymethylpentanal Diethylacetal* [37]

$$
\underset{\text{CH}_2\text{OH}}{\text{HO}-\text{CH}_2-\underset{|}{\text{CH}}-\text{CH}_2\text{CH}_2\text{CH(OC}_2\text{H}_5)_2} \longrightarrow \left[-\text{O}-\text{CH}_2-\text{CH} \underset{\text{CH}_2-\text{CH}_2}{\overset{\text{CH}_2-\text{O}}{\diagup\diagdown}} \text{CH}- \right]_n
$$

(22)

To a small, heavy-walled polymerization tube with a connection to a vacuum system are added 1.0 gm (0.005 mole) of 5-hydroxy-4-hydroxy-methylpentanol diethylacetal (n_D^{20} 1.4530–1.4532) along with a small crystal of p-toluenesulfonic acid monohydrate. The tube is heated to 120°C under a reduced pressure (0.02 mm Hg) for 15–30 min. The reaction mixture is cooled, neutralized with concentrated ammonium sulfate, and extracted with chloroform. The chloroform extract is decolorized with Darco and evaporated to dryness under reduced pressure to give the polymer [average mol. wt. (from bornyl bromide) 1190 ± 100; IR 3440 cm^{-1} (OH) and 1736 cm^{-1} (>C=O); softening point 49°C].

3-4. *Preparation of Polyspiroacetal from the Pentaerythritol– Dipentaerythritol–Glutaraldehyde Condensation Reaction* [7b,41]

To a resin flask are added 480 gm (3.5 moles) of technical grade pentaerythritol, 1383 gm of a 24% aqueous (pH 2.5–4.0) solution of glutaraldehyde (3.3 moles), and 7.4 gm (0.082 mole) of oxalic acid catalyst. The solution is refluxed for 2 hr, cooled, filtered, and the solid washed with water and dried at 55°C to afford 620 gm (94%) of product, m.p. 201°–205°C, mol. wt. 1750, [η] 0.17 (determined at 20°C in cresylic acid).

$$\begin{array}{c} \text{IOCH}_2 \quad \text{CH}_2\text{OH} \\ \diagdown\text{C}\diagup \\ \text{IOCH}_2 \quad \text{CH}_2\text{OH} \end{array} + (\text{HOCH}_2-)_3-\text{C}-\text{CH}_2\text{O}-\text{CH}_2\text{C}(\text{CH}_2\text{OH})_3 + \overset{\displaystyle O}{\overset{\|}{\text{HC}}}-(\text{CH}_2)_3-\overset{\displaystyle O}{\overset{\|}{\text{CH}}} \longrightarrow$$

$$\left[\begin{array}{c} \text{O}-\text{CH}_2 \quad \text{CH}_2-\text{O} \\ -\text{CH}\diagup\diagup\text{C}\diagdown\diagup\text{CH}-(\text{CH}_2)_3- \\ \text{O}-\text{CH}_2 \quad \text{CH}_2-\text{O} \end{array}\right]_x$$

$$\left[\begin{array}{c} \text{O}-\text{CH}_2 \quad \text{CH}_2-\text{O}-\text{CH}_2 \quad \text{CH}_2-\text{O} \\ -\text{CH}\diagup\diagup\text{C}\diagdown\diagup \diagdown\text{C}\diagup \diagdown\text{HC}-(\text{CH}_2)_3- \\ \text{O}-\text{CH}_2 \quad \text{CH}_2 \quad \text{CH}_2 \quad \text{CH}_2-\text{O} \\ \qquad\qquad | \qquad\qquad | \\ \qquad\qquad \text{OH} \qquad\quad \text{OH} \end{array}\right]_y$$

(23)

4. POLYACETALS WITH OTHER FUNCTIONAL GROUPS

Polyacetals with other functional groups are prepared by the reaction of substituted polyols or aldehydes with each other as shown in Eq. (24).

$$\overset{\displaystyle X}{\overset{|}{\text{HO}}}-\text{R}-\text{OH} + \text{Y}-\text{R}'\text{CH}{=}\text{O} \longrightarrow \left[\begin{array}{c} \text{X} \\ | \\ -\text{O}-\text{R}-\text{OCH}- \\ | \\ \text{R}' \\ | \\ \text{Y} \end{array}\right]_n \qquad (24)$$

$\quad$ X or Y = $\text{CH}_2{=}\text{CH}-$, halogen, etc.
$\quad$ R = alkyl, cycloalkyl
$\quad$ R' = $\text{CH}_2{=}\text{CH}$, ester, amide, etc.

Furthermore, methyl azelaaldehydate obtained from the ozonization of methyl oleate [42–44] is a useful starting material for the preparation of cyclic polyacetals and mixed acetal–ester or acetal–amide polymers as shown in Scheme 4.

The polyesters prepared using the adduct of pentaerythritol and methyl-azelaaldehydate (2,4,8,10-tetraoxaspiro[5,5]undecane) can be used to give cross-linked rubbery polymer gels by heating with such catalysts as zinc oxide, zinc acetate, litharge, boric acid, and *p*-toluenesulfonic acid (see Fig. 1).

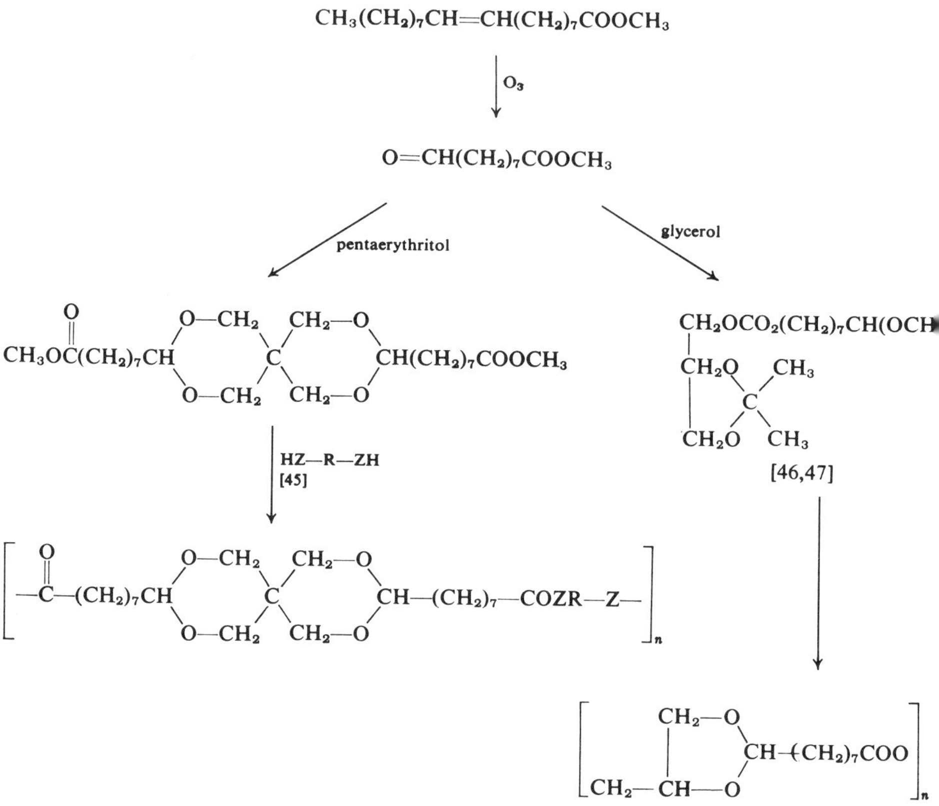

$$CH_3(CH_2)_7CH=CH(CH_2)_7COOCH_3$$

$$\downarrow O_3$$

$$O=CH(CH_2)_7COOCH_3$$

pentaerythritol   glycerol

$$CH_3OC(CH_2)_7CH \quad \text{(O—CH}_2 \quad CH_2\text{—O)} \quad C \quad CH(CH_2)_7COOCH_3$$

O—CH₂   CH₂—O

$$CH_2OCO_2(CH_2)_7CH(OCH$$

CH₂O   CH₃
C
CH₂O   CH₃

HZ—R—ZH
[45]

[46,47]

$$\left[-C-(CH_2)_7CH \begin{array}{c} O-CH_2 \quad CH_2-O \\ C \\ O-CH_2 \quad CH_2-O \end{array} CH-(CH_2)_7-COZR-Z- \right]_n$$

$$\left[\begin{array}{c} CH_2-O \\ | \\ CH_2-CH-O \end{array} CH-(CH_2)_7COO \right]_n$$

Z = O or NH

SCHEME 4. Reactions of methyl azelaaldehydate.

In addition, the reaction of divinyl ethers or diallylidene acetals derived from acrolein with active hydrogen-containing substituted compounds yields acetals containing recurring *gem*-hemiacetal with other functional groups. [See Eq. (25).]

$$HZ-R-ZH + CH_2=CH-O-R'-OCH=CH_2 \longrightarrow$$

$$\left[\begin{array}{c} -R-Z-CH-O-R'-OCH-Z- \\ | \qquad\qquad\qquad | \\ CH_3 \qquad\qquad\qquad CH_3 \end{array} \right]_n \quad (25)$$

$$Z = \overset{O}{\overset{\|}{C}}O, \text{ substituted alkoxy, O, S}$$

R = alkyl, alkylene ether (as in diethylene glycol), or aryl

Some typical examples are described in Table V.

TABLE V

PREPARATION OF POLYACETALS WITH OTHER FUNCTIONAL GROUPS

Diol, diacid, diamine, or diester	Acetal	Reaction conditions				Yield (%)	Mol. wt.	Ref.
		Temp. (°C)	Time (hr)	Catalyst	Solvent			
—	Isopropylideneglyceryl azelaaldehydate dimethyl acetal[a] (34.0 gm)	7–9	1	Conc. HCl (340 ml)	C_6H_6 (340 ml)	77	1530	c
Ethylene glycol (0.022 mole)	3,9-Bis(7-carboxymethoxy-heptyl)-2,4,8,10-tetra-oxaspiro[5,5]undecane[b] (0.02 mole)	210–295	11	—	—	—	5700	d
Hexamethylene diamine (0.008 mole)	3,9-Bis(7-carboxymethoxy-heptyl)-2,4,8,10-tetra-oxaspiro[5,5]undecane[b] (0.008 mole)	290 294	2 2½ at 28 mm Hg	— —	— —	100 —	3030 —	e
Dimethylterephthalate (0.32 mole)	3,9-Bis(7-carboxymethoxy-heptyl)-2,4,8,10-tetra-oxaspiro[5,5]undecane[b] (0.08 mole)	190–260	17½	—	—	—	—	f
Ethylene glycol (0.48 mole)	3,9-Bis(7-carboxymethoxy-heptyl)-2,4,8,10-tetra-oxaspiro[5,5]undecane[b] (0.08 mole)	260	3 at 0.15 mm Hg	—	—	—	—	f
HOC—R—COH (with O=C double bonds)	CH₂=CHO—R—OCH=CH₂	—	—	—	—	—	—	g

[a] Glycerol–methylazelaaldehydehyde adduct.   [b] Pentaerythritol–methyl azelaaldehydate adduct.

[c] W. C. Miller, R. A. Awl, E. H. Pryde, and J. C. Cowan, *J. Polym. Sci., J. Polym. Chem. Ed.* **8**, 415 (1970).

[d] E. H. Pryde, R. A. Awl, H. M. Teeter, and J. C. Cowan, *J. Polym. Sci.* **59**, 1 (1962).

[e] E. H. Pryde, D. J. Moore, H. M. Teeter, and J. C. Cowan, *J. Polym. Sci.* **58**, 611 (1962); E. H. Pryde, U.S. Patent 3,223,683 (1965).

[f] E. H. Pryde, U.S. Patent 3,183,215 (1965).   [g] Deutsche Solvay-Werke G.m.b.H., British Patent 768,305 (1957).

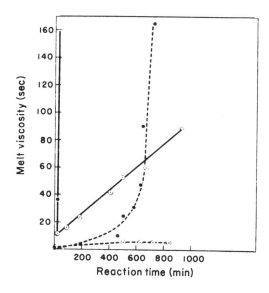

4-1. *Preparation of Poly(ester–acetal) by the Reaction of the Pentaerythritol Acetal of Methyl Azelaaldehydate with Dimethyl Terephthalate and Ethylene Glycol* [48]

a. Preparation of the Pentaerythritol Acetal of Methyl Azelaaldehydate [3,9-Bis(7-carbomethoxyheptyl)-2,4,8,10-tetraoxaspiro[5,5]undecane]

$$C(CH_2OH)_4 + (CH_3O)_2C(CH_2)_7COOCH_3 \xrightarrow{-CH_3OH}$$

$$CH_3O-\overset{\overset{O}{\|}}{C}-(CH_2)_7-\underset{O-CH_2}{\overset{O-CH_2}{CH}} \underset{CH_2-O}{\overset{CH_2-O}{C}} CH-(CH_2)_7\overset{\overset{O}{\|}}{C}OCH_3 \quad (26)$$

To a three-necked, 500-ml, round-bottomed flask equipped with a ther-mometer, a capillary tube for nitrogen, a mechanical stirrer, and a 4-inch Vigreux column with distilling head are added 100.7 gm (0.348 mole) of methyl azelaaldehydate dimethyl acetal (80.2% purity), 29.3 gm (0.215 mole) of pentaerythritol, and 0.1 gm of potassium bisulfate. The flask and contents are heated at a pot temperature of 127°–134°C for 5 hr during which time 19.0 gm (0.6 mole) of methanol is distilled off and collected. Toluene (100 ml) is added and heating continued for 3 hr with slow distillation of the toluene until the pot temperature reaches 142°C and the vapor temperature is 105°C. The reaction mixture is cooled and the solid product dissolved in 150 ml of methylene chloride. The methylene chloride solution is filtered of unreacted pentaerythritol (4.33 gm) and then washed with water. The methylene chloride is then removed at atmospheric pressure and the product distilled under reduced pressure (0.35 mm Hg) at 241°–260°C. Recrystallization from absolute ethanol affords 69.9 gm (85%), m.p. 65°–67°C.

b. Polymerization

$$\text{CH}_3\text{OC(CH}_2)_7\text{—CH} \overset{\text{O—CH}_2}{\underset{\text{O—CH}_2}{\diagup}} \text{C} \overset{\text{CH}_2\text{—O}}{\underset{\text{CH}_2\text{—O}}{\diagdown}} \text{HC—(CH}_2)_7\text{COCH}_3 \; + \; \text{CH}_3\text{OC—}\langle\text{arene}\rangle\text{—C—OCH}_3$$

$$+ \quad \overset{\text{CH}_2\text{—CH}_2}{\underset{\text{OH} \quad \text{OH}}{|\qquad|}}$$

$$\swarrow$$

$$\text{—O—C—(CH}_2)_7\text{—CH} \overset{\text{O—CH}_2}{\underset{\text{O—CH}_2}{\diagup}} \text{C} \overset{\text{O—CH}_2}{\underset{\text{O—CH}_2}{\diagdown}} \text{HC(CH}_2)_7\text{COCH}_2\text{CH}_2\text{OC—}\langle\text{arene}\rangle\text{—C—O—}\underset{\text{CH}_2}{\underset{\text{CH}_2\text{—}}{|}} \Big]_n$$

(27)

To a 250-ml distilling flask equipped with a thermometer and capillary inlet for nitrogen are added 62.1 gm (0.32 mole) of dimethyl terephthalate, 37.8 gm (0.08 mole) of the acetal (see 4-1,a, above), 30.0 gm (0.48 mole) of ethylene glycol (20% excess), and 0.01 gm of calcium oxide. The flask is heated gradually and at 190°C methanol begins to distill out. Heating is continued first at 240°–260°C for $7\frac{1}{2}$ hr at atmospheric pressure, than for 1 hr at 200 mm Hg, and finally at 260°C and 0.15 mm Hg for 3 hr. During the entire reaction nitrogen is bubbled through the reaction mixture to provide good agitation. The resulting polymer is an opaque white solid, m.p. 174°–182°C.

Cross-linked polymers are prepared as described in Preparation 4-2.

4-2. Preparation of Cross-Linked Poly(ester–acetals) [42]*

$$
CH_3O\overset{O}{\overset{\|}{C}}(CH_2)_7CH\underset{O-CH}{\overset{O-CH}{<}}\underset{CH-O}{\overset{CH-O}{>}}CH(CH_2)_7COOCH_3 \;+\; CH_3O\overset{O}{\overset{\|}{C}}-\!\!\!\bigcirc\!\!\!-\overset{O}{\overset{\|}{C}}OCH_3
$$

$$
+\;\; \underset{OH}{\overset{CH_2}{|}}-\underset{OH}{\overset{CH_2}{|}}
$$

$$
H\!\!\left[OCH_2CH_2-O\overset{O}{\underset{\|}{C}}-(CH_2)_7-CH\underset{O-CH_2}{\overset{O-CH_2}{<}}\underset{CH_2-O}{\overset{CH_2-O}{>}}CH(CH_2)_7COOCH_2CH_2O-\overset{O}{\overset{\|}{C}}-C_6H_4-\overset{O}{\overset{\|}{C}}\right]_n\!\!OCH
$$

(28)

Dimethyl terephthalate (13.59 gm, 0.07 mole), 3,9-bis(7-carbomethoxy-heptyl)-2,4,8,10-tetraoxaspiro[5,5]undecane (14.16 gm, 0.03 mole), ethylene glycol (7.45 gm, 0.12 mole), and calcium oxide (0.005 gm) are placed in a 50-ml modified distilling flask equipped with a thermometer and a fine capillary inlet for nitrogen. The flask and its contents are heated gradually, and at 180°C methanol begins distilling from the melt. The temperature is gradually increased to 235°C over a 10-hr period. Heating is continued at 110 mm Hg and 233°–258°C for 2 hr, and finally at 0.1 mm Hg 234°–280°C for an additional $2\frac{1}{4}$ hr. The flask is then immersed in a silicone oil bath at 280°C and fitted with a 1-ml graduated pipet from which the tip has been removed. The time required for flow between two 0.1-ml marks is 0.4 sec. Boric acid (0.015 gm) is added, and as heating is continued, the melt viscosity increases gradually at first, then more rapidly (as shown in Fig. 1) until the polymers have set to a rubbery, insoluble gel. At a melt viscosity of 90 sec the polymer can readily be drawn out into a fiber which shows considerable elasticity. The gel undergoes a phase change at 160°C but does not melt at 300°C. It adheres strongly to the glass of the reaction flask and can be separated only by chilling the flask and polymer to −60°C. A polymer prepared in a similar manner from dimethyl terephthalate and ethylene glycol but containing no ester–acetal has a melting point of 246°–248°C.

In a similar experiment, litharge is added to a soluble poly(ester–acetal) which has a melt viscosity of about 10 sec at 270°C and which has been heated 12 hr with lime. Within 40 min the polymer sets to an insoluble gel. Similarly,

---

* Reprinted from E. H. Pryde, R. A. Awl, H. M. Teeter, and J. C. Cowan, *J. Org. Chem.* **25**, 2260 (1960). Copyright 1960 by the Am. Chem. Soc. Reprinted by permission of the copyright owner.

zinc oxide and zinc acetate give gelled polymers. With 10 mole % of the ester–acetal, polymers at the borderline of gelation were obtained which had melting points of 226°C (after a reaction period of 20 min at 280°C with litharge) and of 221–225C° (after 173 min at 250°–280°C with boric acid). These polymers were soluble in 50% phenol in tetrachloroethane. They were light colored, hard solids which adhered strongly to the glass walls of the reaction flask.

4-3. Preparation of Poly(acetal–amide) by the Reaction of the Pentaerythritol Acetal of Methyl Azelaaldehydate with Hexamethylenediamine [49]

$$
\text{CH}_3\text{OC(CH}_2)_7\text{CH}
\begin{array}{c}
\text{O—CH}_2 \quad \text{CH}_2\text{—O} \\
\diagdown \quad \diagup \\
\text{C} \\
\diagup \quad \diagdown \\
\text{O—CH}_2 \quad \text{CH}_2\text{—O}
\end{array}
\text{CH(CH}_2)_7\text{C—OCH}_3 + \text{H}_2\text{N—(CH}_2)_6\text{—NH}_2
$$

$$
\left[-\overset{\text{O}}{\underset{\|}{\text{C}}}-(\text{CH}_2)_7-\text{CH}
\begin{array}{c}
\text{O—CH}_2 \quad \text{CH}_2\text{—O} \\
\diagdown \quad \diagup \\
\text{C} \\
\diagup \quad \diagdown \\
\text{O—CH}_2 \quad \text{CH}_2\text{—O}
\end{array}
\text{CH(CH}_2)_7\overset{\text{O}}{\underset{\|}{\text{C}}}\text{—NH(CH}_2)_6\text{NH—} \right]_n
\tag{29}
$$

To a flask as described in Preparation 4-2 are added 4.00 gm (0.008 mole) of the pentaerythritol acetal of methyl azelaaldehydate and 0.93 gm (0.008 mole) of hexamethylenediamine. The mixture is first heated for 2 hr at 290°C with capillary nitrogen agitation and then for $2\frac{1}{2}$ hr at 290°C at 28 mm Hg to give 3.65 gm of a white brittle solid, m.p. 160°–163°C, mol. wt. 3030. Heating for 20 min at 260°C with *p*-toluenesulfonic acid catalyst (3–10 wt.%) results in a transparent, infusible, brownish solid with good adhesion to glass. Using no catalyst, the polymer gels with continued heating at 290°C for 10 hr.

4-4. Preparation of a Polyacetal by the Condensation of the Divinyl Ether of Diethylene Glycol with Diethylene Glycol [33]

$$
\text{HOCH}_2\text{CH}_2\text{O—CH}_2\text{CH}_2\text{OH} + \text{H}_2\text{C}{=}\text{CH—O—CH}_2\text{CH}_2\text{OCH}_2\text{CH}_2\text{OCH}{=}\text{CH}_2 \longrightarrow
$$

$$
\left[-\text{O—CH}_2\text{CH}_2\text{—O—CH}_2\text{CH}_2\text{O}\underset{\underset{\text{CH}_3}{|}}{\text{CH}}\text{—O—CH}_2\text{CH}_2\text{—O—CH}_2\text{CH}_2\text{—O}\underset{\underset{\text{CH}_3}{|}}{\text{CH}}\text{—} \right]_n
\tag{30}
$$

To a resin flask equipped with a mechanical stirrer and a nitrogen inlet is added 26.5 gm (0.25 mole) of diethylene glycol and the flask is cooled to 20°C while 41 gm (0.25 mole) of the divinyl ether of diethylene glycol is added. Then 2.0 gm (0.012 mole) of trichloroacetic acid is added over a 15-min period while the temperature is maintained at 20°–30°C. The viscous liquid is diluted with 100 gm of methylene chloride while powdered sodium hydroxide is added to neutralize the acidity. The reaction is filtered and the methylene chloride concentrated to give a residue. The residue is heated for 2 hr at $100°-120°C$ at 1 mm Hg to give the product, mol. wt. 3950, n_D^{20} 1.4630.

5. POLYKETALS

The reaction of a ketone with diols or diketones with polyols yields polyacetals by the direct reaction or by the transketalization using methyl or ethyl ketals with polyols.

$$\begin{array}{c}\text{HOCH}_2\\ \qquad\qquad\text{C}\\ \text{HOCH}_2\end{array}\begin{array}{c}\text{CH}_2\text{OH}\\ \\ \text{CH}_2\text{OH}\end{array} + \text{O=C-R-C=O} \longrightarrow \left[\begin{array}{c}\text{-O-CH}_2\\ \qquad\qquad\text{C}\\ \text{-O-CH}_2\end{array}\begin{array}{c}\text{CH}_2\text{-O}\\ \\ \text{CH}_2\text{-O}\end{array}\text{C-R}\right]_n$$

$$+ 2\text{H}_2\text{O} \qquad (31)$$

$$\text{HO-R'-OH} + (\text{R})_2\text{C(OR')}_2 \xrightarrow{\text{H}^+} \left[\begin{array}{c}\text{-O-R'-O-C-}\\ \qquad\qquad\diagup\diagdown\\ \qquad\quad\text{R}\quad\text{R}\end{array}\right]_n + 2\text{R'OH} \qquad (32)$$

A. Polyketals by Transketalization Reactions

High molecular weight polyacetals have been reported to be prepared by the reaction of 2,2-dimethoxypropane with 1,4-cyclohexanedimethanol using PTSA (0.03%) as a catalyst [50].

$$(\text{CH}_3\text{O})_2\text{C(CH}_3)_2 + \text{HOCH}_2-\bigcirc-\text{CH}_2\text{OH} \xrightarrow[\text{C}_6\text{H}_{14}]{\text{PTSA}}$$

$$\left[\begin{array}{c}\qquad\qquad\qquad\qquad\qquad\qquad\text{CH}_3\\ \text{-O-CH}_2-\bigcirc-\text{CH}_2\text{OC-}\\ \qquad\qquad\qquad\qquad\qquad\qquad\text{CH}_3\end{array}\right]_n \qquad (33)$$

m.p. $140°-170°C$
mol. wt. 26,000

The polymer is produced by heating the reactants and slowly (approximately 20 hr) distilling off the methanol and replacing from time to time the hexane solvent.

B. Polyspiroketals

In 1951, Orth [6] prepared polyketals using 1,4-cyclohexanedione and, in 1962, Cohen and Laven [7] reported on the ketal prepared by the transketalization reaction of the ethyl ketal of 2,5-hexanedione with pentaerythritol [Eq. (34)].

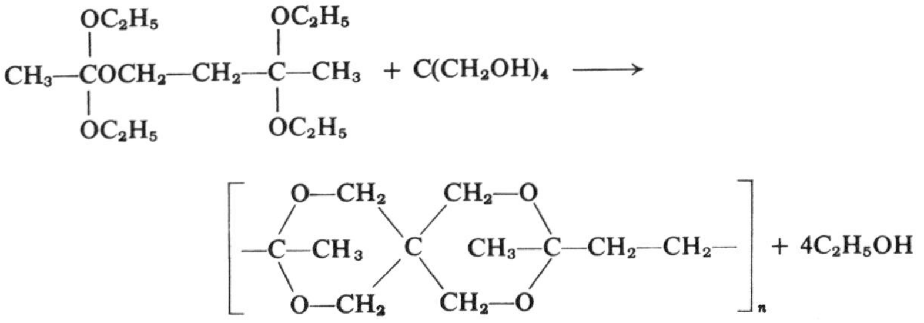

$$
\begin{array}{cc}
OC_2H_5 & OC_2H_5 \\
| & | \\
CH_3\!-\!COCH_2\!-\!CH_2\!-\!C\!-\!CH_3 & + \; C(CH_2OH)_4 \longrightarrow \\
| & | \\
OC_2H_5 & OC_2H_5
\end{array}
$$

$$
\left[\begin{array}{cc}
O\!-\!CH_2 \quad CH_2\!-\!O \\
\diagup \qquad \diagup \\
-C\!-\!CH_3 \quad C \quad CH_3\!-\!C\!-\!CH_2\!-\!CH_2- \\
\diagdown \qquad \diagdown \\
O\!-\!CH_2 \quad CH_2\!-\!O
\end{array}\right]_n + \; 4C_2H_5OH \quad (34)
$$

mol. wt. 8250

Bailey and Volpe [50a] reported on the preparation of a series of thermally stable spiro polymers by reacting various diketones with pentaerythritol [Eq. (35)].

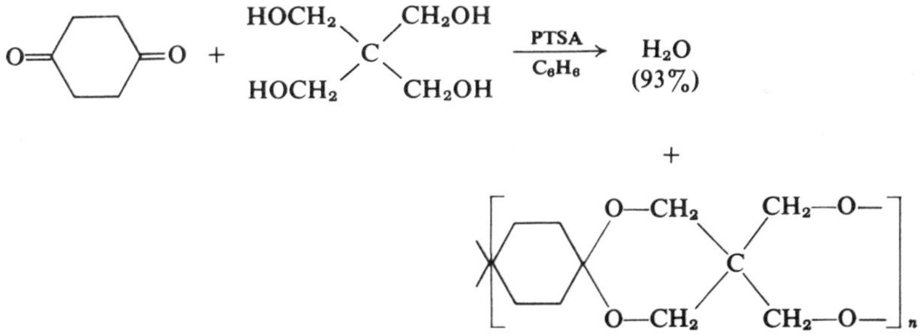

$$
O\!=\!\bigcirc\!=\!O \; + \;
\begin{array}{cc}
HOCH_2 & CH_2OH \\
\diagdown \quad \diagup \\
C \\
\diagup \quad \diagdown \\
HOCH_2 & CH_2OH
\end{array}
\xrightarrow[C_6H_6]{PTSA}
\begin{array}{c}
H_2O \\
(93\%)
\end{array}
$$

$$
+
$$

$$
\left[\bigcirc\!\!\!\times\!\!\!\times
\begin{array}{cc}
O\!-\!CH_2 \quad CH_2\!-\!O- \\
\diagup \qquad \diagup \\
C \\
\diagdown \qquad \diagdown \\
O\!-\!CH_2 \quad CH_2\!-\!O-
\end{array}\right]_n \quad (35)
$$

m.p. 350°C (decomp.)

The polymer from 1,4-cyclohexanedione did not melt but started to decompose at 250°C. The polymer is insoluble in most common solvents for polymers (DMF, DMSO) but did dissolve in concentrated sulfuric acid with decomposition. X-Ray data indicated that the polymer was linear with at least 95% crystallinity. Attempts to reduce the crystallinity by copolymerization of pentaerythritol with two diketones still gave very crystalline polymers. The thermal stability and intrinsic viscosities in hexafluoropropanol of some of the polymers are shown in Table VI.

PHYSICAL PROPERTIES OF POLYKETALS FROM PENTAERYTHRITOL[a]

Polymer from pentaerythritol and	Intrinsic viscosity in hexafluoroisopropanol at 25°C	Thermal gravimetric data (50% wt. loss; °C)
1,4-Cyclohexanedione	0.049	450
1,10-Cyclooctadecanedione	0.092	365
Mixture of 1,4-cyclohexanedione and 1,10-cyclooctadecanedione	0.045	—
95% 1,4-Cyclohexanedione and 5% cyclohexanone	—	425
90% 1,4-Cyclohexanedione and 10% cyclohexanone	—	414
80% 1,4-Cyclohexanedione and 20% cyclohexanone	—	407
50% 1,4-Cyclohexanedione and 50% cyclohexanone	—	360
1,10-Cyclooctadecanedione (dimethyl sulfoxide-insoluble fraction)	—	400
1,10-Cyclooctadecanedione (dimethyl sulfoxide-soluble fraction)	—	400
50% 1,4-Cyclohexanedione and 50% 1,10-cyclooctadecanedione	—	438

[a] Data taken from W. J. Bailey and A. A. Volpe, *Polym. Prepr., Am. Chem. Soc., Div. Polym. Chem.* **8**, 292 (1967).

5-1. Preparation of a Polyketal by Reaction of Pentaerythritol with 1,4-Cyclohexanedione [50a]*

$$C(CH_2OH)_4 \; + \; O=\langle\!\!\langle \;\rangle\!\!\rangle=O \; \longrightarrow \; \left[\begin{array}{c} O-CH_2 \quad CH_2-O \\ \diagdown\,C\,\diagup \\ O-CH_2 \quad CH_2-O \end{array} \right]_n \quad (36)$$

A mixture of 6.8 gm (0.05 mole) of pentaerythritol, 5.6 gm (0.05 mole) of 1,4-cyclohexanedione and 0.5 gm of *p*-toluenesulfonic acid monohydrate is heated under reflux in 100 ml of benzene. After the mixture has been heated for 3 hr, 1.71 ml (97% of the theoretical amount) of water is collected in a Bidwell–Sterling trap. The mixture is cooled to room temperature and allowed to stand overnight. The product is a white precipitate which is collected by

* Preparation 5-1 is reprinted from W. J. Bailey and A. A. Volpe, *Polym. Prepr., Am. Chem. Soc., Div. Polym. Chem.* **8**, 293 (1967). Copyright 1967 by the American Chemical Society Reprinted by permission of the copyright owner.

filtration, washed several times with benzene, and dried in air. About 11.2 gm (90%) of a white, powdery polymer is obtained which does not melt below 310°C and which does not hydrolyze in water but does hydrolyze in concentrated sulfuric acid to give the starting material. Since a small amount of pentaerythritol is recovered, the reaction, based on unrecovered starting material, seems to be essentially quantitative.

Analysis of the Spiro Polymer

Polymer from Pentaerythritol and 1,4-Cyclohexanedione. An infrared spectrum of the spiro polymer indicates that this material contains a very small number of hydroxyl groups (3400 cm^{-1}) which are due either to entrapped starting materials or to end groups on the polymer. For the most part, however, the polymer seems to consist of a polyspiroketal structure since its spectrum is similar to that of pentaerythritol dibutyral.

X-Ray analysis of the polymer gives a diffraction pattern that contains very sharp peaks with no amorphous background, indicating that it is very highly crystalline (probably greater than 95% crystalline). The interplanar species to which all of its peaks corresponds and their intensities are shown in the following tabulation.

Location of X-ray reflections (Å)	Intensity
5.50	Low
5.10	Very high (dominates pattern completely)
4.67	Low
4.35	Medium
3.97	Very low
3.53	Very low
3.07	Very low
2.81	Very low
2.60	Very low
2.50	Very low

Bailey also extended this work to the preparation of the polythioketals using pentaerythritol tetrathiol and cyclic diketones, and he reported the use of a number of tetrols with cyclohexanedione to give thermally stable, insoluble polyspiroketals as described in Table VII.

TABLE VII
PREPARATION OF POLYSPIROKETALS BY THE REACTION OF TETROLS WITH
1,4-CYCLOHEXANEDIONE [51]

Tetrol	Polymer yield (%)	M.p. (°C)	Decomposition point (°C)	Hexafluoro-isopropanol (dl/gm 25°C)
1,1,4,4-Tetrakis(hydroxymethyl)-cyclohexanone	70	370	410–450	—
2,2,5,5-Tetrakis(hydroxymethyl)-cyclopentanone	74	370–395	—	—
1,2,4,5-Cyclohexanetetrol	—	—	—	0.056
1,2,4,5-Tetrakis(hydroxymethyl)-cyclohexane	99	—	—	0.04

5-2. Preparation of Polyspiroketal by the Reaction of 1,4-Cyclohexanedione with 1,1,4,4-Tetrakis(hydroxymethyl)cyclohexane [51]*

$$O = \left\langle \text{cyclohexane} \right\rangle = O \; + \; \begin{array}{c} HOCH_2 \\ HOCH_2 \end{array} \left\langle \text{cyclohexane} \right\rangle \begin{array}{c} CH_2OH \\ CH_2OH \end{array} \longrightarrow$$

$$\left[\left\langle \text{cyclohexane} \right\rangle \begin{array}{c} O-CH_2 \\ O-CH_2 \end{array} \left\langle \text{cyclohexane} \right\rangle \begin{array}{c} CH_2-O \\ CH_2-O \end{array} \right]_n \quad (37)$$

A mixture of 10.2 gm (0.05 mole) of 1,1,4,4-tetrakis(hydroxymethyl)-cyclohexane, 5.6 gm (0.05 mole) of 1,4-cyclohexanedione, and 0.5 gm of p-toluenesulfonic acid is heated under reflux in 200 ml of benzene (reagent grade). After the mixture has been heated with rapid stirring for 5 hr, another 150 ml of benzene is added and the mixture heated until no further water is collected in the Dean & Stark trap. The hot mixture is filtered through a Büchner funnel, and the solid is washed sequentially with 400-ml portions of boiling benzene, methanol, carbon tetrachloride, chloroform, and N,N-dimethylformamide. Finally the polymer was washed with 200 ml of benzene, air dried, and then vacuum desiccated for 6 hr to give 10.0 gm (70%) of polymer which shows no loss of water at 100°C and no change of

* Preparation 5-2 is reprinted from W. J. Bailey, C. F. Bean, Jr., and I. Haddad, Polym. Prepr., Am. Chem. Soc., Div. Polym. Chem. 12, 169 (1971). Copyright 1971 by the American Chemical Society. Reprinted by permission of the copyright owner.

color until 400°C. Some shrinkage is observed to start at 370°C, and melting with decomposition results at 410°–450°C.

Analysis

Calculation for $(C_{16}H_{24}O_4)_n$: C, 68.54   H, 8.63
Found: C, 68.38   H, 8.37

After a 0.2-gm sample is added to 40 ml of hexafluoroisopropanol, the mixture is stirred overnight and filtered. A residue of approximately 2 mg remains undissolved. An IR spectrum of the vacuum-desiccated sample shows trace absorptions at 3420 and 1699 cm^{-1} which are attributed to entrapped starting material or endgroups on the polymer chain. An IR spectrum of the same sample although nondesiccated shows a pronounced increase in the hydroxyl absorption at 3420 cm^{-1}, indicative of the hydroscopic nature of the polymer. A detailed IR (Nujol) spectrum contains the following absorptions (cm^{-1}): 1360 (m), 1345 (m), 1270 (m), 1240 (w), 1169 (m), 1144 (s), 1132 (m), 1110 (s), 1101 (s), 1079 (m), 1059 (m), 1050 (m), 1038 (m), 1025 (m), 988 (w), 977 (m), 948 (m), 929 (w), 908 (s), and 891 (m).

6. POLYTHIOACETALS AND POLYTHIOKETALS

The reaction of a polymercaptan with dialdehydes [52] or diketones [53] affords polymers as shown in Eqs. (38) and (39).

CAUTION: Alkylene dimercaptans have been reported to cause severe skin rashes [54].

Autenrieth, Geyer, and Beuttel [55] reported that pentamethylenedithiol reacts with acetone and diethylene ketone to give crystalline products. However, the same dithiol reacts with either acetaldehyde, benzaldehyde, or benzophenone to give noncrystalline products. The latter are probably polymeric but were not recognized at that time. Fisher and Wiley [56] prepared the spiran-type polymer shown in Eq. (39) by reaction of the tetramercaptan derived from pentaerythritol and 1,4-cyclohexanedione.

TABLE VIII

PROPERTIES OF POLYMERCAPTALS AND POLYMERCAPTOLS[a]

Dithiol	Grams	Carbonyl compound	Grams	Time (min) passing HCl	Reaction detd. by unused –SH (%)	M.p. (°C)	Inherent viscosity	Mol. wt. (calc.)	Equiv. wt. by amperometric titration
Decamethylene	0.9444	Acetaldehyde	0.2078	30	95.39	35	0.042	1100	—
Hexamethylene	0.9800	Acetaldehyde	0.3685	11½	—	60	0.146	5100	3930
Decamethylene	0.9370	Butyraldehyde	0.4587	17	—	—	0.102	3300	—
Hexamethylene	0.97	Butyraldehyde	0.8170	15	—	40	0.049	1400	—
Decamethylene	0.98	Benzaldehyde	1.0504	12 (hr)[b]	—	135	0.428	18,600	—
Decamethylene	0.98	Benzaldehyde	1.0504	15[c]	—	120	0.452	19,900	—
Hexamethylene	0.97	Benzaldehyde	1.0504	7	—	115	1.374	76,000	—
The above polymer after a month in corked tube						—	0.864	43,000	25,030
Hexamethylene	0.97	Benzaldehyde	1.0504	10	—	110	0.98	50,000	—
The above polymer after a month in vacuum						—	0.93	47,000	29,500
Hexamethylene	1.5329	Vanillin	1.0178	25	—	40	0.065	1900	528

Hexamethylene	0.9810	Vanillin	10	—	150	Insoluble	—	—
Hexamethylene	0.9840	Anisaldehyde	10	99.8	130	0.44	19,000	—
Decamethylene	1.8847	m-Nitrobenzaldehyde	60	99.75	75	0.16	5700	—
Hexamethylene	0.9872	p-Nitrobenzaldehyde	22	99.71	80	Insoluble	—	—
Hexamethylene	0.9818	p-Bromobenzaldehyde	17	99.16	72	0.067	2000	2380
Hexamethylene	0.9823	2-Butanone	1¾ (hr)	70.87	Low	0.088	2700	—
Hexamethylene	0.9886	3-Methyl-2-pentanone	60	45.83	Low	0.048	1300	1360
Hexamethylene	0.9779	4-Methyl-2-pentanone	60	48.45	Low	0.053	1500	—
Hexamethylene	0.9920	Acetophenone	12	71.40	30	0.088	2700	—
Hexamethylene	0.9797	Cyclohexanone	10	90.00	—	—	—	—
Hexamethylene	0.9815	Cyclohexanone	10^d	95.93	75	0.34	14,300	—
Decamethylene	0.9502	Vanillin	10	97.18	140	Insoluble	—	—

[a] Reprinted in part from C. S. Marvel, E. H. H. Shen, and R. R. Chambers, *J. Am. Chem. Soc.* **72**, 2106 (1950). Copyright 1950 by the American Chemical Society. Reprinted by permission of the copyright owner.

[b] Hydrogen chloride gas was passed very slowly. This reaction was done without solvent.

[c] After passing hydrogen chloride into the mixture for 15 min, the reaction was left in contact with air for 6 hr. This reaction was carried out in dioxane solution (10 ml).

[d] The polymer solution was exposed to air for 10 hr before precipitating by methanol.

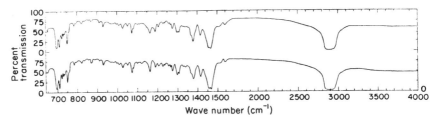

Fɪɢ. 2. Infrared absorption of benzaldehyde decamethylenedithiol polymer. Upper, polymer prepared in absence of solvent; lower, polymer prepared in solvent. [Reprinted from C. S. Marvel, E. H. H. Shen, and R. R. Chambers, *J. Am. Chem. Soc.* **72**, 2106 (1950). Copyright 1950 by the American Chemical Society. Reprinted by permission of the copyright owner.]

Marvel, Shen, and Chambers extended this work further by studying a series of different aldehydes with either decamethylenedithiol or hexamethylenedithiol in cold dioxane using hydrogen chloride gas as a catalyst. The results are shown in Table VIII and a typical infrared spectrum of these polymers is shown in Fig. 2.

The polymers are stable under basic conditions but are degraded in acid as evidenced by a lowering of the solution's viscosity over a 24-hr period [57].

6-1. General Procedure for the Preparation of Polythioacetals and Polythioketals Reported in Table VIII [57]

To a reaction flask are added the carbonyl compound and an equivalent amount of dithiol along with 10 ml of freshly distilled dry dioxane. The solution is placed in an ice bath while dry hydrogen chloride is bubbled through for varying lengths of time. In all cases the reaction is exothermic and a pink color develops. The mixture is cooled and 20 ml of cold methanol added to precipitate the polymer. The polymer is dried under reduced pressure (approximately 1 mm Hg) at room temperature and the results are shown in Table VIII.

In contrast to the results in Table VIII and Fig. 2, which indicate that only a polymer is obtained from the reaction of benzaldehyde with decamethylenedithiol, it was later reported that this reaction also proceeds to give a 50% yield of a crystalline product with the following structure [58]:

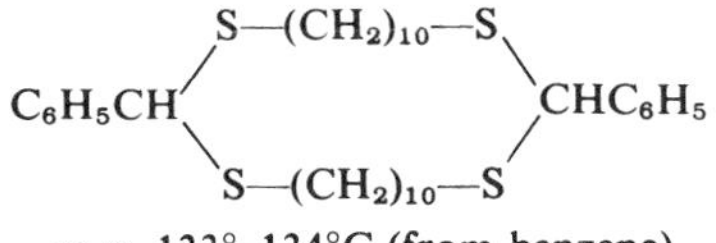

$$
\begin{array}{c}
\mathrm{S-(CH_2)_{10}-S} \\
\mathrm{C_6H_5CH} \qquad\qquad \mathrm{CHC_6H_5} \\
\mathrm{S-(CH_2)_{10}-S}
\end{array}
$$

m.p. 133°–134°C (from benzene)

Similar results are obtained with benzaldehyde diethyl acetal, where a 30% yield of the cyclic product is obtained as described in Table IX.

TABLE IX

REACTIONS OF DECAMETHYLENE DITHIOL AT 30°C[a,b]

	Benzaldehyde diethyl acetal	Vanillin	Acetone	p-Chloro-benzaldehyde diethyl acetal
Dithiol (gm)	19.28	12.9	19.3	19.3
Carbonyl compound	Benzaldehyde diethyl acetal	Vanillin	Acetone	p-Chloro-benzaldehyde diethyl acetal
Amount (gm)	16.83	9.52	5.425	20.052
Dioxane (ml)	80	80	70	80
Catalyst (ml) dioxane sat. with dry HCl	9	20	15	9
Reaction time (days)	12	5	5	44
Isolation method	A	B	B	A
Polymer (gm)[c]	19.1(69.5%)	0	0	20.1 (60.0%)
Inherent viscosity	0.21	—	—	0.18
Cyclic product (gm)	8.23 (29.9%)	21[c] (98.9%)	19.3 (83.9%)	12.08 (39.4%)
M.p. (°C)	133–134	163–164	120–121	144–145
Analyses (%)				
Calc. C		63.48	63.35	62.06
Obs. C		63.51	63.29	62.28
Calc. H		8.29	10.63	7.66
Obs. H		8.45	10.42	7.35
Calc. S		18.83	26.02	19.50
Obs. S		18.93	26.61	19.39[d]
Calc. mol. wt.		681	493	658
Obs. mol. wt.[e]		659	463	623

[a] Reprinted from C. S. Marvel and R. C. Farrar, Jr., *J. Am. Chem. Soc.* **79**, 986 (1957). Copyright 1957 by the American Chemical Society. Reprinted by permission of the copyright owner.

[b] Polymer isolated by addition of dioxane (free from crystalline cyclic product) to methanol.

[c] At the end of five days stirring this product precipitated suddenly from the solution.

[d] Anal. calc.: Cl, 10.78; found: Cl, 10.50.

[e] Average of 5 by boiling point elevation method of A. W. C. Menzies and S. L. Wright, Jr., *J. Am. Chem. Soc.* **43**, 2315 (1921).

**6-2. Preparation of the Polythioketal by Reaction of
Tetrakis(mercaptomethyl)methane with 1,4-Cyclohexanedione** [56]

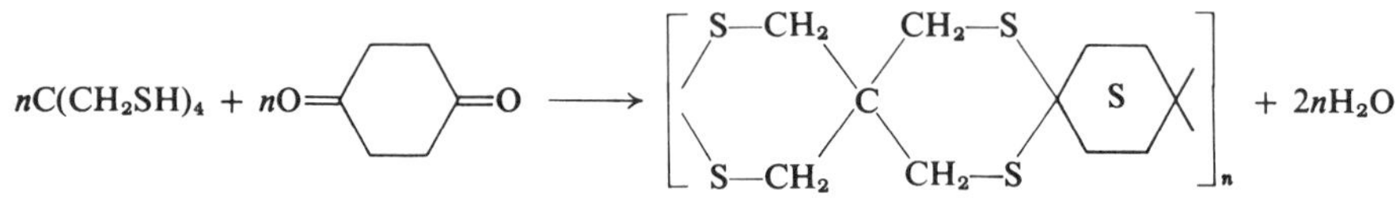

$$nC(CH_2SH)_4 + nO=\langle\ \rangle=O \longrightarrow \left[\begin{array}{c} S-CH_2 \quad CH_2-S \\ \diagdown \quad \diagup \\ C \quad \quad S \\ \diagup \quad \diagdown \\ S-CH_2 \quad CH_2-S \end{array}\right]_n + 2nH_2O$$

(40)

To a reaction flask are added 2.0 gm (0.01 mole) of tetrakis(mercapto-methyl)methane (m.p. 72°–73°C) and 1.12 gm (0.01 mole) of 1,4-cyclo-hexanedione (m.p. 76°–78°C) along with 10 gm of purified dioxane. The solution is cooled and hydrogen chloride gas is introduced. As the mixture warms to room temperature, a polymeric spiroketal precipitates, m.p. 314°–318°C. The polymer is insoluble in *m*-cresol, DMF, butyl Cellosolve, phenol, furfural, tetralin, chloroform, camphor, C_6H_5Cl, C_6H_5—$N(CH_3)_2$, CS_2, and cyclohexanediol. The polymer is slightly soluble in boiling naph-thalene and fairly soluble in boiling anthracene, in which it has a molecular weight of 613.

7. POLY(VINYL ACETALS)

Poly(vinyl alcohol) reacts as a 1,3-diol with aldehydes to form cyclic acetals. The acetalization reaction is sometimes between polymer chains, and the resultant cross-linking reaction renders the polymer insoluble.

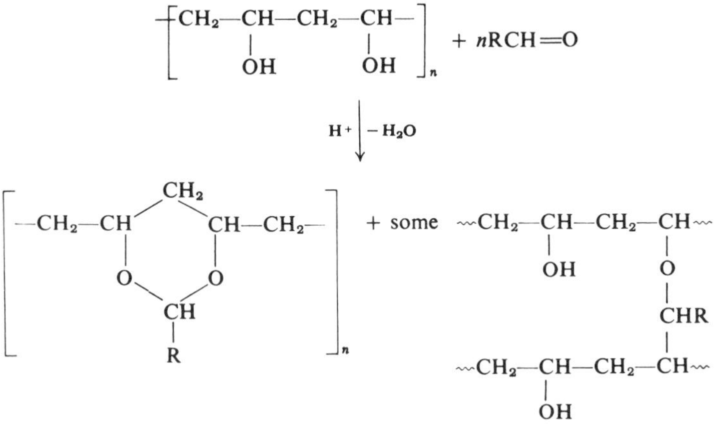

$$\left[\begin{array}{c} CH_2-CH-CH_2-CH- \\ \quad | \quad \quad \quad | \\ \quad OH \quad \quad OH \end{array}\right]_n + nRCH=O$$

$$H^+ \downarrow -H_2O$$

(41)

$$\left[\begin{array}{c} \quad \quad CH_2 \\ \quad \diagup \quad \diagdown \\ -CH_2-CH \quad \quad CH-CH_2- \\ \quad \diagdown \quad \diagup \\ \quad O \quad \quad O \\ \quad \diagdown \quad \diagup \\ \quad \quad CH \\ \quad \quad | \\ \quad \quad R \end{array}\right]_n + \text{some} \quad \begin{array}{c} \sim CH_2-CH-CH_2-CH\sim \\ \quad | \quad \quad \quad | \\ \quad OH \quad \quad O \\ \quad \quad \quad \quad | \\ \quad \quad \quad \quad CHR \\ \quad \quad \quad \quad | \\ \sim CH_2-CH-CH_2-CH\sim \\ \quad \quad | \\ \quad \quad OH \end{array}$$

R = H, alkyl, or aryl

In most cases poly(vinyl acetal) resins contain 5–25% hydroxyl groups, on a weight basis, calculated as poly(vinyl alcohol) (see Table X).

Poly(vinyl acetals) were first prepared by Hermann and Haehnel [59] in 1924 by a reaction of benzaldehyde with poly(vinyl alcohol). The commercialization of poly(vinyl acetals) began in the 1930's when poly(vinyl alcohol)

TABLE X

PREPARATION OF POLY(VINYL ACETALS)[a]

Aldehyde	Grams per 100 gm PVOH	Reaction conditions		Analysis	
		Time (hr)	Temp. (°C)	% Volatile	% PVOH
Paraldehyde	50	69	60	0.75	11.8
	67	40+	60	—	9.1
	90	69	60	1.38	7.1
Propionaldehyde	57.5	4	75	—	19.7
	67	4	75	—	16.7
	82	4	75	—	13.7
	93	4	75	—	10.3
Isobutyraldehyde	66	$4\frac{1}{2}$	75	1.3	25.3
	62	5	75	1.5	19.8
	100	5	73	3.0	12.9
Butyraldehyde	53	7	75	0.8	25.4
	65	5	77	1.5	18.0
	76	5	75	1.4	15.8
	100	5	75	1.2	12.0
2-Ethyl butyraldehyde	95	18	75	0.4	21.7
	100	16	75	1.1	17.5
	93	17	75	0.8	15.4
n-Hexaldehyde	79	16	75	0.5	24.5
	87	16	75	1.65	19.3
	90	16	75	1.16	15.9
	106	16	75	1.0	12.1
n-Heptaldehyde	90	16	75	0.9	20.0
	100	16	75	1.2	19.3
	104	16	75	1.08	15.25
	109	16	75	1.0	13.2
2-Ethyl hexaldehyde	102	17	75	2.2	19.3
	123	18	75	2.3	14.1
	135	17	75	1.1	12.7
	165	17	70	3.8	6.3

[a] Reprinted from A. F. Fitzhugh and R. N. Crozier, *J. Polym. Sci.* **8**, 2251 (1952). Copyright 1952 by the Journal of Polymer Science. Reprinted by permission of the copyright owner.

was reacted in solution with aldehydes (formaldehyde, acetaldehyde, butyraldehyde) in the presence of mineral acids to give products useful for electrical insulation [60,60a]. Large-scale developments followed in the United States as a result of efforts by Shawinigan, DuPont, Monsanto, and Carbide [61,61a].

The five major methods of preparing poly(vinyl acetals) involve the following:

1. Conversion of poly(vinyl acetate) to poly(vinyl alcohol) in acid solution followed by acetalization [62,62a].

2. Reaction of an aqueous poly(vinyl alcohol) solution with the aldehyde until precipitation of the acetal occurs [63].

3. A method similar to 2 except that there is added a solvent for the acetal which is also miscible with water, thereby preventing precipitation.

4. Heterogeneous reaction of film or fibre of poly(vinyl alcohol) with the aldehyde to form the acetal.

5. Poly(vinyl alcohol) is suspended in a suitable nonsolvent which dissolves the aldehyde and the final product.

The first two methods are more widely used for commercial manufacture of poly(vinyl acetals).

Some of the aldehydes used to acetalize poly(vinyl alcohol) are formaldehyde [64], acetaldehyde [65], propionaldehyde [65], butyraldehyde [65], heptanol [65], palmitic and stearyl aldehydes [66], glyoxal [67], benzaldehyde [59], p-tolualdehyde [68], 2-naphthaldehyde [68], vinyl benzaldehyde [69], and glyoxalic acid [70]. Ketones such as cyclohexanone [71] and methyl ethyl ketone have been used to give ketals [72].

The hydrolysis of various poly(vinyl acetals) in ethanolic 85% phosphoric acid at reflux temperatures has been measured by the percentage of aldehyde liberated in a given length of time [73]. The order of resistance to hydrolysis of the polymers examined is poly(vinyl formal) > poly(vinyl propional) > poly(vinyl acetal) > poly(vinyl butyral).

$$+ \text{RCHO} + \text{H}^+ \quad (42)$$

Poly(vinyl acetals) find use in the areas of coatings [74] and adhesives [75] but their high cost has limited more widespread use. Poly(vinyl formals) are used as wire enamels [76] and are also combined with phenolics to increase toughness. Other resins have also been combined with poly(vinyl formal) to increase overload resistance [77]. Poly(vinyl butyral) is mainly used as an interlayer material in automobile safety glass and approximately 30 million lb was used in 1970 in the United States for this application [78].

Poly(vinyl acetals) have been reviewed elsewhere and these references should be consulted for further details [79].

A. Poly(vinyl formal)

Poly(vinyl formal) was developed by General Electric as a wire coating and the early process involved direct reaction of a mixture of poly(vinyl acetate), glacial acetic acid, formalin and concentrated sulfuric acid at 70°C [77,80]. After the desired degree of polymerization is obtained, ammonium hydroxide is added to bring the pH to about 9–10. The reaction mixture is poured into water to precipitate the poly(vinyl formal). Poly(vinyl formals) have also been used as fibers [81].

B. Poly(vinyl acetaldehyde acetal)

Poly(vinyl acetates) can be condensed directly with acetaldehyde in the presence of acidic catalysts [60a,82] or by using pure poly(vinyl alcohol) and acetaldehyde [83].

C. Poly(vinyl butyral)

Poly(vinyl butyral) may be prepared by dissolving poly(vinyl alcohol) (PVA) in a mixture of water and methanol and adding butyraldehyde and a small amount of sulfuric acid as a condensation agent and heating to 60°–70°C [61,84]. The best quality poly(vinyl butyral) is stabilized by slurrying an alkali in order to prevent viscosity changes and discolorization. Urea is also used to stabilize the pH to about 1.5 [85].

Some processes involve saponifying poly(vinyl acetate) (PVAc) to PVA with aqueous hydrochloric acid; butyraldehyde is then added directly [61,84,86]. The resulting poly(vinyl butyral) solution is neutralized and the product precipitated with water, washed, and dried.

7-1. General Procedure for the Preparation of Poly(vinyl acetals) [62]

Acetalization (see Note) was carried out in a three-necked resin flask equipped with a mechanical stirrer and condenser. The reaction times vary from 4 to 40 hr at 60°–75°C. At the end of this time water is slowly added to a vigorously stirred solution to precipitate the polymer in a granular form.

The heat stability is enhanced by steeping for approximately 4 hr with a 0.02 N potassium hydroxide solution at 50°C. The resin is finally washed with water and dried at 55°–60°C. The hydroxyl content is determined by the acetalation.

NOTE: Acetals of acetaldehyde, propionaldehyde, and n- and isobutyraldehyde are prepared by condensation with PVA in the presence of sulfuric acid in ethanol. Acetals of 2-ethyl butyraldehyde, n-hexaldehyde, n-heptaldehyde, and 2-ethyl hexaldehyde are prepared in a mixture of dioxane and ethanol. Some typical results are shown in Table X.

7-2. Preparation of Poly(vinyl formal) [87]

$$-CH_2-CH-CH_2-CH- \longrightarrow \left[-CH_2-CH-CH_2-CH- \right]_n + nCH_2=O$$

with OAc, OAc substituents on the left and OH, OH on the right product.

$$\left[-CH_2-CH \begin{array}{c} CH_2 \\ \diagup \quad \diagdown \\ \quad \quad CH- \\ O \diagdown \quad \diagup O \\ CH_2 \end{array} \right]_n \quad (43)$$

To a round-bottomed flask equipped with a reflux condenser, mechanical stirrer, and a heating mantle are added 12.0 gm (0.138 mole) of poly(vinyl acetate) of high viscosity and 40 gm (0.66 mole) of glacial acetic acid. The mixture is heated with stirring for 1 hr or until complete solution is obtained.

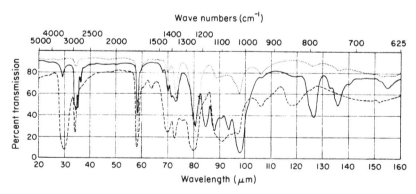

FIG. 3. Infrared spectra: solid line, poly(vinyl formal); dashed line, poly(vinyl alcohol); dotted line, poly(vinyl acetate). [Reprinted from H. C. Beachell, P. Fotis, and J. Hacks, *J. Polym. Sci.*, 7, 353 (1951). Copyright 1951 by the Journal of Polymer Science. Reprinted by permission of the copyright owner.]

To the solution are added 4.1 gm of concentrated sulfuric acid, 22.0 gm (0.27 mole) of 37% formaldehyde solution (formalin), and 8.4 gm of water. The reaction mixture is stirred for 24 hr and then the polymer is precipitated as a fine powder by the slow addition of water to give about 50% concentration of acetic acid. The dilute solution is poured into a vigorously agitated water bath to wash the polymer. The polymer is then dried to constant weight.

Figure 3 shows the infrared spectra of poly(vinyl formal) and compares it with that of the starting poly(vinyl acetate) and intermediate poly(vinyl alcohol).

7-3. Preparation of Poly(vinyl acetaldehyde acetal) [88]

$$\left[\begin{array}{c} -CH_2-CH-CH_2-CH \sim \\ \quad | \qquad \qquad | \\ \quad OH \qquad \quad OH \end{array} \right]_n + nCH_3CH{=}O \xrightarrow{HCl} \left[\begin{array}{c} CH_2 \\ -CH \quad CH-CH_2 \sim \\ | \qquad | \\ O \quad\quad O \\ CH \\ | \\ CH_3 \end{array} \right]_n$$

(44)

To a resin flask equipped with a mechanical stirrer, dropping funnel, and condenser are added 24 gm (0.54 mole) of poly(vinyl alcohol) (100% hydrolyzed grade), 250 ml of water, and 60 ml of concentrated hydrochloric acid. Then dropwise is slowly added a solution of 10 gm (0.23 mole) of acetaldehyde in 320 ml of isopropanol. The reaction mixture is allowed to stand at room temperature for 1 hr and the polymer is precipitated with the aid of 320 ml of water. The polymer is purified by kneading several times with fresh water, cut into small pieces, and dried. Further purification is effected by adding the dried polymer to dioxane and filtering the resulting solution to leave behind one-fourth of the product. The clear solution is freeze dried to remove the dioxane and other volatiles and give a product with about 27% bound acetaldehyde.

7-4. Preparation of Poly(vinyl butyral) [89]

$$\left[\begin{array}{c} -CH_2-CH-CH_2-CH- \\ \quad | \qquad \qquad | \\ \quad OH \qquad \quad OH \end{array} \right]_n + nC_3H_7CH{=}O \longrightarrow \left[\begin{array}{c} CH_2 \\ -CH_2-CH \quad C- \\ | \qquad | \\ O \quad\quad O \\ CH \\ | \\ C_3H_7 \end{array} \right]_n$$

$$+ H_2O \quad (45)$$

To a resin flask equipped with a mechanical stirrer and condenser are added 400 gm (9.0 moles) of poly(vinyl alcohol), 2210 gm of ethanol, 250 gm

(3.47 moles) of *n*-butyraldehyde and 10 gm of sulfuric acid. The reaction mixture is stirred and warmed at 75°–78°C for 2 hr. After this time water is added to precipitate the acetal. The product is washed first with water, then with a warm (80°C), dilute aqueous potassium hydroxide solution to neutralize [89–91] the residual acid, and finally dried.

8. MISCELLANEOUS PREPARATIONS

1. Acetals from formaldehyde–diethyleneglycol for textile applications [27].

2. Dicyclic acetals from the reaction of 2-alkoxy-3,4-dihydropyrans and diols [92].

3. Polyacetals of formaldehyde with methylsilanols [93].

4. Cross-linked polyacetals of diallyl formal and diallyl butyral [94].

5. Poly(vinyl alcohol) reacted with crotonaldehyde to give a poly(vinyl acetal) that can be cross-linked with sulfur or peroxides [95].

6. Poly(vinyl methoxyacetals) [96].

7. Poly(vinyl acetals) based on chloral, cyclohexanone, sulfur-containing dialdehydes, heterocyclic aldehydes, and aldehydes with carboxylic and sulfuric acid groups [97].

8. Poly(vinyl acetal) from β-cyanopropionaldehyde [98].

9. Poly(vinyl acetal) from polyacrolein [99].

10. Acetalization of poly(vinyl alcohol) with substituted benzaldehydes [100].

11. A process for producing homo- and copolymers of cyclic acetals with the aid of perfluoroalkyl sulfuric acid catalysts [101].

REFERENCES

1. S. R. Sandler and W. Karo, "Organic Functional Group Preparations," Vol. 3, Chapter 1. Academic Press, New York, 1972.

2. S. R. Sandler and W. Karo, "Polymer Syntheses," Vol. 1, Chapter 5. Academic Press, New York, 1974.

3. J. W. Hill and W. H. Carothers, *J. Am. Chem. Soc.* **57**, 925 (1935).

4. W. H. Carothers, U.S. Patent 2,071,252 (1937).

5. J. Read, *J. Chem. Soc.* **101**, 2090 (1912).

6. H. Orth, *Kunstoffe* **41**, 454 (1951).

7. S. M. Cohen and E. Lavin, *J. Appl. Polym. Sci.* **23**, 503 (1962).

7a. S. M. Cohen, C. F. Hunt, R. E. Kass, and A. H. Markhart, *J. Appl. Polym. Sci.* **6**, 508 (1962).

7b. S. M. Cohen and E. Lavin, U.S. Patent 2,963,464 (1960); British Patent 896,254 (1962).

8. H. Schulz and H. Wagner, *Angew. Chem.* **62**, 105 (1950); H. S. Rothrock, U.S. Patent 2,401,776 (1946); H. Wagner, German Patent 870,032 (1950).
9. T. N. Shakhtakhtinskii and L. V. Endreev, *Dokl. Akad. Nauk. Az. SSR* **18**, 17 (1962); W. M. Kraft, U.S. Patent 2,870,121 (1959); B. H. Silverman and E. Balgley, U.S. Patent 3,109,830 (1963).
10. H. Orth, German Patent 852,301 (1952); U.S. Patent 2,687,407 (1954).
10a. F. Brown, D. E. Hudgin, and R. J. Kray, *J. Chem. Eng. Data* **4**, 182 (1959).
11. H. S. Hill and H. Hibbert, *J. Am. Chem. Soc.* **45**, 3117 and 3124 (1923).
12. R. L. Adelman, U.S. Patent 2,682,532 (1954).
13. J. M. Stellman, *Chem. & Eng. News* **52**, 3 (1974).
14. N. G. Gaylord, *Encycl. Polym. Technol.* **10**, 319 (1969); *in* "Polyethers" (N. G. Gaylord, ed.), Part I, Vol. XIII, Chapter VII. Wiley (Interscience), New York, 1963; R. W. Lenz, "Organic Chemistry of Synthetic High Polymers," pp. 125–128. Wiley, New York, 1967; H. Gibellow, *Off. Matieres Plast.* **6**, 1258 (1959); D. Sek, *Wiad. Chem.* **26**, 677 (1972).
15. E. Schonfeld, *J. Polym. Sci.* **59**, 87 (1962).
15a. E. Schonfeld, *J. Polym. Sci.* **49**, 277 (1961).
16. B. H. Kress, U.S. Patent 2,785,949 (1957).
17. B. H. Kress, U.S. Patent 2,785,995 (1957); C. B. Hayworth, U.S. Patent 3,862,916 (1975).
18. A. H. Mataszak and H. R. Ready, U.S. Patent 2,838,573 (1958).
19. R. E. Kass, A. H. Markhart, and E. Lavin, U.S. Patent 3,063,955 (1962).
20. R. M. Kullman, H. B. Moore, R. M. Reinhardt, and J. D. Reid, *Am. Dyest. Rep.* **52**, 26 (1963).
21. S. M. Cohen, M. D. Kellert, and J. A. Snelgrove, *Off. Dig. J. Paint Technol.* **37**, 1215 (1965).
22. K. Hori, M. Nakamura, and Y. Hobo, Japanese Patent 74/31,300 (1974).
23. R. S. Kholodovskaya, M. B. Fromberg, G. Y. Gordo, L. P. Kofman, and D. B. Demina, *Lakokras. Mater. Ikh Primen.* No. 6, p. 15 (1965); *Chem. Abstr.* **64**, 9952g (1966).
24. D. B. Pattison, *J. Org. Chem.* **22**, 662 (1957).
25. Hudson Foam Plastics Corp., British Patent 850,178 (1960).
26. B. H. Kress, U.S. Patent 2,786,081 (1957).
27. B. H. Kress and E. Abrams, U.S. Patent 2,785,947 (1957).
28. H. R. Musser and W. J. Jackson, U.S. Patent 3,875,257 (1975).
29. H. R. Musser and W. J. Jackson, U.S. Patent 3,809,681 (1974).
30. E. Schonfeld, *J. Polym. Sci., Part A* **2**, 2489 (1964).
31. E. Hammerschmid, German Patent 889,336 (1953).
32. H. Orth, German Patent 838,827 (1952).
33. Deutsche Solvay-Werke G.m.b.H., British Patent 789,458 (1958).
34. *Chem. Week* **87**, No. 16, 105 (1960).
35. L. S. Abbott, D. Faulkner, and C. E. Hollis, U.S. Patent 2,739,972 (1956).
36. D. B. Capps, U.S. Patent 2,889,290 (1959).
37. C. S. Marvel and J. J. Drysdale, *J. Am. Chem. Soc.* **75**, 4601 (1953).
38. E. L. Kropa and W. M. Thomas, U.S. Patent 2,643,236 (1953).
39. B. H. Kress, U.S. Patent 2,785,996 (1957).
40. European Research Assoc., Belgian Patent 568,181 (1958).
41. A. H. Markhart, C. F. Hunt, and E. Lavin, German Patent 1,210,964 (1966).
42. E. H. Pryde, D. E. Anders, H. M. Teeter, and J. C. Cowan, *J. Org. Chem.* **25**, 618 (1960).

43. R. W. Lenz, W. R. Miller, and E. H. Pryde, *J. Polym. Sci.* **8**, 429 (1970).
44. E. H. Pryde, R. A. Awl, H. M. Teeter, and J. C. Cowan, *J. Org. Chem.* **25**, 2260 (1960).
45. E. H. Pryde, D. J. Moore, H. M. Teeter, and J. C. Cowan, *J. Polym. Sci.* **58**, 611 (1962); E. H. Pryde, R. A. Awl, H. M. Teeter, and J. C. Cowan, *ibid.* **59**, 1 (1962).
46. W. R. Miller, E. H. Pryde, and J. C. Cowan, *Polym. Lett.* **3**, 131 (1965).
47. W. R. Miller, J. C. Cowan, and E. H. Pryde, U.S. Patent 3,285,880 (1966).
48. E. H. Pryde, U.S. Patent 3,183,215 (1965).
49. E. H. Pryde, U.S. Patent 3,223,683 (1965).
50. Chemische Werke Huels A.-G., French Patent 1,361,204 (1964); *Chem. Abstr.* **62**, 6591 (1965).
50a. W. J. Bailey and A. A. Volpe, *Polym. Prepr., Am. Chem. Soc., Div. Polym. Chem.* **8**, 292 (1967).
51. W. J. Bailey, C. F. Beam, Jr., and I. Haddad, *Polym. Prepr., Am. Chem. Soc., Div. Polym. Chem.* **12**, 169 (1971).
52. J. C. Patrick, U.S. Patent 2,527,377 (1950).
53. Imperial Chemical Industries, Ltd., British Patent 577,205 (1946).
54. C. S. Marvel and R. C. Farrar, Jr., *J. Am. Chem. Soc.* **79**, 986 (1957).
55. W. Autenreith and A. Geyer, *Ber.* **41**, 4249 (1908); W. Autenreith and F. Beuttel, *ibid.* **42**, 4346 and 4357 (1909).
56. N. G. Fisher and R. H. Wiley, U.S. Patent 2,389,662 (1945); *Chem. Abstr.* **40**, 1062 (1946).
57. C. S. Marvel, E. H. H. Shen, and R. C. Chambers, *J. Am. Chem. Soc.* **72**, 2106 (1950).
58. C. S. Marvel, E. A. Sienicki, M. Passer, and C. N. Robinson, *J. Am. Chem. Soc.* **76**, 933 (1954).
59. W. O. Herrmann and W. Haehnel, German Patent 480,866 (1924).
60. H. Hopff, U.S. Patent 1,955,068 (1934); H. Bauer, J. Heckmaier, R. Reinecke, and and E. Bergmeister, German Patent 929,643 (1952).
60a. G. O. Morrison, F. W. Skirrow, and K. B. Blackie, U.S. Patent 2,036,092 (1935).
61. G. S. Stamstoff, U.S. Patent 2,400,957 (1946).
61a. G. S. Stamstoff, U.S. Patent 2,422,754 (1947); T. S. Carswell, U.S. Patent 2,378,619 (1945).
62. A. F. Fitzhugh and R. N. Crozier, *J. Polym. Sci.* **8**, 225 (1952).
62a. A. F. Fitzhugh and R. N. Crozier, *J. Polym. Sci.* **9**, 96 (1952).
63. K. Rosenbusch, W. Pense, and F. Winkler, German Patent 1,069,385 (1959).
64. R. D. Dunlop, Fiat Final Report No. 1109 (1947).
65. S. Okamura and T. Motoyama, *Kogyo Kagaku Zasshi* **55**, 774 (1952).
66. K. Noma and T. Sone, *Kobunshi Kagaku* **4**, 50 (1947).
67. S. Okamura, T. Motoyama, and K. Uno, *Kogyo Kagaku Zasshi* **55**, 776 (1952).
68. E. T. Cline and H. B. Stevenson, U.S. Patent 2,606,803 (1953).
69. E. L. Martin, U.S. Patent 2,929,710 (1954).
70. G. Kranzlein and U. Campert, German Patent 729,774 (1937).
71. G. Kranzlein, A. Voss, and W. Starck, German Patent 661,968 (1930).
72. J. D. Ryan, U.S. Patent 2,425,568 (1947).
73. R. A. Barnes, *J. Polym. Sci.* **27**, 285 (1958).
74. A. F. Fitzhugh, E. Lavin, and G. O. Morrison, *J. Electrochem. Soc.* **100**, 351 (1953).
75. *Mod. Plast.* **40**, 90 (1962).
76. R. W. Hall, U.S. Patent 2,114,877 (1938).
77. R. G. Flowers and C. A. Winter, U.S. Patent 3,442,834 (1969).

78. W. O. Herrmann, German Patent 690,332 (1940); J. D. Ryan, U.S. Patent 2,232,806 (1937); D. S. Plumb, *Ind. Eng. Chem.* **36**, 1035 (1944); F. T. Buckley, G. R. Mason, and R. F. Riek, U.S. Patent 3,388,033 (1968).
79. M. K. Lindemann, *Encycl. Polym. Sci. Technol.* **14**, 208 (1971); N. Platzer, *Mod. Plast.* **28**, 142 (1951); G. O. Morrison, *Kirk-Othmer, Encycl. Chem. Technol. 2nd Ed.* Vol. 21, p. 304 (1970).
80. B. N. Norlander and R. E. Burnett, U.S. Patent 2,216,020 (1941).
81. G. Kranzlein and H. Reis, German Patent 765,265 (1954).
82. W. H. Mueller, *Can. Chem. Process Ind.* **29**, 395 (1945).
83. E. Kuehn and H. Hopff, U.S. Patent 2,044,734 (1936).
84. A. Koshita, H. Sakata, H. Asahara, and N. Suyama, Japanese Patent 74/26,953 (1974); *Chem. Abstr.* **82**, 86865 (1975).
85. J. H. Hopkins and G. H. Wilder, U.S. Patent 2,282,057 (1942); B. C. Bren, U.S. Patent 2,122,277 (1938).
86. Société Nobel Française, French Patent 849,460 (1939); A. Voss and W. Starck, German Patent 737,630 (1930); *Chem. Eng.* **61**, 122 (1954).
87. G. Kranzlein and H. Reis, German Patent 765,265 (1954); H. C. Beachell, P. Fotis, and J. Hucks, *J. Polym. Sci.* **7**, 353 (1951); S. N. Ushakov, *J. Appl. Chem. USSR (Engl. Transl.)* **19**, 553 (1946).
88. R. A. Barnes, *J. Polym. Sci.* **27**, 285 (1958).
89. E. Lavin, A. T. Marinaro, and W. R. Richard, U.S. Patent 2,496,480 (1950).
90. J. H. Hopkins and G. H. Wilder, U.S. Patent 2,282,057 (1942).
91. B. C. Bren, J. H. Hopkins, and G. H. Wilder, U.S. Patent 2,282,026 (1942).
92. H. F. Reinhardt, U.S. Patent 3,232,907 (1966).
93. F. A. Henglein and P. Schmulder, *Makromol. Chem.* **13**, 53 (1954).
94. D. E. Adelson and H. F. Gray, Jr., U.S. Patent 2,469,288 (1949).
95. A. F. Fitzhugh, U.S. Patent 2,527,495 (1950).
96. M. Dole and I. L. Faller, *J. Am. Chem. Soc.* **72**, 414 (1950); K. F. Beal and C. J. B. Thor, *J. Polym. Sci.* **1**, 540 (1946).
97. C. E. Schildknecht, "Vinyl and Related Polymers," p. 360. Wiley, New York, 1952.
98. Y. Oshira, M. Miyasaka, Y. Shirota, and H. Mikawa, *J. Polym.* **6**, 12 (1974); *Chem. Abstr.* **82**, 86785m (1975).
99. R. C. Schulz, *Kunstst.-Plast.* **6**, 32 (1959).
100. M. Kato, E. Tanaka, and S. Konotsune, Japan Kokai B 76,873 (1973); *Chem. Abstr.* **80**, 3260 (1974).
101. K. Burg and H. Schlaf, U.S. Patent 3,883,450 (1975).

POLY(VINYL ETHERS)

I. INTRODUCTION

The first vinyl ether to be polymerized was reported 94 years ago by Wislicenus [1] who treated ethyl vinyl ether with iodine and obtained a violent reaction giving a resinous material. More recent studies [2,2a] on this reaction indicate that I^+ is the active initiator and that carbonium ions are involved.

Reppe [3–3c] and co-workers described the polymerization of a wide variety of vinyl ethers by acidic reagents used in small amounts. Some typical acidic reagents or catalysts used by Reppe are $SnCl_4$, $AlCl_3$, BF_3, BF_3

complexes, $FeCl_3$, $ZnCl_2$, $SnCl_4$, H_2SO_4, SO_2, H_3PO_4, etc. The polymerizations were reported to be violent at room temperature and above and afforded low molecular weight resins. Friedel–Crafts catalysts were used by Favorski and Shostakovskii [4,5] to polymerize methyl, ethyl, isopropyl, butyl, amyl, and other vinyl ethers.

Eley and Pepper [6] polymerized vinyl *n*-butyl ether in bulk and in petroleum ether over the temperature range of 20°–94°C using a 2% solution of $SnCl_4$ in petroleum ether as the catalyst [6]. The polymerization kinetics of vinyl 2-ethylhexyl ether was studied similarly [2a].

Anionic catalysts fail to initiate polymerization and the mechanism involving Grignard agent catalysts is not clear at this time.

Vinyl ethers are also polymerized by free-radical initiators in bulk, solution, or emulsion to give low molecular weight polymers.

Vinyl ethers copolymerize well with a wide variety of monomers (olefins, haloolefins, alkoxybutadienes, acrylates, acrylonitrile, maleic anhydride, maleates, fumarates, allyl, vinyl pyrrole and vinyl carbazole, etc.) using free-radical ionic initiators or coordination-type catalysts [7].

Some of the major companies producing vinyl ether polymers are GAF Corporation, Union Carbide Corporation, and Badische Anilin- und Soda-Fabrik A.G.

Vinyl ether polymers are useful in lacquer resins, plasticizers, adhesives, paints, copolymer compositions such as those for fire retardants, marine coatings, anticorrosion agents, thickening agents, and other uses [8,8a].

2. CATIONIC POLYMERIZATION

Vinyl octadecyl ether was the first vinyl ether to be polymerized commercially and boron trifluoride was used as the catalyst [9]. The polymerization was carried out at 90°–95°C and low molecular weight polymers were obtained with K values of 20–30, m.p. 50°C (white, brittle solid). Other similar products were suggested in patents by I. G. Farbenindustrie and also the copolymerization with maleic anhydride was mentioned.

Vinyl methyl ether was commercially polymerized in bulk using a 3% solution of boron trifluoride:water (1:2) in dioxane as the initiator [10]. The catalyst was added portionwise at 5°–12°C with cooling, and the polymerization temperature was then allowed to rise to 100°C. The polymer had a balsamlike consistency and was soluble in cold water.

Schildknecht and co-workers [11] were the first to report the preparation of crystalline, isotactic poly(isobutyl vinyl ether).

Methyl vinyl ether yields a crystalline polymer only when methylene chloride is present as a solvent. However, ethyl, isopropyl, and *n*-butyl vinyl ethers do not yield crystalline polymers. Branched alkyl vinyl ethers, other than isobutyl vinyl ether, and benzyl ether also yield crystalline polymers [12]. The crystallinity of the polymers (isotactic) is similar in soluble and insoluble catalyst systems [13].

The boron trifluoride catalyst requires water in order to generate a proton to start initiation by the carbonium ion. The carbonium ion is formed by reaction of the vinyl ether with the proton. A variety of alkyl vinyl ethers has been reported to be prepared by similar cationic catalysts [14]. Other catalysts which have been found useful as initiators are $AlCl_3$, $FeCl_3$, $SnCl_4$, $MgCl_2$–dioxane–ether or tetrahydrofuran complexes, $ZnCl_2$–*tert*-BuCl, and sulfonated polystyrene cation exchange resins. The use of metal fluorides with titanium alkoxides [15] or titanium trifluoride [16] gave stereoregular or crystalline poly(vinyl alkyl ethers). Since most ionic initiators are catalysts for the hydrolysis of vinyl ethers the homopolymerization is usually not carried out in aqueous systems.

Some representative examples of the methods and cationic catalysts used in the preparation of either atactic or isotactic poly(vinyl ethers) are shown in Table I.

Alkyl vinyl ethers can also be copolymerized cationically with such monomers as styrene using $SnCl_4/AlCl_3$ catalyst and in ethyl chloride solution [17–17b].

Care should generally be exercised when acidic agents are used to initiate vinyl ether polymerization. Some polymerizations may occur violently, especially in the case of bulk polymerizations. In order to avoid this situation low temperatures and diluents or solvents should be used and the reaction carried out on a small scale using adequate precautions.

Vinyl ethers will hydrolyze to acetaldehyde and alcohols if the pH is not maintained above 7. For this reason emulsion techniques are rarely used with cationic initiators.

2-1. Preparation of Poly(vinyl ethyl ether) Using Aluminum Chloride Catalyst [3]

$$C_2H_5-OCH=CH_2 \xrightarrow[C_6H_6]{AlCl_3} \left[\begin{array}{c} -CH_2-CH- \\ | \\ OC_2H_5 \end{array} \right]_n \qquad (1)$$

To a flask containing 200 gm of benzene is added 1.0 gm of aluminum chloride. The mixture is stirred vigorously to suspend the catalyst and then the gradual addition of 200 gm (2.78 moles) of vinyl ether is begun. The reaction is exothermic and when the temperature rises to 60°–80°C the addition is slowed so that only 1–2 gm of vinyl ethyl ether is added at one time. Polymerization is detected by the onset of turbidity and a slight brown color. When the brown color disappears, the next increment of vinyl ethyl ether is added.

TABLE I

CATIONIC POLYMERIZATION OF VINYL ETHERS

Polymerization type	Solvent	Catalyst	Tempera- ture (°C)	Stereo- regularity	Ref.
Bulk	—	$BF_3 \cdot 2H_2O$	3–5	Atactic	a
	—	$BF_3 \cdot Et_2O$	−78	Isotactic	b
	—	$SnCl_2$	12	Atactic	c
Solution	Hexane–CH_2Cl_2	$BF_3 \cdot Et_2O$	30	Isotactic	d
	Propane	$BF_3 \cdot Et_2O$	−70	Atactic	e
	Propane–butane	$BF_3 \cdot Et_2O$	−25	Atactic	f
	Petroleum ether	$BF_3 \cdot Et_2O/AlR_3$	0	Isotactic	g
	Hexane–CH_2Cl_2	TiF_3	60	Isotactic	h
	Toluene	R_2AlCl	−78	Isotactic	i

ᵃ J. W. Copenhaver and M. H. Bigelow, "Acetylene and Carbon Monoxide Chemistry," p. 49. Van Nostrand-Reinhold, Princeton, New Jersey, 1949.

ᵇ A. D. Ketley, *J. Polym. Sci.* **62**, S81 (1962).

ᶜ I. P. Losev, O. Y. Fedotova, and M. F. Shostakovskii, *J. Gen. Chem. USSR (Engl. Transl.)* **13**, 428 (1943).

ᵈ S. Okamura *et al., J. Chem. Soc. Jap.* **61**, 1636 (1958).

ᵉ A. O. Zoss, U.S. Patent 2,609,364 (1952); C. E. Schildknecht, U.S. Patent 2,513,820 (1950); A. O. Zoss, U.S. Patent 2,616,879 (1952); C. E. Schildknecht and P. H. Dunn, *J. Polym. Sci.* **20**, 597 (1956).

ᶠ L. Fishbein and F. A. Magnotta, U.S. Patent 3,030,352 (1963); L. Fishbein and B. F. Crowe, *Makromol. Chem.* **48**, 221 (1961).

ᵍ S. Nakano, K. Iwasaki, and H. Fukutani, British Patent 896,981 (1962); Mitsubishi Chem. Co., Japanese Patent 15,942 (1961).

ʰ S. Okamura, T. Higashimura, and H. Yamamoto, *J. Polym. Sci.* **33**, 510 (1958).

ⁱ G. Dall'Asta and N. Oddo, *Chim. Ind. (Milan)* **42**, 1234 (1960).

After the complete addition the reaction mixture is heated to 80°C. The benzene and volatiles are steam distilled off and the residue amounts to about 144 gm (72%) of a yellow balsamlike product. Additional examples of cationic polymerizations of vinyl ethers are shown in Table II.

2-2. Preparation of a Poly(vinyl n-butyl ether) Using Stannous Chloride Catalyst [3b]

$$n\text{-}C_4H_9\text{—}OCH\text{=}CH_2 \xrightarrow{\ SnCl_2\ } \left[\begin{array}{c} \text{—}CH_2\text{—}CH\text{—} \\ | \\ OC_4H_9 \end{array} \right]_n \qquad (2)$$

To 50 gm (0.50 mole) of vinyl 2-butyl ether is added 0.25 gm of stannous chloride. After a few minutes the temperature begins to rise to 45°–50°C and then more rapidly to 140°C. The product obtained is a yellowish brown,

TABLE II

CATIONIC POLYMERIZATION OF VINYL ETHERS[a]

Vinyl alkyl ether (gm)	Solvent (ml)	Catalyst (gm)	Reaction conditions		
			Temp. (°C)	Time (hr)	Product (gm)
$C_4H_9OCH{=}CH_2$ 50	—	$SnCl_4$ 0.25	45–140	$\frac{1}{2}$–1	35
$C_2H_5OCH{=}CH_2$ 200	C_6H_6 200	$AlCl_3$ 1.0	60–80	1–2	144
$C_6H_5OCH{=}CH{-}CH_3$ 20	C_6H_6 80	$SnCl_4$ 1.0	80	1–2	b
$C_2H_5O{-}C({CH_3}){=}CH_2$ 50	C_6H_6 150	$AlCl_3$ 1.0	80	1–2	b
2-methylphenyl $OCH{=}CH_2$ (o-cresyl vinyl ether) 140	C_6H_6 40	$AlCl_3$ 0.3	80	1–2	b

CH_3 — benzene ring — $OCH=CH_2$

Substrate		Catalyst	Temp		
50.0	—	ZnCl$_2$ 5.0	25	48	b
50	—	5.0	80–95	1–2	b
C$_2$H$_5$OCH=CH$_2$ 2100	—	2% BF$_3$—O(C$_4$H$_9$)$_2$ 4.0	35–50	2–3	b
C$_4$H$_9$O—CH=CH$_2$ 2100	—	2% BF$_3$—O(C$_2$H$_5$)$_2$ 4.0	40–50	2–3	b
CH$_3$OCH=CH$_2$ 1100	—	2% BF$_3$—O(C$_4$H$_9$)$_2$ 2.0	–5–30	2–3	b
C$_{18}$H$_{37}$OCH=CH$_2$ 2200	—	2% BF$_3$—O(C$_4$H$_9$)$_2$ 3.1	50–60	3–4	b
CH$_2$=CH—O—CH$_2$CH$_2$OCH$_2$CH$_2$OC$_2$H$_5$ 900	—	2% BF$_3$—O(C$_4$H$_9$)$_2$ 0.6	40–50	3–4	b

[a] From refs. 3a, c.
[b] Yield not specified.

219

balsamlike mass. Purification by steam distillation leaves 35 gm (70%) of a viscous, liquid, yellow polymerization product.

Similar results are obtained with stannic chloride or aluminum chloride.

2-3. Preparation of Poly(vinyl n-butyl ether) Using Boron Trifluoride Etherate Catalyst [18]

$$C_4H_9-OCH=CH_2 \xrightarrow{BF_3 \cdot O(C_2H_5)_2} \left[\begin{array}{c} -CH_2-CH- \\ | \\ OC_4H_9 \end{array} \right]_n \qquad (3)$$

To a reactor containing 200 gm (2.0 moles) of vinyl n-butyl ether are added 600 gm of propane and 750 gm of Dry Ice. The reactor is cooled to $-80°C$ by Dry Ice–acetone and then $BF_3 \cdot O(C_2H_5)_2$ maintained at 25°C is added dropwise until 5 parts have been added. The reaction is complete in 1 hr. The catalyst is deactivated by adding 10 gm of 30% ammonium hydroxide at $-50°C$ followed by 300 gm of propane at $-60°C$. The liquid propane containing the polymer is removed to give 85% (170 gm) of high molecular weight polymer and 1.5% of low molecular weight polymer.

2-4. Preparation of Poly(vinyl ethyl ether) [3c]

$$C_2H_5-OCH=CH_2 \longrightarrow \left[\begin{array}{c} -CH_2-CH- \\ | \\ OC_2H_5 \end{array} \right]_n \qquad (4)$$

One-hundred grams (1.39 moles) of vinyl ethyl ether at $-15°C$ is condensed in 100 gm of sulfur dioxide and then the polymerization commences. After 10–15 hr the viscous solution is heated to expel the SO_2 and then placed under reduced pressure. The final product is a viscous, colorless mass.

3. FREE-RADICAL POLYMERIZATION

Using free-radical initiators affords a slow polymerization of vinyl ethers and only low molecular weights are attainable. The use of heat, light, or radiation [19] gives the same results. Nelson, Banes, and FitzGerald in 1961 reported that 2–10% di-*tert*-butyl peroxide or *tert*-butyl or cumene hydroperoxide gives low molecular weight polymers useful as lubricating oils [20]. Azobis(isobutyronitrile) has been reported to show some ability to effect free-radical polymerization of vinyl ethers [21].

The redox polymerization $[(NH_4)_2S_2O_8 + NaHSO_3]$ at 50°C of 2-chloroethyl vinyl ether has been claimed [22].

Vinyl ethers copolymerize with methyl methacrylate to give alternating copolymer units as well as homopolymer segments of methyl methacrylate [23]. With other monomers this is also true and either solution, emulsion, suspension, or bulk polymerization techniques may be used. The pH should be kept at about 8 or above in aqueous systems to prevent hydrolysis of the vinyl ether. The r_2 (alkyl vinyl ether) values are low and approach zero for bulk polymerization systems [24–24b] utilizing such monomers as acrylonitrile, butyl maleate, methyl acrylate, methyl methacrylate, styrene, vinyl acetate, vinyl chloride, or vinylidene chloride.

Free-radical copolymerization of alkyl vinyl ethers has been carried out with the following typical monomers: acrylic acid (bulk and emulsion) [25,26], acrylonitrile (emulsion) [17a,17b], acrylic esters (emulsion) [27], methyl methacrylate (bulk) [28], maleic anhydride (solution) [29], vinyl acetate (bulk and emulsion) [17b,30,31], and vinyl chloride (emulsion) [17a,24a,32]. The properties of these and other copolymers are described in a technical bulletin by General Aniline & Film Corporation [24b].

Other monomers that copolymerize with alkyl vinyl ethers are vinyl ketones [33], acrolein diacetate [34], acrylamide [35], alkoxy 1,3-butadienes [36], butadiene [37], chloroprene [38], chlorotrifluoroethylene [39], tri- and tetrafluoroethylene [40], cyclopentadiene [41], dimethylaminoethyl acrylate [42], fluoroacrylates [43], fluoroacrylamides [44], *N*-vinyl carbazole [45,46], triallyl cyanurate [45,46], vinyl chloroacetate [47,48], *N*-vinyl lactams [49], *N*-vinyl succinimide [49], vinylidene cyanide [50,51], and others. Copolymerization is especially suitable for monomers having electron-withdrawing groups. Solution, emulsion, and suspension techniques can be used. However, in aqueous systems the pH should be buffered at about pH 8 or above to prevent hydrolysis of the vinyl ether to acetaldehyde. Charge-transfer complexes have been suggested to form between vinyl ethers and maleic anhydride, and these participate in the copolymerization [52]. Examples of the free-radical polymerization of selected vinyl ethers are shown in Table III.

3-1. *Preparation of Poly(vinyl ethyl ether) Using Di-tert-butyl Peroxide Initiator* [20]

$$C_2H_5OCH{=}CH_2 \longrightarrow \left[\begin{array}{c} -CH_2-CH- \\ | \\ OC_2H_5 \end{array} \right]_n \quad (5)$$

To a 1.8-liter stainless steel reactor bomb are added 376 gm (500 ml, 5.22 moles) of vinyl ethyl ether, 400 ml of cyclohexane, and 23.8 gm (0.163 mole) of 6.33 wt.% di-*tert*-butyl peroxide. The reactor bomb is sealed and heated at 159°C for 4 hr with agitation in a rocking apparatus. Then the bomb is cooled and the contents removed. The cyclohexane and other volatiles are removed by stripping at 100°C (overhead temperature) at atmospheric pressure to afford 295.4 gm of product (78%)

TABLE III

FREE-RADICAL POLYMERIZATION OF SELECTED VINYL ETHERS

Vinyl ether(s) (gm)	Comonomer	Solvent	Initiator (gm)	Reaction conditions Temp. (°C)	Time (hr)	Yield (%)	Ref.
2-Aminoisobutyl vinyl ether 20	—	—	Dimethyl azoisobutyronitrile 3	75	16	90	a
2-Aminoisobutyl vinyl ether 20	—	C_6H_6 20	Dimethyl azoisobutyronitrile 3	100 (at 0.2–0.4 mm Hg)	24	44	a
2-Chloroethyl vinyl ether[f] 20	—	H_2O 80	$(NH_4)_2S_2O_8$ 0.8, $NaHSO_3$ 20	50	—	—	b
Vinyl ethyl ether 20	Acrylic acid 30	H_2O 100	$NaHSO_3$ 2, $K_2S_2O_8$ 1	50	16	—	c
Vinyl-n-butyl ether 30	Styrene 70	—	—	130–150	24	—	d
Vinyl ethyl ether 50	Vinyl chloride 300	—	Benzoyl peroxide 1	60	12–24	—	d
Vinyl ethyl ether 376	—	Cyclohexane 600	75% Cumene hydroperoxide 30	125	4	28.5	e
Vinyl isobutyl ether 385	—	Cyclohexane 500	Di-tert-butyl peroxide 23.8	156–158	2	78.7	ν

[a] W. H. Watanabe and S. Melamed, U.S. Patent 2,845,407 (1958).

[b] R. R. Dreisbach and J. L. Lang, U.S. Patent 2,859,209 (1958).

[c] J. L. Lang, U.S. Patent 2,937,163 (1960).

[d] H. Fikentscher, U.S. Patent 2,016,490 (1935).

[e] J. F. Nelson, F. W. Banes, and W. P. FitzGerald, U.S. Patent 2,967,203 (1961).

TABLE IV

Free-Radical Polymerization of Vinyl Ethers [20]

Vinyl ether (gm)	Solvent (ml)	Initiator (gm)	Reaction conditions		
			Temp. (°C)	Time (hr)	Product (gm)
Vinyl ethyl ether 150	Cyclohexane 300	Benzoyl peroxide 4 (2.7%)	95	4	0
376	600	Cumene hydroperoxide 30 (75%)	125	4	107.4
300	600	tert-Butyl hydroperoxide 23.8	125	4	199.6
Vinyl isobutyl ether 385	500	Di-tert-butyl peroxide 23.8	156–158	2	303
Vinyl 2-ethylhexyl ether 162	800	23.8 (14.7%)	153–162	4	145.7
Vinyl dodecyl ether 91	300	3.9	155	4	67

Analysis

Calculated: C, 66.7   H, 11.1   O, 22.2
Found: C, 68.2   H, 11.1   O, 20.7

Additional examples of the free radical polymerization of vinyl ethers is shown in Table IV.

3-2. Preparation of Poly(vinyl isobutyl ether) Using Di-tert-butyl Peroxide Initiator [20]

$$CH_3-CH-CH_2-O-CH=CH_2 \longrightarrow \begin{bmatrix} -CH_2-CH- \\ | \\ O \\ | \\ CH_2 \\ | \\ CH(CH_3)_2 \end{bmatrix}_n \qquad (6)$$
$$\quad\ \ |$$
$$\quad\ CH_3$$

To a reactor bomb are added 385 gm (500 ml, 5.35 moles) of vinyl isobutyl ether, 500 ml of cyclohexane, and 23.8 gm of di-*tert*-butyl peroxide. The bomb is sealed and heated at 156°–158°C for 2 hr with agitation. The bomb is cooled, opened, and then stripped at 100°C to afford 303 gm (78.7%) of product. About 65% of the product is reported to boil in the lubrication oil range.

4. COORDINATION-CATALYZED POLYMERIZATION

Ziegler catalysts [Al(Et)$_3$ plus TiCl$_4$] and related catalysts, such as Al(i-Bu)$_3$ plus the chlorides of Fe, Cr, V, Mo or Ti, have been used to polymerize vinyl ethers to crystalline polymers, as shown in Table V [53]. The polymerizations are usually carried out at low temperatures and give some stereospecificity. The mechanism of polymerization with these catalysts may also be of a cationic type. The use of TiCl$_4$ alone gives amorphous polymers and Al(Et)$_3$ does not cause any polymerization. However, the combination of both catalysts gives some degree of stereospecific polymerization.

Isotactic polymers can be obtained by the use of (Et)$_2$AlCl, and other mixed alkylaluminum halides give crystalline polymers [54]. One report asserts that the latter catalysts are effective only in the presence of a proton-active cocatalyst such as water or hydrochloric acid [55]. Sulfuric acid has been used as a cocatalyst for Al(i-Pr)$_3$ to give stereoregular poly(vinyl *sec*-butyl ether) and poly(vinyl 2-methylbutyl ethers) [56]. Furukawa has also reported the use of diethylzinc with either oxygen, water, or alcohols as cocatalysts to give stereoregular poly(vinyl ethers) [57].

TABLE V
CRYSTALLINE POLY(VINYL ETHERS)[a]

Vinyl ether	M.p. (°C)[b]	Solubility at 25°C
Methyl	144	Insol. water, methanol, heptane
Ethyl	86	Insol. methanol, heptane
n-Propyl	76	Insol. heptane, acetone
Isopropyl	190	Insol. heptane, methanol, acetone
n-Butyl	64	Insol. heptane
Isobutyl	165	Insol. heptane, benzene
tert-Butyl	240–260	Insol. heptane, benzene
Neopentyl	216	Insol. heptane, benzene
Benzyl	162	Insol. acetone, ether
2-Chloroethyl	150	Insol. acetone
2-Methoxyethyl	73	Sol. water; insol. ether
2,2,2-Trifluoroethyl	128	Insol. heptane, benzene, dioxane

[a] Reprinted from E. J. Vandenberg, R. F. Heck, and D. S. Breslow, *J. Polym. Sci.* **41**, 138 (1959). Copyright 1959 by the Journal of Polymer Science. Reprinted by permission of the copyright owner.

[b] By loss of birefringence under a polarizing microscope.

Grignard compounds (*tert*-BuMgBr) are similar effective catalysts [58]. Oxygen has been found to be a necessary cocatalyst for Grignard compounds [59].

Many catalysts have been studied and those with tetrahedral structures and one negative edge give stereoregular polymers at ambient temperatures [60].

Copolymers are also easily prepared using other vinyl monomers. For example ethylene–vinyl ether block copolymers are prepared and show improved transparency and dyeability compared to propylene homopolymers [61].

In the coordinated cationic mechanism the polymerization is initiated by a proton formed by interaction of the catalyst-active hydrogen-containing compounds. The metal atom coordinates the monomer and the growing polymer chain, thus leading to some degree of stereospecificity during the polymerization.

4-1. Preparation of Poly(vinyl methyl ether) Using Transition Metal Catalysts [53e]*

$$CH_3O-CH=CH_2 \longrightarrow \left[\begin{array}{c} -CH_2-CH- \\ | \\ OCH_3 \end{array} \right]_n \qquad (7)$$

* Preparation 4-1 reprinted from E. J. Vandenberg, *J. Polym. Sci., Part C* **1**, 207 (1963). Copyright 1963 by the Journal of Polymer Science. Reprinted by permission of the copyright owner.

a. Polymerization Procedure

Polymerization and catalyst preparations are carried out under nitrogen using capped pressure vessels fitted with a Buna N rubber, self-sealing liner which has been extracted with benzene for three days and dried. Hypodermic equipment is used for evacuations and nitrogen reagent addition. Polymerizations are run with 10 gm of monomer. In general, nonvolatile components such as diluent are charged into the pressure bottle, the free space swept out with nitrogen, and the bottle capped using a self-sealing liner. Air is further removed by evacuating the system through a 20-gauge needle with an oil pump for 1 min (for a 250 ml vessel), the system nitrogen-pressured to 15 psi, evacuated again for 1 min, and then either repressured with nitrogen to 15 psi of charged with a volatile monomer such as vinyl methyl ether. Other low-boiling ingredients are then injected. Next, an organoaluminum compound, referred to as activator, is added and the pressure bottle is placed on a rotating rack in a 30°C water bath for about $\frac{1}{2}$ to 1 hr. At this time the second catalyst component, usually the transition metal component, is added to initiate the polymerization.

b. Pretreated Stoichiometric Vanadium (PSV) Catalyst

The PSV catalyst is the Ziegler-type transition metal catalyst which gives the most stereoregular polymerization. It is prepared by reacting SV catalyst with $(i\text{-Bu})_3\text{Al}$–tetrahydrofuran (THF) complex (1 M in n-heptane) at a mole ratio of 2:1 $(i\text{-Bu})_3\text{Al}$:V (0.1 M vanadium concentration) for 20 hr at room temperature. When $(i\text{-Bu})_3\text{Al}$ alone is used as the pretreating agent, then the PSV catalyst is aged only 0.08 hr before using. Modifications of the PSV catalyst and other catalysts are prepared in a similar manner in n-heptane.

The PSV catalyst, prepared with $(i\text{-Bu})_3\text{Al}$–THF pretreatment, is a chocolate brown dispersion. It can be separated by centrifuging into heptane-insoluble and heptane-soluble components. The heptane-insoluble component, which is also completely insoluble in ether, contains 99% of the vanadium in the PSV catalyst [88% as V(II) and 12% as V(III)] and 0.21 mole of aluminum per mole vanadium. It is highly crystalline by X-ray, with a diffraction pattern different from VCl_3, VCl_2, and $Al^{19}Cl_3$ and resembling that of the SV catalyst with about one-half the lines missing (Table VI). There are isobutyl–metal bonds present, 1.4 isobutyl per aluminum, as determined by measuring the gas evolved after acid hydrolysis. These alkyl groups are probably attached to the aluminum and not vanadium since they are stable to heat treatment at 100°C (5 hr). The heptane-soluble fraction contains only 1% of the vanadium in the PSV catalyst and, based on its chlorine and aluminum analysis, contains 2.35 isobutyls per aluminum.

Excessive heat treatment (20 hr, 90°C) of the PSV catalyst, which reduces

TABLE VI

EFFECT OF VARIOUS ORGANOMETALLICS AS ACTIVATORS AND PRETREATING AGENTS FOR PSV CATALYST IN ETHER DILUENT[a,b]

Activator			PSV catalyst[c]		Total conv. (%)	PVME Alcohol insoluble		
(i-Bu)₃Al (mmole)	Other organometallics Name	mmole	Pretreating agent	Aging time (hr)		Conv. (%)	$(\ln \eta_r)/C$	% of total
1	—	—	(i-Bu)₃Al	0.08	75	25	7.7	33
—	(Et)₃Al	1	(i-Bu)₃Al	0.08	82	27	6.2	33
1	(C₆H₅)₃Al	—	(Et)₃Al	0.08	37	16	6.5	43
—	(C₆H₅)₃Al	0.8	(i-Bu)₃Al	0.08	89	30	6.6	34
1	—	—	(i-Bu)₃Al:THF	0.08	90	27	4.3	29
1	—	—	(i-Bu)₃Al:THF	1	88	34	5.5	39
1	—	—	(i-Bu)₃Al:THF	20	83	34	6.1	41
—	(i-Bu)₃Al:THF	1	(i-Bu)₃Al:THF	20	79	41	6.5	52
—	(i-Bu)₃Al:THF	1	(i-Bu)₃Al:THF	1 at 90°C	57	13	9.4	24
—	(i-Bu)₃Al:THF	1	(i-Bu)₃Al:THF	20 at 90°C	70	6	8.2	9
—	(Et)₃Al:THF	1	(Et)₃Al:THF	0.08	62	27	8.3	44
1	—	—	(i-Bu)₃Al:TEA	0.08	<15	Trace	—	—

[a] 10 gm VME, 25 ml diluent (70–78% ether plus n-heptane), 0.5 mmole V as PSV catalyst, polymerized 21 hr at 30°C.

[b] Reprinted from E. J. Vandenberg, J. Polym. Sci., Part C 1, 207 (1963). Copyright 1963 by the Journal of Polymer Science. Reprinted by permission of the copyright owner.

[c] 2:1 Mole ratio of pretreating organometallic to SV catalyst. Here TEA is triethylamine.

its catalytic activity, changes the color to a light gray, decreases the V(III) content somewhat (down to 4%), and destroys its crystallinity.

c. Polymer Isolation

If the nature of the reaction product permits, a total solids method is used to determine the total percent conversion to polymer. This method consists in taking an aliquot (3–5 gm) of the reaction mixture, adding 2 ml of a 1% sodium hydroxide solution in 99% ethanol as shortstop, and drying (1 hr, 80°C, 15 mm Hg). At the end of the polymerization a shortstop was generally used, preferably 5 ml of 1 M ammonia in ethanol (prepared from 28% aqueous ammonia).

The preferred procedure for isolating PVME consists of evacuating through a 13G hypodermic needle for 20 min with a water pump in order to remove the unreacted monomer. Then 100 ml of anhydrous ethanol is added under nitrogen and the product agitated for at least 16 hr at 30°C. Next 5 ml of 10% methanolic hydrochloric acid is added and the entire product, with sufficient anhydrous ethanol to give a total volume of about 200 ml, is agitated in a Waring Blendor if large lumps are present. The PVME is isolated by centrifuging. It is washed twice with anhydrous ethanol, once with 0.1% sodium hydroxide in methanol, once with methanol, and once with 0.05% Santonex [4,4'-thiobis(6-*tert*-butyl-*m*-cresol); Monsanto Chemical Co.] in methanol. It is then dried in, partly on a steam bath under nitrogen and then for 16 hr at 50°C in a vacuum oven. The amount of alcohol-soluble polymer is determined by combining the alcohol washes prior to the alkali treatment and then determining the total solids (no shortstop) on an aliquot. The PVME is of reasonably low ash, usually less than 0.1%.

In PSV catalyst runs under the best conditions (Table VI), the alcohol-soluble polymer corresponds to 50–60% conversion to a nontacky, tough rubber. It is isolated and purified by combining the ethanol washes (prior to sodium hydroxide wash), concentrating to about one-fifth the original volume, adding water to the original volume, heating to 50°C, and collecting the insoluble residue. This crude polymer is purified by two similar precipitations from hot water. It is then dissolved in methanol, stabilized with about 0.5% Santonex, and dried to constant weight at 50°C *in vacuo*. The tough rubbery product, recovered in 20% conversion and largely soluble in methanol, had $(\ln \eta_r)/C = 2.5$, low crystallinity by X-ray, and 2% crystallinity by infrared.

Crystalline vinyl ethyl ether polymer is isolated by the same procedure used for PVME.

5. MISCELLANEOUS METHODS AND PREPARATIONS

1. Emulsion copolymerization of isobutyl vinyl ether and vinyl chloride [62].

2. Polymerization of vinyl ethers by Grignard compounds [63].

3. Nickel cyanide as a polymerization catalyst for vinyl ethers [64].

4. Preparation of poly(vinyl ethers) starting with poly(vinyl alcohol) [65].

5. Addition of maleic anhydride to poly(alkyl vinyl ethers) [66].

6. Polymerization of vinyl ethers with α-olefins [67].

7. Octadecyl vinyl ether–maleic anhydride copolymer [68].

8. Cotelormerization of vinyl ethers with chloroalkanes [69].

9. Polymerization of α-naphthylmethyl vinyl ether and 3-pyrenylmethyl vinyl ether [70].

10. 1,3-Butadiene–vinyl ether copolymer using cobalt–diene complexes as catalysts [71].

11. Octadecyl vinyl ether–maleic anhydride copolymer as mold release agent [68].

12. Polymerization of isobutyl vinyl ether by diethyl aluminum chloride-oxygen systems [72].

REFERENCES

1. J. Wislicenus, *Justus Liebigs Ann. Chem.* **192**, 106 (1878).

2. W. Chalmers, *Can. J. Res.* **7**, 113 and 472 (1932); *J. Am. Chem. Soc.* **56**, 912 (1934); D. D. Eley and J. Saunders, *J. Chem. Soc.* p. 4167 (1952); *ibid.* pp. 1668, 1672, and 1677 (1954); D. D. Eley and A. Seabrooke, *ibid.* p. 2226 (1964).

2a. D. D. Eley and A. W. Richards, *Trans. Faraday Soc.* **45**, 425 (1949).

3. W. Reppe and O. Schlichting, U.S. Patent 2,104,000 (1937).

3a. W. Reppe and O. Schlichting, U.S. Patent 2,104,001 (1937).

3b. W. Reppe and O. Schlichting, U.S. Patent 2,104,002 (1937).

3c. W. Reppe and E. Kuehn, U.S. Patent 2,098,108 (1937).

4. A. E. Favorski and M. F. Shostakovskii, *J. Gen. Chem. USSR (Engl. Transl.)* **13**, 1 428 (1943).

5. M. F. Shostakovskii and I. F. Bogdanov, *J. Gen. Chem. USSR (Engl. Transl.)* **15**, 249 (1942); M. F. Shostakovskii, *ibid.* **20**, 609 (1950).

6. D. D. Eley and D. C. Pepper, *Trans. Faraday Soc.* **43**, 112 (1947).

7. C. E. Schildknecht, "Vinyl and Related Polymers," pp. 593–634. Wiley, New York, 1952; N. M. Bikales, *Encycl. Polym. Sci. Technol.* **14**, 511 (1971).

8. S. A. Miller, "Acetylene," Vol. 2, pp. 242–244. Academic Press, New York, 1966.

8a. "Alkyl Vinyl Ethers," Tech. Bull. 7543-055, pp. 18–19. Commer. Dev. Dept., GAF Corporation, New York, 1966.

9. FIAT 856.B105 1602.

10. BIOS 742 and 1292. FIAT 1602.

11. C. E. Schildknecht, A. O. Zoss, and C. McKinley, *Ind. Eng. Chem.* **39**, 180 (1947); C. E. Schildknecht, S. T. Gorss, H. R. Davidson, J. M. Lambert, and A. O. Zoss, *ibid.* **40**, 2104 (1948); C. E. Schildknecht, S. T. Gorss, and A. O. Zoss, *ibid.* **41**, 1998 (1949); C. E. Schildknecht, A. O. Zoss, and F. Grosser, *ibid.* p. 2891.

12. C. E. Schildknecht, *Ind. Eng. Chem.* **50**, 107 (1958).

13. G. Natta, I. Bassi, and P. Corradini, *Makromol. Chem.* **18–19**, 455 (1955); S. Okamura, T. Higashimura, and I. Sakurada, *J. Polym. Sci.* **39**, 507 (1959).

14. D. D. Coffman, G. H. Kalb, and A. B. Ness, *J. Org. Chem.* **13**, 223 (1948); J. D. Coombes and D. D. Eley, *J. Chem. Soc.* p. 3700 (1957); L. Fishbein and B. F. Crowe, *Makromol. Chem.* **48**, 221 (1961).
15. R. F. Heck, U.S. Patent 3,157,626 (1964).
16. E. J. Vandenberg, U.S. Patent 3,159,613 (1964).
17. H. Fikentscher, German Patent 634,408 (1936); H. G. Hammon, R. A. Clark, and J. W. Uttley, Jr., U.S. Patent 2,994,681 (1961); R. E. Florin, *J. Am. Chem. Soc.* **71**, 1867 (1949).
17a. G. A. Richter, Jr. and G. L. Brown, U.S. Patent 2,869,977 (1959).
17b. H. Fikentscher, U.S. Patent 2,016,490 (1935).
18. C. E. Schildknecht, U.S. Patent 2,513,820 (1950).
19. S. H. Pinner and R. J. Worrel, *J. Appl. Polym. Sci.* **2**, 122 (1959); Imperial Chemical Industries, Ltd., British Patent 585,179 (1947); E. I. du Pont de Nemours, British Patent 586,297 (1947).
20. J. F. Nelson, F. W. Banes, and W. P. FitzGerald, U.S. Patent 2,967,203 (1961).
21. M. F. Shostakovskii and A. V. Borgdanova, *Izv. Akad. Nauk SSSR, Otd. Khim. Nauk* p. 919 (1954); p. 387 (1957); N. M. Bortnick and S. Melamed, U.S. Patent 2,734,890 (1956).
22. R. R. Dreisbach and J. L. Lang, U.S. Patent 2,859,209 (1958).
23. A. M. Khomutov, *Izv. Akad. Nauk SSSR, Otd. Khim. Nauk* p. 116 (1961).
24. G. Akazome, *Chem. High Polym.* **17**, 449, 452, 478, 482, 558, and 620 (1960); *Chem. High Polym., Ind. Chem. Sect.* **62**, 1247 (1959); J. Alfrey, J. J. Bohrer, and H. Mark, "Copolymerization." Wiley (Interscience), New York, 1952.
24a. E. C. Chapin, G. E. Ham, and C. L. Mills, *J. Polym. Sci.* **4**, 597 (1949).
24b. "Alkyl Vinyl Ethers," Tech. Bull. 7543-055. GAF Corporation, New York, 1966.
25. J. L. Lang, U.S. Patent 2,937,163 (1960).
26. R. R. Dreisbach and J. F. Malloy, U.S. Patent 2,778,812 (1957).
27. Nitto Electrical Industries, Japanese Patent 21,891 (1963); H. Fikentscher and R. Gäth, German Patent 745,424 (1943).
28. P. J. Stedry, U.S. Patent 2,811,501 (1957).
29. M. F. Shostakovskii and A. M. Khomutov, *Bull. Acad. Sci. USSR* p. 931 (1953); F. Grosser, U.S. Patent 2,694,697 (1954); R. S. Towne, U.S. Patent 2,744,098 (1956); J. J. Giammana, U.S. Patent 2,698,316 (1954); E. Knopf and H. Scholz, German Patent 707,321 (1941).
30. A. M. Khomutov and M. F. Shostakovskii, *Izv. Akad. Nauk SSSR, Otd. Khim. Nauk* p. 2017 (1959); A. M. Khomutov, *ibid.* p. 352 (1961); *Polym. Sci. USSR (Engl. Transl.)* **5**, 181 (1964).
31. S. Imada, Japanese Patent 548 (1955).
32. M. F. Shostakovskii, *Zh. Prikl. Khim.* **28**, 1123 (1953).
33. J. M. Wilkinson, Jr. and J. P. Barker, U.S. Patent 2,655,267 (1954).
34. L. M. Minsk and C. C. Unruh, U.S. Patent 2,417,404 (1947).
35. W. Zerweck and W. Kanze, German Patent 948,282 (1956).
36. R. F. Heck, U.S. Patent 3,025,276 (1962); J. Lal, U.S. Patent 3,038,889 (1962).
37. S. N. Ushakov, S. P. Mitsengendler, and V. N. Krasulina, *Izv. Akad. Nauk SSSR, Otd. Khim. Nauk* p. 490 (1957).
38. S. P. Mitsengendler, *Izv. Akad. Nauk SSSR, Otd. Khim. Nauk* p. 1120 (1956).
39. J. J. Robertson, U.S. Patent 2,905,660 (1959).
40. F. Grosser, U.S. Patent 2,547,819 (1951).
41. Mitsubishi Chem. Industrie, British Patent 863,237 (1961).
42. C. S. Scanley, F. H. Siegele, and R. L. Webb, U.S. Patent 3,088,931 (1958).

43. F. W. Knobloch and H. C. Hamlen, P. B. Report 131,998 from *U.S. Gov. Res. Rep.* **31**, 159 (1959).
44. F. W. Knobloch, *J. Polym. Sci.* **25**, 453 (1957).
45. K. Takakura, *Polym. Lett.* p. 565 (1965).
46. D. E. Jefferson, French Patent 1,350,905 (1964).
47. M. F. Shostakovskii, A. M. Khomutov, and P. Alimov, *Izv. Akad. Nauk SSSR, Ser. Khim.* p. 1839 (1963).
48. F. P. Sidel'kovskaya, M. F. Shostakovskii, F. Ibinginov, and M. A. Askarov, *Vysokomol. Soedin.* **6**, 1585 (1964).
49. J. Furukawa, T. Tsuruta, H. Fukutani, and N. Yamamoto, *Kogyo Kagaku Zasshi* **60**, 1085 (1957).
50. B. F. Goodrich, British Patent 756,839 (1956).
51. F. F. Miller and H. Gilbert, German Patent 953,660 (1956).
52. S. Iwatsuki and Y. Yamashita, *J. Polym. Sci., Polym. Chem. Ed.* **5**, 1753 (1967).
53. E. J. Vandenberg, R. F. Heck, and D. S. Breslow, *J. Polym. Sci.* **41**, 519 (1959); E. J. Vandenberg, German Patent 1,033,413 (1958); *J. Polym. Sci., Part C* **1**, 207 (1963); Goodyear Tire & Rubber Co., British Patent 841,238 (1960); E. J. Vandenberg, French Patent 1,381,326 (1964).
54. G. Natta, G. Dall'Asta, G. Mazzenti, U. Giannini, and S. Cesca, *Angew. Chem.* **71**, 205 (1959); G. Dall'Asta and I. W. Bassi, *Chim. Ind. (Milan)* **43**, 999 (1961).
55. H. Sinn, H. Winter, and W. von Tirpitz, *Angew. Chem.* **72**, 522 (1960); *Makromol. Chem.* **48**, 59 (1961).
56. G. P. Lorenzi, E. Benedetti, and E. Chiellini, *Chim. Ind. (Milan)* **46**, 1474 (1964).
57. J. Furukawa, *Makromol. Chem.* **32**, 90 (1959).
58. R. J. Kray, *J. Polym. Sci.* **44**, 265 (1960).
59. M. Bruce and D. W. Farrow, *Polym.* **4**, 407 (1963); G. J. Blake and A. M. Carlson, *J. Polym. Sci.* **4**, 1813 (1966).
60. S. Nakano, K. Iwasaki, and H. Fukutani, *J. Polym. Sci., Part A* **1**, 3277 (1963).
61. Toyo Rayon, British Patent 1,063,040 (1967); E. W. Glueselkamp, U.S. Patent 3,026,290 (1962).
62. W. E. Daniels, U.S. Patent 3,741,946 (1973).
63. R. J. Kray, *J. Polym. Sci.* **44**, 264 (1960).
64. G. F. Walker and D. G. Hawthorne, *J. Catal.* **15**, 83 (1969).
65. H. Ukihashi and K. Nakamura, Japanese Patent 70/33,906 (1970).
66. Cassela Farbwerke Mainkur A.-G., British Patent 1,059,255 (1967).
67. E. W. Duck and B. J. Ridgewell, British Patent 1,107,898 (1968).
68. T. S. Mestetsky, German Patent 2,250,731 (1973).
69. C. E. Schildknecht, D. W. Kent, and K. Williams, *Polym. Prepr., Am. Chem. Soc., Div. Polym. Chem.* **12**, 117 (1971).
70. S. Yoshimoto, K. Okamoto, H. Hirata, S. Kusabyashi, and H. Mikawa, *Bull. Chem. Soc. Jpn.* **46**, 358 (1973).
71. H. Kawatsura, T. Ohmori, and H. Hudo, Japan Kokai 72/42,985 (1972).
72. H. Hirotaka, J. Araki, and H. Tani, *Polym. J.* **4**, 279 (1973).

POLY(*N*-VINYLPYRROLIDONE)

I. NOMENCLATURE
AND MOLECULAR WEIGHT RELATIONSHIPS

Since much of the available literature, particularly commercial data sheets, use the term *N*-vinylpyrrolidone for the monomer of structure **I**, this name will also be used throughout this chapter. The term 1-vinyl-2-pyrrolidone offers a more precise description of the molecule. Older volumes of *Chemical*

$$H_2C\underset{\underset{N}{|}}{\overset{\overset{\displaystyle CH_2}{|}}{\underset{\underset{CH=CH_2}{|}}{\overset{|}{C}}}}$$

$$H_2C\text{------}CH_2$$
$$H_2C\diagdown\underset{N}{}\diagup C{=}O$$
$$|$$
$$CH{=}CH_2$$

I

Abstracts use the term 1-vinyl-2-pyrrolidinone; the current indexes use 1-ethenyl-2-pyrrolidinone (i.e., for indexing purposes, "2-pyrrolidinone, 1-ethenyl").

In this chapter, the addition polymer derived from this compound will be termed poly(*N*-vinylpyrrolidone), sometimes abbreviated "PVP."

In discussions of PVP, the viscosity measure called *Fikentscher's K value* has survived. This term is calculated from Eqs. (1) and (2):

$$\text{Fikentscher's } K \text{ value} = 1000k \tag{1}$$

in the equation:

$$\frac{75k^2}{1 + 1.5kC} + k = \frac{\log \eta_{rel}}{C} \tag{2}$$

where C is the concentration of polymer in a solvent measured in gm/100 ml of solution, and η_{rel} is the ratio of the viscosity of this solution to the viscosity of the pure solvent [1].

The approximate relationship between intrinsic viscosity $[\eta]$ and Fikentscher's K value/1000 (i.e., k) is given by Eq. (3),

$$\frac{[\eta]}{10} = 0.23(75k^2 + k) \tag{3}$$

where the concentration for the determination of $[\eta]$ is given in liters/gm rather than the conventional dl/gm [2].

Table I is a guide to the relationships between Fikentscher's K value, intrinsic viscosity (calculated), and the weight-average molecular weight, $\overline{M}_w$, calculated from Eq. (4).

$$[\eta] = 1.4 \times 10^{-5} \, \overline{M}_w^{0.7} \qquad \text{[Reference 2]} \tag{4}$$

This equation is not to be taken as definitive, since Reppe, himself [2] mentions the expression

$$[\eta] = 1.6 \times 10^{-4} \, \overline{M}_w^{0.68} \tag{5}$$

as one which had been published in 1952–1953, and more recently we have found the *viscosity* molecular weight, $\overline{M}_v$, to be given by the following expression:

$$[\eta] = 6.75 \times 10^{-4} \, \overline{M}_v^{0.55} \text{ (dl/gm)} \qquad \text{[Reference 3]} \tag{6}$$

TABLE I

APPROXIMATE RELATIONSHIPS
BETWEEN VISCOSITY MEASURE-
MENTS AND WEIGHT-AVERAGE
MOLECULAR WEIGHT OF PVP [2]

Fikentscher's K value	$[\eta]^a$	$\overline{M}_w{}^b$
16	—	10,000
18.9	0.11	14,000
22.5	0.14	20,000
24.9	0.17	25,000
26	—	28,000
27.9	0.020	32,000
30	—	38,000
30.4	0.023	39,000
33	—	47,000
34.7	0.029	52,500
35.7	0.031	56,000
36	—	60,000
46.1	0.048	106,000
60	—	220,000
90	—	750,000

^{a} Calculated, probably in units of liters/gm.
^{b} Calculated, based on $[\eta] = 1.4 \times 10^{-5} M^{0.7}$; see text.

Another expression is

$$[\eta] = 6.75 \times 10^{-4} \overline{M}_w^{0.58} \quad ([\eta] \text{ in ml/gm}) \qquad [\text{Reference 4}] \qquad (7)$$

The reasons for the variation in the constants are not clear. To some extent the cause may be variations in the molecular weight distribution found in the fractionated polymers, the nature of the solvents, the degree of hydration of the polymer coil [4–6], etc. Variations due to the molecular weight range indicated were also noted [5,6].

2. INTRODUCTORY REMARKS

The polymerization of the *N*-vinylpyrrolidone received its greatest impetus during World War II, when the Germans made use of saline solutions of the polymer as a blood plasma extender or substitute for their troops.

In view of the current American drug regulations, it is not entirely clear whether the use of PVP blood extenders is still permitted. Despite this, these polymers, and many copolymers, are considered to have a low order of toxicity. They are used in cosmetics, toiletries, and a variety of pharmaceuticals, including the manufacture of tablets and microencapsulated materials. Other applications are in such diverse fields as in adhesives, the textile and dyeing industries, suspending agents, protective colloids, pigment dispersants, leveling agents, desalination membranes, surfactants, flexible contact lenses, and fiberglass sizing. Since the polymer forms complexes with many materials, such as phenolic compounds, cross-linked copolymers are used in beverage clarification. A PVP complex with iodine is used as a household antiseptic [7].

The literature on poly(N-vinylpyrrolidone) is voluminous. References 1, 2, and 7–22 represent a selection of the reviews available. Linke's article [21] is of particular interest for its discussion of PVP and its copolymers in hairsprays and some of the problems associated with packaging in aerosol spray cans.

N-Vinylpyrrolidone is soluble in water and in a large variety of other solvents. The anhydrous polymer is also soluble in water. With highly concentrated warm alkalies or with sodium chloride solutions, the polymer may be salted out of aqueous solutions. When a polymer solution is boiled with concentrated alkali, an insoluble product forms [1]. Heating an aqueous polymer solution with mineral acids is said to cause partial rupture of the pyrrolidone rings to form poly(N-vinyl-γ-aminobutyric acid) [Eq. (8)] [2].

$$
\begin{bmatrix} -CH_2-CH- \\ \quad | \\ \quad N \\ \end{bmatrix}_n \xrightarrow{H_3O^+} \begin{bmatrix} -CH_2-CH- \\ \quad | \\ \quad NH \\ \quad | \\ (CH_2)_3 \\ \quad | \\ CO_2H \end{bmatrix}_n \qquad (8)
$$

During polymerization in acidified aqueous solution, however, partial hydrolysis of the monomer takes place to produce appreciable quantities of acetaldehyde [1]. This is one reason for carrying out aqueous solution polymerizations of this monomer in the presence of a slightly basic buffer.

The anhydrous polymer is also soluble in the lower molecular weight alcohols, ketones, tetrahydrofuran, chlorinated hydrocarbons, pyridine, pyrrolidone, butyrolactone, triethanolamine, dimethylformamide, and glyoxal [1,2]. It is insoluble in ether, aliphatic, and cycloaliphatic hydrocarbons [2]. The polymer is swollen by esters and aromatic hydrocarbons [1]. It is also soluble in many mixed solvents.

Depending on the molecular weight of PVP, in PVP–water-acetone three-component systems, there are compositions from which the polymer may precipitate [1,2]. By use of this observation, molecular weight fractionation of PVP may be accomplished by the judicious addition of acetone to aqueous solutions of poly(*N*-vinylpyrrolidone) [1].

The polymerization of *N*-vinylpyrrolidone has been initiated by many of the systems which are conventional for the formation of high polymers of vinyl compounds. The notable exception to conventional initiations is that dibenzoyl peroxide is not a satisfactory initiator for this monomer. Oxidative side reactions may interfere with the formation of a high molecular weight product. On the other hand, hydrogen peroxide (particularly in the presence of ammonia), hydroperoxides, 2,2′-azobis(isobutyronitrile), persulfates, sodium peroxide, sodium sulfite, etc. have been used as polymerization initiators. Ultraviolet radiation, electron beams, γ-radiation, heat (in the presence of oxygen), titanium tetrachloride, boron trifluoride etherate, halides of mercury, antimony, and bismuth, and $Al(C_2H_5)_{1.5}Cl_{1.5}$ have been used to form PVP. Grafting of *N*-vinylpyrrolidone to various substrates, particularly to textile fibers, has been accomplished.

Bulk polymerization procedures are known. Suspension polymerization in aqueous media loaded with high concentrations of electrolyte are possible. Work has also been carried out on solid-state polymerizations. However, the most important methods of polymerization and copolymerization are solution processes, particularly using water as the solvent. If the dry polymer is desired, it may be extracted with a suitable solvent, precipitated, or isolated by solvent evaporation. Freeze drying or spray drying have also been used.

Copolymerizations and terpolymerizations of *N*-vinylpyrrolidone have been carried out with a large variety of other monomers. Among the comonomers mentioned in the literature are the following:

Acrylamide	Maleic anhydride
Acrylic acid	Maleate esters
Sodium acrylate	Methacrylamide
Allyl alcohol	Methoxystyrene
Biallyl	Methylene diacrylamide
Crotonic acid	Methyl vinyl ketone
Diallyl phthalate	Methyl vinylpyrrolidone
Diethylene glycol	Styrene
bis(allyl carbonate)	Tetramethallyl isocyanurate
Dimethylaminoethyl vinyl sulfide	Trichloroethylene
Dimethylvinyl ethynyl carbinol	Tris(trimethylsiloxy)vinyl silane
Divinylbenzene	Vinyl acetate
Divinyltetrachlorobenzene	Vinylcaprolactam
Fumarate esters	Vinylcarbazole
Isobutyl vinyl ether	Vinyl laurate
Itaconic acid	Vinyl methyl benzimidazole

TABLE II

PHYSICAL PROPERTIES OF *N*-VINYLPYRROLIDONE

Property	Value	Ref.
Molecular weight	111	*a*
Boiling point (°C; mm Hg)	65; 1	*a*
Melting point (°C)	13.5–17.0	*a*
Refractive index, n_D^{25}	1.5120	*a*
Density, d_4^{20} (gm/ml)	1.0458	*a*
Colorless liquid (free from UV-fluorescing impurities)		*b*
Inhibitor	0.1% Sodium hydroxide	*c*
Solubility	Soluble in water and common solvents	*c*
Viscosity at 25°C	2.07 Cp	*d*

a J. P. Schroeder and D. C. Schroeder, *in* "Vinyl and Diene Monomers" (E. C. Leonard, ed.), Part 3, pp. 1362 ff. Wiley, New York, 1971.
b W. Reppe, "Poly(vinylpyrrolidone)." Verlag Chemie, Weinheim, 1954.
c D. H. Lorenz, *Encycl. Polym. Sci. Technol.* **14**, 239 (1971).
d "V-Pyrol, *N*-Vinyl-2-Pyrrolidone," Tech. Bull. 9653-011. GAF Corporation, New York.

Vinyl methyl dichlorosilane	Vinylpyridine
Vinyl methyl oxazolidinone	Vinyl siloxane
Vinyl oxethylurea	Vinylsuccinimide
Vinyl propionate	Vinyl stearate

Graft copolymers of *N*-vinylpyrrolidone with such polymers as dextran, poly(acrylate esters), polyacrylonitrile, polytetrafluoroethylene, poly(methyl methacrylate) films, polyester films, and polyolefins have been reported [22].

Table II gives some of the physical properties of monomer *N*-vinylpyrrolidone.

3. POLYMERIZATION PROCEDURES

A. Bulk Polymerization

As supplied by the manufacturer, monomeric *N*-vinylpyrrolidone is normally inhibited with 0.1% flake sodium hydroxide [22]. This inhibitor may be separated by decantation or filtration. An alternative inhibition system involves the use of gaseous ammonia or organic amines. These inhibitors are said not to interfere with the polymerization process and, in fact, may activate it [23]. This activation by ammonia is not entirely surprising. As will be discussed below in Section 3,B ("Aqueous Solution Polymerization"), aqueous ammonia has been known as an activator in aqueous systems for many years [1]. Certain impurities which have a distinctly inhibitory effect on the polymerization of N-vinylpyrrolidone are γ-butyric acid and γ-butyrolactone [24].

Strictly speaking, early preparative procedures for the polymerization of N-vinylpyrrolidone might be classified as solution processes; however, the level of solvents—usually water—was so low that this work properly belongs in this section. Thus, for example, 50 gm of the monomer was heated with 1 gm of 3% hydrogen peroxide under pressure, with 1 gm of potassium peroxide, with 1 gm of water and 0.1 gm of benzoyl peroxide, or with 1 gm of sodium peroxide [25]. Procedure 3-1 gives more details of a pilot plant-scale operation. This method is given here primarily as an illustration of early work in the field.

3-1. Bulk Polymerization with Hydrogen Peroxide [1]

To a suitable pressure reaction vessel containing 35 kg of N-vinylpyrrolidone is added 150 ml of 30% hydrogen peroxide. The mixture is heated with agitation to 110°C. The heating cycle is discontinued when the temperature of the reaction mixture has risen to 180°–190°C as the polymerization proceeds. When the temperature begins to drop, the molten polymer is poured onto cold plates (proper ventilation and protective equipment for personnel are required). The cooled polymer is chipped off the plates and ground to a fine powder in a ball mill. The resulting powder is somewhat hygroscopic and tends to be discolored. The residual unreacted monomer level is on the order of 10%. (The residual monomer may be extracted with diethyl ether.) The molecular weight of the product is low.

The heat of polymerization is more readily controlled in this process if 1 kg of water is incorporated in the initial reaction charge [1]. In general the solution procedures are of greater importance for the preparation of large quantities of PVP.

For bulk homopolymerization of N-vinylpyrrolidone, benzoyl peroxide appears not to be a satisfactory initiator [26]. In this regard, the monomer differs from N-vinylcaprolactam, which could be polymerized in the presence of this peroxide. The initiator of choice for N-vinylpyrrolidone is 2,2'-azobis(isobutyronitrile) (AIBN) as described in Procedure 3-2.

3-2. Bulk Polymerization with AIBN [26]

In a suitable glass ampoule are placed 5 gm (45 mmoles) of freshly distilled N-vinylpyrrolidone and 0.01 gm (0.062 mmole) of 2,2'-azobis(isobutyronitrile). The ampoule is flushed with nitrogen and sealed. After shaking the reaction to insure complete dissolution of the initiator, a protective sleeve is placed around the ampoule and the assembly is heated in an oil bath at

TABLE III
BULK POLYMERIZATION OF UNINITIATED
N-VINYLPYRROLIDONE [27]

Reaction temperature (°C)	Initial rate of polymerization (wt.% of polymer formed per hr)	Average degree of polymerization
140	0.026	440
160	0.18	300
180	0.60	250

$60° \pm 1°$C for 72 hr. The ampoule is then removed from the bath, cooled, and cautiously opened. The product is dissolved in ethanol and precipitated from solution with ligroin to yield 3.4 gm (68% of theoretical yield).

Breitenbach and Schmidt [27] found that without initiator present, the bulk polymerization of *N*-vinylpyrrolidone proceeded readily at elevated temperatures (Table III).

At lower temperatures, AIBN was used in this study. The initial rate of polymerization was found to be proportional to the square root of the initiator concentration. In the temperature range of $20°$–$50°$C, this rate followed the Arrhenius equation (cf. Table IV):

$$\log (\%/\text{hr}) = 24.5 - 7670/T \tag{9}$$

TABLE IV
BULK POLYMERIZATION OF
N-VINYLPYRROLIDONE IN THE PRESENCE
OF 0.5×10^{-3} MOLES OF AIBN PER
MOLE OF MONOMER [27]

Reaction temperature (°C)	Initial rate of polymerization (wt.% of polymer formed per hr)
20	0.018
25	0.08
30	0.23
35	0.4
40	1.0
45	2.8
50	5.7

At 50°C the decomposition constant for AIBN, K_0, is 0.0106/hr. The average degree of polymerization ($\overline{\text{DP}}$) of the polymer is only slightly dependent on the AIBN concentration. (At 20°C, with 1×10^{-3} moles of AIBN per mole of monomer, $\overline{\text{DP}} = 2500$; with 0.25×10^{-3} moles of AIBN per mole of monomer, $\overline{\text{DP}} = 2600$.) This degree of polymerization does not appear to be identical with the kinetic chain length. It seems to be dependent on the interaction of the propagation step of the growing chain and chain transfer to monomer.

The temperature dependence of the average degree of polymerization is only slight. (At 20°C, with 0.25×10^{-3} moles of AIBN per mole of monomer, $\overline{\text{DP}} = 2600$; at 50°C, $\overline{\text{DP}} = 2300$.) This indicates that the activation energy of the transfer reaction is only slightly greater than that of the propagation step.

In all of this work, the average degree of polymerization was calculated from viscosity measurements in methanol by the equation of H. P. Frank and B. Levy [Eq. (10)]:

$$\log \overline{\text{DP}} = 4.62 + 1.47 \log [\eta] \tag{10}$$

where $[\eta]$ is in units of liter/gm.

Both bulk and solution polymerizations of N-vinylpyrrolidone have been initiated at 0°–100°C by metal salts such as the chloride, bromide, iodide, sulfate, and nitrate of mercury(II); the chloride, bromide, and iodide of bismuth; and the chloride of antimony [28]. The polymerization may be sufficiently exothermic to require external cooling. Procedure 3-3 is given here as an illustration of this patented process.

3-3. Bulk Polymerization with Mercuric Chloride as Catalyst [28]

With due precaution for the handling of mercuric chloride, in a stirred reactor equipped for external cooling, to 100 gm (0.90 mole) of N-vinyl-pyrrolidone, at room temperature, is added slowly 5 gm (0.018 mole) of mercuric chloride. The rapid evolution of heat is controlled by external cooling. After the reaction has subsided, benzene is added to the reaction mass to precipitate the polymer. The product is recovered by filtration.

This process may also be carried out in such solvents as acetone, dioxane, 2,5-dioxahexane, tetrahydrofuran, methanol, and other lower alcohols.

In general, bulk polymerization processes have been used to study the copolymerization of N-vinylpyrrolidone with a variety of monomers such as vinyl laurate [29], styrene [30], methyl methacrylate [30], vinyl acetate [30,31], vinyl chloride [30], crotonaldehyde [32], crotonic acid [33], N-vinylsuccinimide [34], butyl methacrylate [35], N-vinylphthalimide [36], acrylic acid [37], various alkyl acrylates and methacrylates including lauryl methacrylate and stearyl methacrylate [38], and ethylene [39]. Table V lists reactivity ratios of several copolymer systems.

TABLE V
REACTIVITY RATIOS OF VARIOUS MONOMERS (M_2) WITH *N*-VINYLPYRROLIDONE (M_1)

Monomer	Reactivity ratios		Ref.
	r_1	r_2	
Acrylonitrile	0.06 ± 0.07	0.18 ± 0.007	*a*
Allyl alcohol	1.0	0.0	*a*
Allyl acetate	1.6	0.17	*a*
Butyl methacrylate	0.23	1.16	*b*
Crotonic acid	0.85 ± 0.05	0.02 ± 0.02	*c*
Maleic anhydride	0.16 ± 0.03	0.08 ± 0.03	*a*
Methyl methacrylate	0.005 ± 0.05	4.7 ± 0.5	*d*
	0.02 ± 0.02	5	*a*
Styrene	0.045 ± 0.05	15.7 ± 0.5	*d*
Trichloroethylene	0.54 ± 0.04	<0.01	*a*
Tris(trimethylsiloxy)vinyl silane	4.0	0.1	*a*
Vinyl acetate	3.30, 2.7, 2.0, 0.38	0.205, 0.19, 0.24, 0.44	*d,e,a*
Vinyl carbonate	0.4	0.7	*a*
Vinyl chloride	0.38	0.53	*d*
Vinyl cyclohexyl ether	3.84	0	*f*
Vinyl phenyl ether	4.43	0.22	*f*
N-Vinylphthalimide	1.28 ± 0.04	0.35 ± 0.002	*g*

a "V-Pyrol, *N*-Vinyl-2-pyrrolidone," Tech. Bull. 9653-011. GAF Corporation, New York.

b F. Ibragimov, D. Mukhamadaliev, and T. G. Gafurov, *Vysokomol. Soedin., Ser. A* **12** 1475 (1970); *Chem. Abstr.* **73**, 779272x (1970).

c S. N. Ushakov, V. A. Kropachev, L. B. Trukhmanova, R. I. Gruz, and T. M. Markelova, *Vysokomol. Soedin., Ser. A* **9** 1807 (1967); *Polym. Sci. USSR (Engl. Transl.)* **9**, 2042 (1967); *Chem. Abstr.* **67**, 100426 (1967).

d J. F. Bork and L. E. Coleman, *J. Polym. Sci.* **43**, 413 (1960).

e T. B. Efremova, A. I. Meos, L. A. Vol'f, and V. V. Zarutskii, *Zh. Prikl. Khim.* **42**, 1196 (1969); *Chem. Abstr.* **71**, 61790 (1969).

f F. P. Sidel'kovskaya, M. A. Askarov, and F. Ibragimov, *Vysokomol. Soedin.* **6**, 1810 (1964); *Polym. Sci. USSR (Engl. Transl.)* **6**, 2005 (1964); *Chem. Abstr.* **62**, 6563c (1965).

g R. I. Gruz, V. G. Shibalovich, E. F. Panarin, and S. N. Ushakov, *Vysokomol. Soedin., Ser. A* **10**, 2096 (1968); *Chem. Abstr.* **70**, 12013g (1969).

Copolymers of *N*-vinylpyrrolidone have been grafted onto cellulose by treating cellulose with mixtures of monomers containing initiators [35].

Hydrophilic contact lens materials have been prepared by graft polymerizing 2-hydroxyethyl methacrylate onto poly(*N*-vinylpyrrolidone) using *tert*-butyl perbenzoate as an initiator [40]. A more complex cross-linked contact lens material has been prepared by polymerizing a composition of PVP, *N*-vinylpyrrolidone, 2-hydroxyethyl methacrylate, and ethylene dimethacrylate with benzoyl peroxide [41].

A cross-linked hydrogel has also been prepared by terpolymerizing N-vinylpyrrolidone, methyl acrylate, and tetraethylene glycol dimethacrylate in the presence of AIBN [42].

B. Aqueous Solution Polymerization

For commercial production of PVP, aqueous solution processes are most important. Product isolation may be somewhat difficult in the laboratory since facilities for spray drying or film casting are usually not available. However, techniques such as precipitation methods may be applicable. If the product is of sufficiently high molecular weight and the electrolyte concentration is sufficiently high, poly(N-vinylpyrrolidone) may separate from the solvent spontaneously.

One of the early patents for the polymerization of N-vinylpyrrolidone discloses sodium (or potassium) sulfite as polymerization initiators. The aqueous solution polymerization process is carried out in neutral or basic media in order to avoid acetaldehyde formation by decomposition of the monomer. In the more usual oxidative initiation, this may also lead to acetic acid generation. In the more generally used procedures, careful buffering of the reaction medium is usual.

3-4. *Aqueous Solution Polymerization with Potassium Sulfite* [43]

In a suitable reactor fitted with a gas inlet tube and a mechanical stirrer, a solution of 30 gm (0.27 mole) of N-vinylpyrrolidone, 40 gm (0.25 mole) of neutral potassium sulfite, and 200 gm of water is agitated under a nitrogen atmosphere for 24 hr at 35°–40°C. The solution is then allowed to stand at room temperature for 24 hr. The polymer is separated by decantation. Then the polymer is dissolved in an equal volume of distilled water. The solution is dialyzed against running water for 48 hr. The dialyzed solution is filtered and evaporated at room temperature. The product is a clear, hornlike mass which, in aqueous solution, is neutral.

3-5. *Generalized Procedure for Polymerization with Ammonia–Hydrogen Peroxide* [1]

A review article by Fikentscher and Herrle [1] discusses the solution polymerization of this monomer in considerable detail. This material is summarized here so that the reader may develop his own reaction procedures on the basis of the information. The reaction conditions based on this work are reliable for the preparation of copolymers of N-vinylpyrrolidone and vinyl acetate [44].

It is postulated that in water, N-vinylpyrrolidone forms a monohydrate. If hydrogen peroxide is added, the medium becomes acidic—presumably because

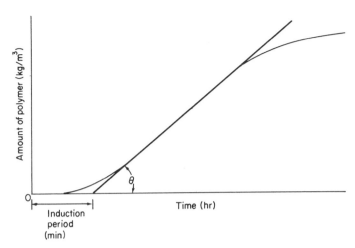

FIG. 1. Polymerization of *N*-vinylpyrrolidone. Tan θ, rate of polymerization in kg/m³/hr.

of the formation of acetic acid from the acetaldehyde which is generated from the monomer. For this reason the reaction medium is usually buffered to maintain the pH between 8 and 7. Ammonia or amines, aside from acting as buffering agents, also have an activating effect.

In the Fikentscher and Herrle treatment, the rate of polymerization is given in units of kilograms of polymer formed per cubic meter of solution per hour. The rate is determined by plotting the amount of polymer formed (in kg/m³) against time (in hr) (cf. Fig. 1). A tangent is drawn at the point where the straight line portion of the curve begins after curving up from the abscissa. The tangent is extended to the time axis. The trigonometric tangent of the angle θ (Fig. 1) is considered to be the rate of polymerization (kg/m³/hr). The intercept on the time axis is the induction period (min). To be noted is that the period from time 0 to the first perceptible evidence of polymer formation is given no special term.

The percentages of initiators or activators are always given as based on the monomer. Ammonia is given in terms of the weight of gaseous ammonia used, although it is actually added as aqueous ammonium hydroxide solution. The quantity of hydrogen peroxide is not precisely defined but is believed to be in terms of the weight of a 30% hydrogen peroxide solution added. The initial pH is always 8 (unless otherwise noted) and is never allowed to drop below 7.

The effect of ammonia on the polymerization characteristics is given in Table VI.

When amines are substituted for ammonia in equivalent amounts, the induction period is increased, the rate of polymerization is decreased, but the K value is essentially unchanged. Substituting equivalent quantities of sodium

TABLE VI

EFFECT OF AMMONIA ON THE POLYMERIZATION OF
N-VINYLPYRROLIDONE[a]

Ammonia added (%)	Induction period (min)	Rate of polymerization (kg/m³/hr)	K value
0[b]	—	No polymerization even after 4 hr	—
0.04[c]	180	200	63
0.1	5	250	53
0.4	0	500	56
1.6	0	850	62

[a] Polymerization conditions: 30% of monomer in distilled water; initial pH = 8 (unless otherwise indicated); 0.5% of 30% H_2O_2; temperature, 50°C [1].

[b] At pH = 7 initially.

[c] Up to 0.5% of sodium bicarbonate had been added to maintain the pH at 7.

hydroxide or sodium bicarbonate for ammonia did not lead to perceptible polymerization even after 4 hr.

The initial pH of the system affects the induction period and the rate of polymerization but does not have a significant effect on the K value. Thus under reaction conditions similar to those given in Table VI, with 0.1% of ammonia, the induction period above a pH of 9 is 5 min or less, the rate of polymerization is above 200 kg/m³/hr, but the K value remains at 55.

As may be anticipated, the effect of temperature on the rate of polymerization is quite significant (Table VII).

TABLE VII

EFFECT OF TEMPERATURE ON THE
POLYMERIZATION OF THE N-VINYLPYRROLIDONE[a]

Temperature (°C)	Induction period (min)	Rate of polymerization (kg/m³/hr)	K value
30	40	100	63
50	5	250	56
70	2	530	54
90	0	1200	52

[a] Polymerization conditions: 30% of monomer in distilled water; 0.5% of 30% H_2O_2; 0.1% of ammonia [1].

TABLE VIII

EFFECT OF HYDROGEN PEROXIDE
CONCENTRATION ON THE POLYMERIZATION
OF *N*-VINYLPYRROLIDONE[a]

Percent of 30% H_2O_2 based on monomer	Induction period (min)	Rate of polymerization (kg/m³/hr)	K value
0.25	10	350	65
0.5	0	400	56
1.0	0	700	45
1.5	0	800	38
2.0	0	1100	33

[a] Polymerization conditions: 50% of monomer in distilled water; 0.3% of ammonia; reaction temperature: 50°C [1].

It is to be noted that the temperature effect on the average molecular weight, as reflected by the K value, is relatively slight.

On the other hand, the amount of hydrogen peroxide used does control the K value. Table VIII summarizes the effect of the level of hydrogen peroxide on the average molecular weight (as expressed in terms of K value).

If the hydrogen peroxide concentration is increased beyond 2%, the K value as well as the rate of polymerization decreases. At a level of 5% of 30% hydrogen peroxide, the K value is approximately 20.

At very low levels of hydrogen peroxide, the K values increase greatly. However, since the initiator seems to be consumed in the process, hydrogen peroxide has to be added portionwise. Table IX summarizes the results.

TABLE IX

EFFECT OF LOW LEVELS OF HYDROGEN PEROXIDE ON THE POLYMERIZATION OF
N-VINYLPYRROLIDONE[a]

Total percent of 30% H_2O_2 based on monomer	No. of H_2O_2 portions added	Induction period (min)	Rate of polymerization (kg/m³/hr)	K value
0.2	2	20	290	80
0.15	3	23	225	90
0.03	3	—	Does not initiate polymerization	—

[a] Polymerization conditions: 30% of monomer in distilled water; 0.3% of ammonia; reaction temperature: 50°C [1].

The relation of hydrogen peroxide concentration to the K values observed is so reproducible that a plot of K values vs. $[H_2O_2]$ can be used to establish the amount of hydrogen peroxide required to produce a polymer of predetermined K value. These data may also be computed from Eq. (11):

$$K = \frac{40}{C^{0.3}} \tag{11}$$

where K is the K value and C is the concentration of hydrogen peroxide.

The concentration of N-vinylpyrrolidone has an effect on the rate of polymerization and the induction period. With increasing monomer concentration, the rate increases up to approximately 30%, then remains constant up to approximately 60%. Beyond this concentration, it decreases sharply. Pure N-vinylpyrrolidone does not seem to polymerize under these reaction conditions.

These phenomena are attributed to the declining formation of the hydrate of the monomer as the monomer concentration increases. The molecular weight of the product remains essentially constant over the concentration range in which polymerization takes place.

If additional monomer is added gradually to a reaction system, the K value does not change significantly. If such a technique is contemplated, small amounts of hydrogen peroxide should be added along with the monomer to replace the initiator which may have decomposed during the reaction.

Table X outlines the effect of changes in the monomer concentration on reaction parameters.

TABLE X

EFFECT OF N-VINYLPYRROLIDONE CONCENTRATION ON THE
POLYMERIZATION OF N-VINYLPYRROLIDONE [a]

Concentration of monomer in distilled water (%)	Induction period (min)	Rate of polymerization (kg/m³/hr)	K value
10	—	Does not polymerize	—
20	0	250	34
40	0	850	39
50	0	850	35
60	0	850	35
80	10	350	35
90	10	150	—
100	—	Does not polymerize	—

[a] Polymerization conditions: 2% of 30% H_2O_2; 0.4% of ammonia; reaction temperature: 50°C [1].

The effect of oxygen on the reaction is particularly significant at lower polymerization temperatures. As oxygen is displaced by nitrogen, the induction period is shortened and the rate of polymerization increases.

For example, a 30% solution of N-vinylpyrrolidone in distilled water, containing 0.5% of 30% hydrogen peroxide and 0.1% of ammonia, does not polymerize at 20°C when exposed to air. When the oxygen is displaced with nitrogen, polymerization is initiated rapidly and the process is virtually complete in 2 hr to give a product with a K value of 56.

Poly(N-vinylpyrrolidone) is normally water soluble but it may be salted out with high concentrations of sodium chloride or warm alkaline solutions. When the polymer is heated to reflux with concentration alkaline solutions, an insoluble material forms.

While the monomer is soluble in virtually all common solvents, the polymer is soluble in water and in many water–solvent systems. Suitable solvents for the polymer are alcohols, ketones, tetrahydrofuran, chlorinated hydrocarbons, pyridine, and lactones. Aromatic solvents and esters swell the polymer and ether and aliphatic hydrocarbons are nonsolvents.

Unreacted monomer may be separated from an aqueous solution of the polymer by extraction with methylene chloride.

The solubility of PVP in water–acetone systems varies considerably with the K value of the polymer. Addition of acetone to an aqueous PVP solution (K value of 100) may be used to fractionate the polymer in order to study the molecular weight distribution. The phase relation of the PVP–acetone–water system is beyond the scope of the present chapter. Interested readers are referred to Fikentscher and Herrle [1] for a more detailed discussion of this topic. Along with this, references 4–6, and 45 should be reviewed for their discussion of molecular weight determinations, viscosity phenomena, and solubility characteristics.

The problem of reducing the level of unreacted monomer in its polymer is very general. In the case of PVP, which has been suggested as a blood plasma substitute, this matter is of particular importance. A patented procedure for reducing the level of residual monomer is given in Procedure 3-6 [46].

3-6. *Method for Reducing Residual Monomer Levels* [46]

One pound of spray-dried poly(N-vinylpyrrolidone) with 3.08% unsaturation, calculated as N-vinylpyrrolidone, is dissolved in 6.6 lb of distilled water. To this solution is added 18 ml of 28% aqueous ammonia followed by 18 ml of 35% hydrogen peroxide. The mixture is immediately spray dried to produce a resin which is analyzed as having 1.1% unsaturation, calculated as N-vinylpyrrolidone. With repeated treatments the residual monomer level is said to be reducible to 0.65%.

TABLE XI

POLYMERIZATION CONDITIONS FOR N-VINYLPYRROLIDONE ACCORDING TO NEY et al. [47]

Expt. no.	Amount of water (gm)[a]	Amount of N-vinylpyrrolidone	Amount of 28% ammonia solution (gm)	Buffer used (gm)	Amount of 38% H_2O_2 (gm)	Heating cycle	Initial pH	Final pH	pH of redissolved polymer	Nature of dry PVP
1	300	100	2	—	2.1	1. NH_3 solution and H_2O_2 added to refluxing monomer solution 2. Exothermic polymerization starts promptly, causing refluxing of the solution for 4–5 min. Continued refluxing with external heating for 20 min	10	8.8	4	Slightly yellow, definite odor
2	300	100	2	—	2.1	As above except that reaction was maintained at 70°C with an increase in reaction time	10	Not reported	Acidic	Objectionable odor
3	30[b]	100	2	—	2.1	As in experiment 1 except that the vapors from the system are distilled out; distillation is continued until reaction subsides	10	5.7	Not reported	Product is spray dried. Product is odorless and colorless
4	300	100	—	—	2.1	Heat at reflux	—	3.8	Not reported	Practically colorless, acetaldehyde odor, no color change on

5	300	100	3.7 gm disodium phosphate	—	2.0	1. H_2O_2 is added to refluxing solution 2. Exothermic polymerization is continued for 5 min 3. Reflux is continued for 20 min	9.4	7.2	6.8[c]	Practically colorless, odorless[d]
6	300	100	2 oz. of boric acid adjusted to pH 9.5 with NaOH	—	2	As in experiment 5	9.5	9.5–9.4	~7.0[c]	Colorless and odorless product[e]
7	300	100	3.7 gm of monosodium phosphate adjusted to pH 7.8 with NaOH	—	2	As in experiment 5	7.8	7.8–7.3	~7.0[c]	Colorless and odorless product,[e] similar to that of experiment 6
8	300	100	3.7 gm of disodium phosphate	—	2	As in experiment 5 except that the vapors from the reacting system are distilled out	4.4	~7.0	7.0[c]	Colorless and odorless

[a] Distilled water boiled in a stream of nitrogen is used throughout.

[b] The patent reads "30 gm" of distilled water. Since all other examples are 300 gm of distilled water, this may be a typographical error in the patent and should read 300 gm.

[c] The product was isolated (1) by removing most of the water by vacuum distillation; (2) precipitating the buffer with methanol; (3) filtration of the precipitating salts; (4) removing methanol by vacuum distillation; (5) redissolving in water to 15% solution; (6) spray drying.

[d] Similar results were observed when the reaction was carried out at 60°C. Similar results were obtained when vapors were permitted to distill out. When sodium acetate was used as buffer, the final reaction solution had a pH of 5.7; when sodium acid phthalate was used the final pH was 4.3.

[e] Similar results were observed when the reaction was carried out at 80°C.

Particularly when PVP is considered for application as a blood plasma extender, trace amounts of extraneous materials may be objectionable. The ammonia–hydrogen peroxide system discussed above may give rise to acetaldehyde–ammonia complexes which may be capable of further condensation reactions. Hydrogen peroxide may also oxidize acetaldehyde to acetic acid. The molecular weight distribution of the polymer also appears to be critical in blood plasma extenders.

One patent which claims to overcome some of the objections resulting from the initiation systems seems to reverse the previous teaching completely in that it recommends the avoidance of ammonia. Instead, disodium phosphate, which may be used in isotonic solutions, is used as the buffer. It also recommends distilling low boilers out of the reaction system along with some of the aqueous solvent [47]. Table XI outlines the procedures found in this patent. To be noted in this table are not only the color and odor of the dried polymer, but also the difference in the pH of the polymer solution after reaction and the pH after the dried polymer is redissolved in distilled water.

Low concentrations of 2,2′-azobis(isobutyronitrile) (AIBN) have been used to initiate the polymerization of N-vinylpyrrolidone in aqueous solution without buffering agents. The process is said to be autocatalytic from the start even in fairly dilute systems. If the Trommsdorff effect is involved, the increase in intrinsic viscosity noted with increases in reaction time imply that along with chain-transfer reactions, the chain termination step is also involved [48]. Table XII shows the effect of increasing concentrations of monomer in solution on the percent converted after one hour along with the intrinsic viscosity of the resultant product. It will be noted that in aqueous solution, polymers of higher average molecular weights are formed than when pure monomer is polymerized. An upper limit of the average degree of polymerization seems to be reached at modest concentrations [48]. Table XIII shows observations on polymerizations carried to higher conversions and with increasing levels of AIBN. In general to be noted is that the rate of polymerization is proportional to the square root of the AIBN concentration [48].

Fractionation of commercial poly(N-vinylpyrrolidone) indicates that when hydrogen peroxide is used as the initiator, considerable polymer of low molecular weight is formed ($[\eta] \sim 0.03$ liters/gm). 2,2′-Azobis(isobutyronitrile) leads to polymers with considerably fewer low molecular weight fractions. This is considered an indication that hydrogen peroxide acts both as an initiator and as a chain-transfer agent [48].

Using a viscometric technique to follow the kinetics of polymerization of N-vinylpyrrolidone, Bond and Lee [49,49a,50] confirmed the observation of Breitenbach and Schmidt [48] that the degree of polymerization of PVP is not significantly influenced by the concentration of initiator or by the reaction

TABLE XII

EFFECT OF INCREASING *N*-VINYLPYRROLIDONE
CONCENTRATION IN SOLUTION POLYMERIZATION
[48]^a

Concentration of monomer (*M*/liter)	Percent converted after 1 hr	Intrinsic viscosity of product (liters/gm)^f
1	5.8	0.18
2	5.9	0.18
3	5.6	0.18
4	6.2	0.18
6.8^b	7.1	0.17
7.8^c	7.2	0.14
8.3^d	5.8	0.12
9^e	5.7	0.11

^{a} Polymerization conditions: reactions carried out in aqueous solution for 1 hr, at 50°C, with 4.5 × 10^{-3} *M*/liter of AIBN.

^{b} Equivalent to 2 moles of water per mole of monomer.

^{c} Equivalent to 1 mole of water per mole of monomer.

^{d} Equivalent to 0.5 mole water per mole of monomer.

^{e} Neat *N*-vinylpyrrolidone.

^{f} Determined in methanol solution.

TABLE XIII

AQUEOUS SOLUTION POLYMERIZATION OF *N*-VINYLPYRROLIDONE TO
HIGH CONVERSIONS AT 50°C [48]

Concentration of monomer (*M*/liter)	Concentration of AIBN (*M*/liter)	Polymerization Time (hr)	Percent conversion	Intrinsic viscosity (liters/gm)^a
1	4.51 × 10^{-3}	1	5.8	0.183
		2	12.2	0.186
		4	44	0.225
		10	92	—
		15	91	0.276
3	13.5 × 10^{-3}	0.5	6.25	0.162
		1	27.5	0.172
		2	68.0	0.187
Neat	4.51 × 10^{-3}	1	5.7	0.113
		4	29.6	0.125

^{a} Determined in methanol.

temperature. This is particularly interesting since their initiator system consisted of hydrazine and cupric sulfate in aqueous solution. The formation of PVP proceeds by first-order kinetics with this initiator system. This is attributed to the fact that poly(*N*-vinylpyrrolidone) does not precipitate during the process. It is expected that, if precipitation of product takes place during solution polymerization, heterogeneous polymerization would occur in the precipitate with an appreciable gel effect and a rapid increase in the rate of polymerization. This phenomenon has been observed in the case of the aqueous solution polymerization of methyl methacrylate, for example [49,49a,50].

The separation of the high molecular weight fraction of a PVP polymer is of considerable importance in connection with its use as a blood plasma extender. Evidently this fraction is retained in the kidney [51].

One patented method of separating the undesirable high molecular weight fraction involves the formation of complexes of the polymer with poly(acrylic acid) or with an ethylene–maleic anhydride copolymer. The procedure is outlined here for reference only.

3-7. *Method for Separation of High Molecular Weight PVP Fractions* [51]

To a solution of 10 gm of poly(vinylpyrrolidone) [with $\eta_{rel} = 1.237$ in 1% aqueous solution] in 30 ml of distilled water is added 0.973 gm of a 24% aqueous solution of poly(acrylic acid) [pH of the aqueous poly(acrylic acid) solution is 4.2]. The mixture is stirred while 4.1 ml of 1 *N* hydrochloric acid is added. After stirring for 3 min, the foam, which has formed, is separated. This foam contains approximately 15% of the PVP and all of the poly(acrylic acid) which had been added. The PVP in the form is the high molecular weight fraction of the polymer. The clear solution contains PVP with $\eta_{rel} = 1.174$ in a 1% aqueous solution.

Another method of separating high molecular weight PVP from lower molecular weight fractions depends on a solubility inversion phenomenon in partially miscible solvent systems [52]. Thus, on extraction of a PVP solution in butanol with cold water, the aqueous layer contains polymer of higher average molecular weight than the butanol layer. On the other hand, upon extraction of the butanol solution with hot water, the average molecular weight of the polymer in the aqueous layer is lower than that of the polymer retained in the butanol layer. Continuous extraction systems may be devised by using these observations.

C. Solution Polymerization in Organic Solvents

Since *N*-vinylpyrrolidone is soluble in a great variety of solvents, its polymerization need not be confined to aqueous systems. However, interest in aqueous solution polymerization processes has been so overwhelming that

TABLE XIV

POLYMERIZATION OF *N*-VINYLPYRROLIDONE IN
BENZENE SOLUTION [48][a]

Concentration of monomer (*M*/liter)	Percent converted after 1 hr	Intrinsic viscosity of product (liters/gm)[b]
1	5.7	0.11
4	5.5	0.11

[a] Polymerization conditions: Reactions carried out in benzene solution; reaction time: 1 hr at 50°C, with $4.5 \times 10^{-3}\ M$ AIBN.
[b] Determined in methanol.

few literature references deal with organic solvent systems as an alternative.

One of the earliest examples is found in the patent of Schuster *et al.* [25]. There *N*-vinylpyrrolidone is heated in ethanol solutions with hydrogen peroxide and benzoyl peroxide. The products do not appear to be very satisfactory resins.

Breitenbach and Schmidt [48] briefly studied the polymerization in benzene. Table XIV outlines their results.

To be noted in this work is that both the rate of conversion and the intrinsic viscosity of the polymer are independent of the monomer concentration in the solvent. The rate of polymerization is dependent on the square root of the AIBN concentration.

In acetone solution, at 60°C (presumably under the slight pressure of a dilatometer) 5.77 moles of monomer per liter of acetone also containing 0.037 moles of AIBN per liter gave a yield of 77.1 mole percent of polymer after 240 min [53].

A patent [54] states that batch polymerizations of *N*-vinylpyrrolidone in aqueous solution with hydrogen peroxide often give rise to gel formation. This difficulty can be overcome by replacing at least part of the water with such substances as isopropyl alcohol, thioglycolic acid, dimethylformamide, ethanolamine, methyl ethyl ketone, trichloroacetic acid, or 2-mercaptoethanol. A continuous polymerization procedure is said to be particularly effective. Procedure 3-8 is given here only as an illustration of this patented process.

3-8. *Continuous Polymerization Process* [54]

The reaction is carried out in a 3050-ml glass flask fitted with an efficient stirrer, two inlets for reactants, an outlet for the product solution, a thermometer, and a nitrogen inlet. Means of controlling the flow rates of reactants also are provided.

The following reactants are prepared: reactant 1, consisting of a solution of 11 liters of N-vinylpyrrolidone, 11 liters of isopropanol, and 154 ml of 28% aqueous ammonia solution; and reactant 2, consisting of approximately 9 liters of 0.12% aqueous hydrogen peroxide.

Into the stirred reactor, maintained during the preparation at 70°C, while a slow stream of nitrogen bubbles through the equipment. Reactant 1 is fed in at a rate of 190 ml/hr, while simultaneously, reactant 2 is fed in at a rate of 80 ml/hr. The overflow is collected as the product solution. After 42 hr the product solution has 87% of the monomer converted to polymer with a Fikentscher K value between 45 and 50 and no sign of gel formation. When the preparation is continued for 110 hr similar results are observed.

If water is substituted for isopropanol in reactant 1, gelation interferes with this continuous process.

It is possible to demonstrate the chain-transfer properties of hydrogen peroxide and utilize this reagent to control the product molecular weight distribution by polymerizing N-vinylpyrrolidone in ethanol using AIBN as initiator and hydrogen peroxide as a chain-transfer agent [55]. In dimethylformamide, ferric chloride acts as a polymerization retarder [56].

D. Suspension Polymerization

Since N-vinylpyrrolidone is quite soluble in pure water, obviously suspension polymerization of this monomer is likely only if specialized techniques are used.

Monagle claimed that a nontacky, hard, yet water-soluble bead of a copolymer of 80% acrylamide and 20% N-vinylpyrrolidone can be produced in a mixture containing at least 40% tertiary butanol (a nonsolvent for the polymer) in water along with electrolytes such as KCl, NH_4Cl, NaCl, Na_2CO_3, $Na_2B_4O_7$–H_3BO_3, citric acid–NaH_2PO_4, Na_2CO_3–$NaHCO_3$, NH_4OH, or Na_2SO_4. The polymerization is carried out in a stirred reactor in an inert atmosphere with such initiators as ammonium persulfate, potassium persulfate, hydrogen peroxide, or AIBN [57].

A cross-linked PVP bead polymer has been produced in an aqueous system containing substantial quantities of an inorganic salt [58,58a]. Procedure 3-9 is given here to illustrate a patented procedure [58].

3-9. *Suspension Polymerization of Cross-Linked N-Vinylpyrrolidone* [58]

Under a nitrogen atmosphere, to a solution of 0.46 gm of a 10% aqueous solution of sodium dibasic phosphate (Na_2HPO_4), and 40.0 gm of sodium sulfate in 240 gm of water heated to 50°–65°C is added a solution of 40 gm of N-vinylpyrrolidone, 0.12 gm of AIBN, and 1.6 gm of methylenebisacrylamide with stirring. The mixture is heated between 50° and 65°C for 4 hr.

Then a mixture of 0.04 gm of AIBN, 0.4 gm of methylenebisacrylamide, 10 gm of ethanol, and 10 gm of water is added and heating is continued for 2 hr at 62°–63°C. The mixture is then cooled to room temperature with stirring. The product is filtered off, washed with water, and dried under reduced pressure. Yield: 41.5 gm (white beads). Alternative cross-linking agents are ethylene dimethacrylate, the dimethylacrylates of the higher ethylene glycols, and divinylbenzene [58]. Sodium chloride may be substituted for sodium sulfate [58a].

E. Cationic Polymerization

Procedure 3-10 represents an early cationic polymerization experiment with *N*-vinylpyrrolidone.

3-10. *Cationic Polymerization with Boron Trifluoride Etherate* [59]

With appropriate protective devices, to 10 ml of *N*-vinylpyrrolidone dissolved in 30 ml of petroleum ether at room temperature is added 2 ml of boron trifluoride etherate. Within a short time polymerization takes place with a rise of the temperature in the reaction mixture to 40°C. The polymer is recovered as a sticky solid at room temperature.

A somewhat more elaborate procedure involves the use of triethylboron for the copolymerization of *N*-vinylpyrrolidone with methyl methacrylate or lauryl methacrylate. The procedure, which is patented and given here only for reference, involves the additional use of hydroperoxide and an amine [60].

3-11. *Copolymerization of N-Vinylpyrrolidone and Methyl Methacrylate with Triethylboron* [50]

With suitable safety precautions for the handling of triethylboron, a small three-necked flask, under nitrogen, is charged with 5.0 gm (0.05 mole) of methyl methacrylate and 4.4 gm (0.04 mole) of *N*-vinylpyrrolidone. With vigorous stirring, the solution is cooled to 0°C. Then, while the reaction temperature is maintained between 0° and 2°C, 0.53 ml of triethylboron, 0.22 ml of cumene hydroperoxide, and 0.28 ml of pyridine are added. The reaction mixture is stirred for 4 hr at a temperature between 0° and 2°C and then is added cautiously to an excess of methanol. The product is filtered off, dissolved in benzene, and reprecipitated with methanol. After filtration, the product is dried overnight under reduced pressure at 60°C. Yield: 8.7 gm (93%); specific viscosity of a 1% solution in benzene at 25°C is 3.152.

An alternating copolymer of *N*-vinylpyrrolidone and acrylonitrile was prepared by reacting *N*-vinylpyrrolidone and acrylonitrile in heptane solution at $-10°C$ with an initiator which had the composition $AlEt_{1.5}Cl_{1.5}$. The polymer yield was not particularly high [61].

Another catalyst system found in the patent literature involves the deposition of halides such as zirconium tetrachloride, vanadium tetrachloride, titanium tetraiodide, or oxyhalides such as chromium oxychloride or vanadium oxychloride on a finely divided particulate inorganic substrate having surface hydroxyl groups. Among such solids are alumina, zirconia, silica (particularly a pyrogenic silica such as "Cab-O-Sil"), or a carbon black such as channel black or furnace black. A toluene slurry of this material is added, under dry nitrogen, to a toluene solution of N-vinylpyrrolidone containing a small amount of triisobutylaluminum. After 24 hr at 80°C, a 25% yield of polymer is produced [62].

F. Radiation and Solid State Polymerization

Since the polymerization of liquid monomers which have been cooled below their melting point is usually induced by radiation, the discussion of radiation and solid-state polymerizations has been combined in this section.

Since poly(vinylpyrrolidone) is water soluble, the application of the monomer to the radiation-cured coatings field is limited. On the other hand, printing plates can be prepared based on a cross-linkable monomer system. In this case, a coated substrate is exposed to ultraviolet radiation through a negative. Where the monomer system is exposed, it becomes cross-linked and insoluble. Where it is not exposed, the coating remains water or solvent soluble. Judging by one patent, intermediate steps of gray in a negative are reproduced by more-or-less proportional degrees of cross-linking to permit reproduction of intermediate shades of gray [63].

In a patent, a simple procedure for ultraviolet-induced polymerization of the monomer is indicated. This procedure is given here for reference.

3-12. Ultraviolet-Induced Polymerization in Aqueous Solution [64]

To a solution of 30 gm of N-vinylpyrrolidone in 70 gm of distilled water is added 0.075 gm of 2,2'-azobis(isobutyronitrile) (AIBN). The solution is exposed to ultraviolet light with wavelength in the range of 310–410 nm. Polymerization takes place in 2–3 hr. Analysis of the solution shows a residual monomer content of only 1.7%. The Fikentscher K value of the product may range up to the unusually high value of 110.

A source of radiation which may be used in simple laboratory experiments is a 275-watt Westinghouse RS sunlamp, which has emission maxima at 365, 405, and 436 nm [65]. This bulb (with proper shielding to protect laboratory personnel) may be placed about 10 cm from the object to be irradiated. In using this type of bulb some precautions are necessary in its proper positioning.

In another connection, we found that the intensity of radiation is not always concentric with the outer shape of the bulb and that the change in intensity from the center of the spot projected by the bulb measured to its outer edge does not follow the expected inverse square relationship from the bulb to various locations on the spot. Also, with use, the intensity and probably also the emission maxima will change [44].

Among sensitizers used in this field are monosodium riboflavin 5'-phosphate in the presence of oxygen [65], a copper–amine catalyst and a photosensitizer [66], zinc chloride and air [67], anthraquinone, α-chloroanthraquinone, benzil, benzoin, benzophenones, the methyl ether of benzoin [68], various metal perfluoroalkane sulfonates [69], di(4-[N,N-dimethylamino]phenyl) ketone, a (2-chlorophenyl)-4,5-bis(3-methoxyphenyl)imidazole dimer, 2-mercapto-benzoxazole, 1,3-bis(4-[N,N-dimethylamino]benzal)acetone [70], and azido-pentaaminecobalt(III) chloride [71].

At least initially, the γ-radiation-induced polymerization in the solid state is quite different from that of liquid N-vinylpyrrolidone. The rate in the solid state is said to increase after several hours until it reaches values comparable to those found in the liquid phase. This autoacceleration may be related to the regularity of the crystal structures of the frozen monomer, the temperature, and the quantity of preformed polymer present [72]. The activation energy for the polymerization of liquid N-vinylpyrrolidone was found to be 9 ± 0.5 kcal/mole while that for the solid state polymerization is 18.0 kcal/mole [73].

A procedure for the continuous polymerization of N-vinylpyrrolidone by chemical initiation has been described in Procedure 2-8. The same authors also claimed the polymerization of a 20% aqueous solution of the monomer, containing 0.1% of thioglycolic acid passed into a stirred reactor having a hold-up volume of 100 ml at a rate of 52 ml/hr, while the reactor was irradiated with a Machlett OE G-50 X-ray tube, 50,000 volts at 50 ma. During a 12-hr period, conversion was 88% of a gel-free polymer having an average Fikentscher K value of 62.5 [54].

In a bulk copolymerization study of the vinyl acetate–N-vinylpyrrolidone system using ^{60}Co γ-radiation it was found that the copolymer composition depended on the monomer concentration. The copolymer was always enriched in N-vinylpyrrolidone [74].

N-Vinylpyrrolidone spread on various substrates, upon irradiation with a Van de Graff electron accelerator, formed adhesive bonds between adjacent surfaces [75].

G. Charge-Transfer Polymerization

The photo copolymerization of N-vinylpyrrolidone with methyl methacrylate in the presence of zinc chloride was mentioned in passing in Section

TABLE XV

PHOTOPOLYMERIZATION OF *N*-VINYLPYRROLIDONE, MALEIC ANHYDRIDE, AND
METHYL METHACRYLATE [76][a]

Conc. *N*-vinyl- pyrrolidone (M/liter)	Conc. maleic anhydride (M/liter)	Conc. methyl methacrylate (M/liter)	Solvent	Initiator	Polymer yield (gm)
3.75	0.41	5.62	—	2×10^{-2} M/liter AIBN, 2 hr	0.290
3.75	0.41	5.62	—	Dark	Negligible
3.75	0.41	5.62	—	UV	0.421
3.75	0.41	0	Benzene[b]	UV	0.120
3.75	0	5.62	—	UV	0.047
0	0.41	5.62	Benzene[b]	UV	0

[a] Polymerization conditions: Volume of reactants was maintained at 5 ml; UV source emitted at 360–370 nm; reaction time, 3 hr at 30°C; monomer exposed to air.
[b] Enough benzene added to produce 5 ml volume in reactor.

3,F above [67]. While the thermal copolymerization of these two monomers does not take place *in vacuo*, it does take place in the presence of oxygen, after an induction period. Thus oxygen seems to participate in the formation of an active species. The rate is increased in the presence of zinc chloride. It has been postulated that a charge-transfer polymerization process is involved here [67].

In the presence of maleic anhydride, the ultraviolet-induced terpolymerization of *N*-vinylpyrrolidone, methyl methacrylate, and maleic anhydride exhibits charge-transfer polymerization characteristics without any zinc chloride being present [76]. The mixture of *N*-vinylpyrrolidone and maleic anhydride, on exposure to ultraviolet radiation, changes color from yellowish through pink to scarlet. The ultraviolet spectrum indicates the formation of a one-to-one complex.

In air, the polymerization proceeds to a significant extent only when all three monomers are present in the system. Table XV illustrates this very well.

Table XV shows that the best yield of product is produced when all three components are present. While the *N*-vinylpyrrolidone–maleic anhydride complex did polymerize, only a small amount of a copolymer formed when *N*-vinylpyrrolidone and methyl methacrylate were exposed to ultraviolet radiation, presumably because of complex formation with oxygen.

A determination of the reactivity ratios for the *N*-vinylpyrrolidone–methyl methacrylate system illustrates the effect of charge-transfer complexes [77].

In air:

$M_1 = N$-vinylpyrrolidone; $r_1 = 4.6 \pm 0.04$

$M_2 = $ methyl methacrylate; $r_2 = 0.02 \pm 0.02$

In the presence of zinc chloride (in air or *in vacuo*):

$M_1 = N$-vinylpyrrolidone; $r_1 = 1.6 \pm 0.8$

$M_2 = $ methyl methacrylate; $r_2 = 0.6 \pm 0.9$

4. MISCELLANEOUS PREPARATIONS

1. Preparation, properties and applications of PVP in the blood field and in other branches of medicine [78,79].

2. Polymerization of 1-vinyl-5-methylpyrrolidone with hydrogen peroxide, organic hydroperoxides, *tert*-butyl perbenzoate, or boron trifluoride etherate [80].

3. Polymerization of N-vinylthiopyrrolidone [73].

4. Copolymerization of N-vinylpyrrolidone with vinyl chloride [53].

5. Copolymerization of N-vinylpyrrolidone with vinyl acetate [53].

6. Copolymerization of N-vinylpyrrolidone with ethylene [81].

7. Copolymerization of N-vinylpyrrolidone with acrylates, methacrylates, or mixed acrylic monomers, initiated by *tert*-butyl perbenzoate [82].

8. Copolymerization of N-vinylpyrrolidone with sodium vinylsulfonate [83].

9. Copolymerization of N-vinylpyrrolidone with vinyl chloride, vinyl acetate, or styrene in acetic or monochloroacetic acid [84].

10. Copolymerization of N-vinylpyrrolidone with N-vinylphthalimide and conversion to a N-vinylpyrrolidone–vinylamine copolymer [85].

11. Copolymerization of N-vinylpyrrolidone with acrylamide. The effect of additions of glycerol in increasing the reactivity of acrylamide [86].

12. Simultaneous polymerization of N-vinylpyrrolidone and alkylation with reagents such as α-eicosenes or 1,1,2,-trifluoro-2-chloroethene [87].

13. Copolymerization of N-vinylcaprolactam and ethylene [88].

14. Cross-linking of PVP with α,ω-diolefins such as 1,7-octadiene [89].

15. Graft copolymerization of N-vinylpyrrolidone onto cellulose [90–92].

16. Graft copolymerization of N-vinylpyrrolidone onto polyethylene terephthalate (Dacron) [93].

17. Graft copolymerization of N-vinylpyrrolidone onto keratin, fibroin, or collagen [65].

18. Effect of PVP on the polymerization of acrylic acid [94].

19. Use of PVP in the encapsulation of hydrophobic materials [95].

20. Graft copolymerization of 2-hydroxyethyl methacrylate onto PVP using di-*sec*-butyl peroxydicarbonate as initiator [96].

21. Graft copolymerization of 2-hydroxyethyl methacrylate onto PVP using benzoyl peroxide followed by radiation cross-linking for plastic contact lenses [97].

22. Preparation of semipermeable membranes for reverse osmosis by cross-linking PVP films with vaporized cross-linking agents such as toluene diisocyanate [98].

23. Graft copolymerization of a mixture of methyl methacrylate and *N*-vinylpyrrolidone onto cellulose [99].

24. Effect of solvents on the polymerization kinetics of *N*-vinylpyrrolidone [100].

REFERENCES

1. H. F. Fikentscher and K. Herrle, *Mod. Plast.* **23**(3), November, 157 (1945).
2. W. Reppe, "Polyvinylpyrrolidone." Verlag Chemie, Weinheim, 1954.
3. H. H. G. Jellinek and B. Blom, *J. Appl. Polym. Sci.* **16**, 527 (1972).
4. K. Dialer and K. Vogler, *Makromol. Chem.* **6**, 191 (1951).
5. H. Scholtan, *Makromol. Chem.* **7**, 104 and 209 (1951).
6. J. Hengstenberg and E. Schuch, *Makromol. Chem.* **7**, 105 and 236 (1951).
7. D. H. Lorenz, *Encycl. Polym. Sci. Technol.* **14**, 239 (1971).
8. J. M. DeBell, W. C. Goggin, and W. E. Gloor, "German Plastics Practice," pp. 1 and 154. DeBell & Richardson, Springfield, Massachusetts, 1946.
9. W. Reppe, "Acetylene Chemistry," P. B. Rep. 18852-S. Charles A. Meyer and Co., Inc., New York, 1949.
10. C. E. Schildknecht, "Vinyl and Related Polymers," p. 662. Wiley, New York, 1952.
11. I. Greenfield, *Ind. Chem.* **32**, 11 (1956).
12. J. Remond, *Rev. Prod. Chim.* **59**, 127 and 260 (1956).
13. E. Ferraris, *Mater. Plast. (Milan)* **25**, 208 (1959).
14. W. W. Myddleton, *Manuf. Chem.* **34**, (7), 316 (1963).
15. "Polyvinylpyrrolidone," Tech. Bulls. 7543-113, 7543-031, 7543-029, and 7543-066. GAF Corporation, New York.
16. F. A. Wagner, *Kirk-Othmer Encycl. Chem. Technol.*, 2nd Ed. Vol. 17, p. 391 (1968).
17. R. H. Kaplan, Thesis, Cornell University, Ithaca, New York, 1969; University Microfilms, Ann Arbor, Michigan, Order No. 70-5801, *Diss. Abstr. Int. B* **31**, 162 (1970).
18. F. P. Sidel'kovskaya, "Chemistry of *N*-Vinylpyrrolidone and Its Polymers." Nauka, Moscow, 1970; *Chem. Abstr.* **74**, 100, 438v (1971).
19. J. P. Schroeder and D. C. Schroeder, *in* "Vinyl and Diene Monomers" (E. C. Leonard, ed.), Part 3, p. 1362 ff. Wiley, New York, 1971.
20. H. Warson, *Polym., Paint Color J.* **161**, 637 and 643 (1972).
21. W. Linke, *Chemtech.* **4**, 288 (1974).
22. "V-Pyrol, N-Vinyl-2-Pyrrolidone," Tech. Bull. 9653-011. GAF Corporation, New York.

23. C. R. Stahl and S. Siggia, U.S. Patent 3,454,558 (1959); *Chem. Abstr.* **71**, 61883 (1969).
24. N. V. Grigor'eva, E. K. Podoval'naya, and J. S. Stogov, *Zh. Prikl. Khim.* (*Leningrad*) **45**, 692 (1972); *Chem. Abstr.* **77**, 20168 (1972).
25. C. Schuster, R. Sauerbier, and H. Fikentscher, U.S. Patent 2,335,454 (1943).
26. F. P. Sidel'kovskaya, M. A. Askarov, and F. Ibragimov, *Vysokomol. Soedin.* **6**, 1810 (1964); *Polym. Sci. USSR* (*Engl. Transl.*) **6**, 2005 (1964); *Chem. Abstr.* **62**, 6563c (1965).
27. J. W. Breitenbach and A. Schmidt, *Monatsh. Chem.* **83**, 833 (1952).
28. B. J. Luberoff and W. D. Gersumky, U.S. Patent 3,162,625 (1964).
29. J. H. Werntz, U.S. Patent 2,497,705 (1950); *Chem. Abstr.* **44**, 6440g (1950).
30. J. F. Bork and L. E. Coleman, *J. Polym. Sci.* **43**, 413 (1960).
31. T. B. Efremova, A. I. Meos, L. A. Vol'f, and V. V. Zarutskii, *Zh. Prikl. Khim.* (*Leningrad*) **42**, 1196 (1969); *Chem. Abstr.* **71**, 61790 (1969).
32. S. N. Ushakov, L. B. Trukhmanova, T. M. Markelova, and V. A. Kropachev, *Vysokomol. Soedin.*, *Ser. A* **9**, 999 (1967); *Polym. Sci. USSR* (*Engl. Transl.*) **9**, 113 (1967); *Chem. Abstr.* **67**, 54489t (1967).
33. S. N. Ushakov, V. A. Kropachev, L. B. Trukhmanova, R. I. Gruz, and T. M. Markelova, *Vysokomol. Soedin.*, *Ser. A* **9**, 1807 (1967); *Polym. Sci. USSR* (*Engl. Transl.*) **9**, 2042 (1967); *Chem. Abstr.* **67**, 100426 (1967).
34. A. F. Nikolaev, G. P. Toreschchenko, N. Ya. Salivon, F. O. Pozdnyakova, and Z. G. Proskuryakova, *Vysokomol. Soedin.*, *Ser. A* **14**, 2368 (1972); *Chem. Abstr.* **78**, 30309k (1973).
35. F. Ibragimov, D. Mukhamadaliev, and T. G. Gafurov, *Vysokomol. Soedin.*, *Ser. A* **12**, 1475 (1970); *Chem. Abstr.* **73**, 77927x (1970).
36. R. I. Gruz, V. G. Shibalovich, E. F. Panarin, and S. N. Ushakov, *Vysokomol. Soedin.*, *Ser. A* **10**, 2096 (1968); *Chem. Abstr.* **70**, 12013q (1969).
37. I. Scondac, A. Roman, and M. Dima, *Rev. Roum. Chim.* **11**, 1229 and 1333 (1966).
38. Rohm and Haas Co., Netherlands Patent Appl. 6,605,998 (1966); *Chem. Abstr.* **66**, 97273g (1967).
39. J. E. McKeon and P. S. Starcher, U.S. Patent 3,318,906 (1967); *Chem. Abstr.* **68**, 39466k (1968).
40. D. G. Ewell, U.S. Patent 3,647,736 (1972); *Chem. Abstr.* **76**, 154737k (1972).
41. M. Seiderman, U.S. Patent 3,639,524 (1972); *Chem. Abstr.* **76**, 141762f (1972).
42. R. Steckler, U.S. Patent 3,532,679 (1970).
43. W. Reppe, C. Schuster, and A. Hartman, U.S. Patent 2,265,450 (1941).
44. W. Karo, Author's laboratory.
45. B. Jirgensons, *Makromol. Chem.* **6**, 30 (1951).
46. H. Beller, U.S. Patent 2,665,271 (1954).
47. W. O. Ney, Jr., W. R. Nummy, and C. E. Barnes, U.S. Patent 2,634,259 (1953).
48. J. W. Breitenbach and A. Schmidt, *Monatsh. Chem.* **83**, 1288 (1952).
49. J. Bond and P. I. Lee, *J. Polym. Sci.*, *Polym. Chem. Ed.* **9**, 1775 (1971).
49a. J. Bond and P. I. Lee, *J. Polym. Sci.*, *Polym. Chem. Ed.* **7**, 379 (1969).
50. J. Bond and P. I. Lee, *J. Polym. Sci.*, *Polym. Chem. Ed.* **9**, 1777 (1971).
51. B. Hargitay, German Patent 1,943,412 (1970); *Chem. Abstr.* **72**, 133579d (1970).
52. I. A. Parshikov, *Zh. Prikl. Khim.* (*Leningrad*) **45**, 318 (1972).
53. K. Hayashi and G. Smets, *J. Polym. Sci.* **27**, 275 (1958).
54. J. F. Volks and T. G. Traylor, U.S. Patent 2,982,762 (1961).
55. General Aniline and Film Corp., British Patent 1,021,121 (1966); *Chem. Abstr.* **64**, 19821h (1966).

56. A. V. Cernobai, *Vysokomol. Soedin., Ser. A* **10**, 1716 (1968); *Polym. Sci. USSR* (*Engl. Transl.*) **10**, 1986 (1968); *Chem. Abstr.* **69**, 97212f (1968).
57. D. J. Monagle, U.S. Patent 3,336,270 (1967).
58. N. D. Field and E. P. Williams, German Patent 1,929,501 (1970).
58a. N. D. Field and E. P. Williams, French Patent 2,010,746 (1970).
59. C. E. Schildknecht, A. O. Zoss, and F. Grosser, *Ind. Eng. Chem.* **41**, 2891 (1949).
60. E. H. Mattus and J. E. Fields, U.S. Patent 3,275,611 (1966).
61. K. Nakaguchi, S. Kawasumi, M. Hirooka, H. Yabunchi, and H. Takao, Japanese Patent 70/09,952 (1970); *Chem. Abstr.* **73**, 35965c (1970).
62. J. C. Mackenzie and A. Orzechowski, U.S. Patent 3,285,892 (1966).
63. E. Wainer, J. M. Lewis, and J. E. Shirey, German Patent 2,112,416 (1971); *Chem. Abstr.* **76**, 8964g (1972).
64. General Aniline and Film Corp., British Patent, 725,674 (1955).
65. H. L. Needles, *J. Appl. Polym. Sci.* **12**, 1557 (1968).
66. A. C. Schoenthaler, U.S. Patent 3,418,295 (1968).
67. H. Tamura, M. Tanaka, and N. Murata, *Bull. Chem. Soc. Jpn.* **42**, 3042 (1969).
68. N. K. Frunze, *Rev. Roum. Chim.* **14**, 1309 (1969); *Chem. Abstr.* **72**, 112201j (1970).
69. J. E. Kropp, German Patent 2,016,018 (1970); *Chem. Abstr.* **74**, 42836k (1971).
70. W. R. Hertler, German Patent 2,133,460 (1972); *Chem. Abstr.* **76**, 106465q (1972).
71. H. Kothandaraman, K. S. V. Srinivasan, and M. Santappa, *J. Polym. Sci., Polym. Chem. Ed.* **10**, 3685 (1972).
72. S. Munari, J. Tealdo, F. Vigo, and G. Bonta, *Proc. Tihany Symp. Radiat. Chem., 2nd, 1966* p. 573 (1967); *Chem. Abstr.* **68**, 30142u (1968).
73. Gy. Hardy and M. A. Morsi, *Kinet. Mech. Polyreactions, Int. Symp. Macromol. Chem., Prepr., 1969* Vol. 4, p. 75 (1969); *Chem. Abstr.* **75**, 64348a (1971).
74. S. A. Taschmukhamedov, A. Sh. Karabaev, and R. S. Tillaev, *Uzb. Khim. Zh.* **16**, 54 (1972); *Chem. Abstr.* **77**, 88912x (1972).
75. N. S. Marans, U.S. Patent 3,424,638 (1969).
76. H. Tamura, M. Tanaka, and N. Murata, *Bull. Chem. Soc. Jpn.* **42**, 3041 (1969).
77. H. Tamura, M. Tanaka, and N. Murata, *Kobunshi Kagaku* **27**, 652 (1970); *Chem. Abstr.* **75**, 6461w (1971); *Kobunshi Kagaku* **27**, 736 (1970); *Chem. Abstr.* **75**, 6459b (1971).
78. M. M. Flannery, ed., "PVP, Polyvinylpyrrolidone, An Annotated Bibiliography to 1950," Commer. Dev. Dep., General Aniline and Film Corporation (GAF), New York, 1951.
79. H. Weese, G. Hecht, and W. Reppe, German Patent 738,994 (1943).
80. R. Bácskai, *Magy. Kem. Foly.* **61**, 97 (1955); *Chem. Abstr.* **50**, 1360f (1956).
81. R. Resz and H. Bartl, French Patent 1,392,354 (1965); *Chem. Abstr.* **63**, 13503a (1965).
82. Rohm and Haas, G.m.b.H., French Patent 1,437,012 (1966); *Chem. Abstr.* **66**, 12777j (1967).
83. K. Sakaguchi and K. Nagase, *Kogyo Kagaku Zasshi* **69**, 1204 (1966); *Chem. Abstr.* **66**, 116330z (1967).
84. Yu. D. Semichikov, A. V. Ryabov, and V. N. Kashaeva, *Vysokomol. Soedin., Ser. B* **12**, 381 (1970); *Chem. Abstr.* **73**, 35882y (1970).
85. R. I. Gruz, T. Yu. Verkhoglyadova, E. F. Panarin, and S. N. Ushakov, *Vysokomol. Soedin., Ser. A* **13**, 647 (1971); *Polym. Sci. USSR* (*Engl. Transl.*) **13**, 736 (1971).
86. A. M. Chatterjee and C. M. Burns, *Can. J. Chem.* **49**, 3249 (1971).
87. General Aniline and Film Corp., British Patent 1,101,163 (1968); *Chem. Abstr.* **68**, 60237w (1968).

88. Farbwerke Hoechst, A.-G., Belgian Patent 655,504 (1965); *Chem. Abstr.* **64**, 19898h (1966).
89. A. Merijan, U.S. Patent 3,350,366 (1967).
90. B. P. Morin, Yu. G. Krazhev, and Z. A. Rogovin, *Vysokomol. Soedin.* **7**, 1463 (1965); *Chem. Abstr.* **63**, 18444e (1965).
91. F. Ibragimov, A. D. Virnik, F. P. Sidel'kovskaya, M. A. Askarov, and Z. A. Rogovin, *Zh. Vses. Khim. Ova.* **11**, 119 (1966); *Chem. Abstr.* **64**, 17767e (1966).
92. T. A. Mal'tseva, A. D. Virnik, G. D. Pestereva, and Z. A. Rogovin, *Izv. Vyssh. Uchebn. Zaved., Teckhnol. Tekst. Promsti.* No. 4, p. 92 (1966); *Chem. Abstr.* **66**, 56484r (1967).
93. G. W. Stanton and T. G. Traylor, U.S. Patent 3,274,294 (corrected patent number) (1966); *Chem. Abstr.* **66**, 3732p (1967).
94. J. Ferguson and S. A. O. Shah, *Eur. Polym. J.* **4**, 343 (1968).
95. T. Kobayashi, French Patent 1,568,500 (1969); *Chem. Abstr.* **72**, 6188a (1970).
96. H. R. Leeds, German Patent 1,952,514 (1970); *Chem. Abstr.* **73**, 110470t (1970).
97. K. F. O'Driscoll and A. A. Isen, French Patent 2,077,538 (1971); *Chem. Abstr.* **77**, 76211e (1972).
98. L. Nicolas and R. Dick, German Patent 2,118,625 (1971); *Chem. Abstr.* **76**, 73284z (1972).
99. D. Mukhamadaliev, A. Adylov, F. Ibragimov, Yu. T. Tashpulatov, T. G. Gafurov, and Kh. U. Usmanov, *Vysokomol. Soedin., Ser. A* **13**, 2572 (1971); *Polym. Sci. USSR (Engl. Transl.)* **13**, 2893 (1971).
100. E. Senogles and R. Thomas, *J. Polym. Sci., Polym. Symp.* **49**, 203 (1974).

Chapter 9

POLYMERIZATION OF ACRYLIC ACIDS AND RELATED COMPOUNDS

I. INTRODUCTION

In this chapter the methods of polymerization of acrylic and of methacrylic acid are discussed along with supplementary notes on the utilization of the closely related itaconic acid and of the salts of the acrylic acids.

The primary applications of acrylic, methacrylic, and itaconic acids are in the field of copolymerization with a host of other monomers. For example, incorporation of up to 2% of methacrylic acid is said to improve the freeze-thaw as well as the mechanical stability of polymer latexes, particularly at a high pH. Resistance of such latexes to coagulation by calcium ions is said to be improved. Since these acids contribute a degree of hydrophilicity to a copolymer, oil resistance is increased. The carboxylic acid groups in copolymers are capable of reacting with alkalis to form salts. Thus a degree of alkali solubility may be conferred to a resin. On the other hand, a copolymer may be cross-linked by the use of di- or polyvalent ions. For example, elastomers which have been prepared with a small percentage of methacrylic acid may be vulcanized with zinc oxide. Other methods of cross-linking copolymers containing acrylic acids include reactions which lead to ester, amide, or anhydride formation by reaction with appropriate reactants, such as epoxy compounds, diols, diamines, amino resins, phenolic resins, hydroxyl terminated polyesters, etc. Copolymers containing carboxylic acid units as well as glycidyl groups may be cross-linked thermally in a second stage of the formation of a plastic object.

The presence of the acrylic acids in a copolymer may enhance the adhesion of the polymer to another substrate, particularly to a substrate which is subject to corrosion by acids such as aluminum. Coatings containing methacrylic acid may have improved abrasion and wet- or dry-rub resistance. The hardness and the softening temperature may be increased because of the polarity conferred by these monomers. Pigments and fillers may be dispersed more readily in copolymers containing carboxylic acid moieties. Amphoteric copolymers of vinyl pyridine and methacrylic acid are said to be compatible with gelatin.

The homopolymers of the acrylic acids are water soluble. This solubility varies considerably with the molecular weight, the degree of cross-linking and branching, the method of preparation, the presence of other solutes, and the pH. While the solubility of poly(acrylic acid) increases as the temperature is raised, poly(methacrylic acid) decreases in solubility with increasing temperature. As the pH of poly(acrylic acid) or poly(methacrylic acid) is increased by the addition of an alkali, the mutual repulsive forces set up between the carboxylate ions leads to an uncoiling of the polymer chains. This manifests itself in an unusual increase in viscosity as the pH is raised until a rigid gel is formed. When more caustic is added to the system beyond the point where all of the carboxylic acid groups have been converted to carboxylate ions, the viscosity of the system decreases again, essentially as expected on the basis of dilution of the mixture. The use of poly(acrylic acid) or poly(methacrylic acid) as thickening agents depends on the viscosity behavior described here. The property is carried over to copolymers containing these acids. Other applications of these polymers and copolymers include their use as textile sizing, leather tanning agents, binders for scrap leather, cationic ion-exchange resins, soil conditioners, drilling muds, flocculating agents, dispersing agents, and retention aids in paper making.

Glacial acrylic and methacrylic acids are usually supplied inhibited with approximately 250 ppm of *p*-methoxyphenol (MEHQ: monomethyl ether of hydroquinone). Other inihibitors which have been used for these compounds are methylene blue, hydroquinone, and *N,N'*-diphenyl-*p*-phenylenediamine (DPPD). (CAUTION: DPPD may be a cancer-causing agent.) The last-mentioned inhibitor is of particular interest. In the acid, it obviously exists as a nonvolatile salt which can act as a still-pot inhibitor when small amounts of acrylic or methacrylic acid have to be distilled under reduced pressure to prepare an inhibitor-free sample of the acids.

CAUTION: Inhibitor-free glacial acrylic acid or glacial methacrylic acid must be utilized within approximately 0.5–1.0 hr since uncontrolled polymerization of these monomers may be extremely violent, if not explosive.

Copper and copper salts have also been suggested as nonvolatile inhibitors of these acids. Nitric oxide may be of value as a volatile inhibitor.

As indicated above, the inhibitor-free acids are temperamental. They may polymerize suddenly with the evolution of considerable heat. Since the polymers are insoluble in their monomers, the possibility of "popcorn" polymers forming with explosive violence is quite strong. Therefore, inhibitor-free monomers should only be prepared in small quantities and utilized very promptly. If the situation permits dilution of the acids with other monomers, a somewhat more stable system can be developed.

As may be seen in Table I, the melting point of acrylic acid is in the range of $13° \pm 0.5°C$ and that of methacrylic acid is in the range of $15° \pm 1°C$. Consequently, these acids should always be stored above their melting points.

TABLE I

PROPERTIES OF MONOMERS

Properties	Acrylic acid	Methacrylic acid	Itaconic acid
Formula	$CH_2=CH-CO_2H$	$CH_2=C-CO_2H$ $\quad\quad\ CH_3$	$CH_2=C-CO_2H$ $\quad\quad\quad CH_2CO_2H$
Molecular weight	72	86	130
Melting point, °C	13, 12.5, 13.5	15, 16, 14	167–168
Heat of fusion (13°C)	2.66 kcal/mole	—	—
Vapor pressure, °C (mm Hg)	27(5)	48.5(5)	—
	66(40)	86.4(40)	—
	103(200)	123.9(200)	—
Boiling point, °C (mm Hg)	141(760)	163.0(760)	—
Kinematic viscosity, centistokes, 25°C	1.1	1.3	—
Water solubility (25°C)	Miscible in all proportions	Miscible in all proportions	9.5 gm/100 ml $H_2O \sim 30\%$ solution at 50°C
pK_a			
Monomer	4.26	4.66, 4.4	—
Polymer	4.75	5.65	—
Dissociation constant(s) (of monomer) [a]	5.50×10^{-5}	3.72×10^{-5}; 2.17×10^{-5}	1.40×10^{-4} (k_1); 3.56×10^{-6} (k_2)
Density, 20°C	1.05 gm/ml	d_4^{20} 1.0153	1.49 ± 0.01 gm/ml
Refractive index, n_D^T, °C	1.4224^{20}, 1.4185^{25}	1.4314^{20}, 1.4288^{25}	—
Heat of polymerization	18.5 kcal/mole	15.84 ± 0.17 kcal/mole	—

[a] Unlike acrylic and methacrylic acids, itaconic acid has two dissociation constants. Two different values have been reported for the dissociation constant of methacrylic acid.

Otherwise, during their crystallization the inhibitors may separate from solution, which may generate regions of uninhibited monomers whose hazardous nature has been mentioned above. Also, on melting of the crystallized monomers, volume increments may form which are essentially uninhibited and therefore susceptible to uncontrolled polymerization. This means that if acrylic or methacrylic acids are to be distilled, the condenser water must be at about 20°C to prevent crystallization in the condensing and receiving system. On the other hand, the condensing water should be at least 20°–30°C below the boiling point of the monomer being distilled for efficient condensation.

For accurate and reproducible polymerization studies involving acrylic acid, distillation of the monomer is mandatory. Unlike methacrylic acid, acrylic acid dimerizes spontaneously around 30°C at a rate of 1.0–1.5% per month [Eq. (1)] [1].

$$2CH_2{=}CH{-}\overset{\overset{\displaystyle O}{\|}}{C}{-}OH \longrightarrow CH_2{=}CH\overset{\overset{\displaystyle O}{\|}}{C}{-}OCH_2CH_2\overset{\overset{\displaystyle O}{\|}}{C}{-}OH \qquad (1)$$

At approximately 20°C, the rate of formation of this acryloxypropionic acid is said to be negligible. The problem appears to be acute primarily during out-of-doors storage of acrylic acid in the summer months. Acryloxypropionic acid does not seem to inhibit or interfere otherwise with the polymerization of acrylic acid, but, of course, its presence represents a variable impurity in the monomer. There appears to be no method of inhibiting the dimerization of acrylic acid.

In the generalized procedure for the distillation of the acrylic acids given here, a distillation column packed with copper is suggested. This packing acts as an inhibitor for the monomeric vapors. Since the acids tend to corrode copper, the top of the distillation column should be left empty and, possibly, fitted with a spray trap to prevent copper salts, which have inhibitory properties, from entering the distillation head. We have found that scouring pads made from woven copper strands (rather than copper or brass coils) constitute a useful column packing. In order to use these, a brass rivet which holds the pad together is removed and the copper "cloth" is carefully unfolded to yield a sleeve which is fitted loosely into a 25-mm-diameter distillation column. Depending on the length of the column, two or more pads may have to be used. For large-scale operations, the copper cloth may be purchased by the yard. It is best not to re-use this packing.

As mentioned before, the inhibitor-free monomers must be used promptly, diluted with other monomers or solvents, or be reinhibited.

In Procedure 1-1 precautions are taken to prevent water from contaminating the acids, since recent work seems to indicate that water may contribute to popcorn formation [2].

1-1. Distillation of Acrylic Acids

In a hood, behind a safety shield, in a 250-ml flask fitted with a thermometer, a fine capillary tube connected to a drying tube for an air bleed, and a distillation column packed partially with copper cloth (see above) topped with a spray trap and a total reflux, partial take-off distillation head are placed 100 ml of glacial acrylic or methacrylic acid and 1 gm of N,N'-diphenyl-*p*-phenylenediamine or phenothiazine.

CAUTION: These inhibitors may have carcinogenic or other physiological properties.

The distillation head and receiver are connected through a pressure regulator, manometer, and suitable traps to a plastic water aspirator. The condenser water is adjusted to deliver condenser water not *lower* than 20°C but is *not* passed through to the condenser until all moist air has been driven from the apparatus. The equipment is protected from light.

The pressure in the distillation head is adjusted to approximately 30 mm Hg and the distillation is started by heating the still pot cautiously. The condenser is turned on after vapors fill the head. The first 25 ml of distillate is reinhibited with approximately 250 ppm of *p*-methoxyphenol and discarded.

CAUTION: In disposing of these acids, OSHA and EPA-approved procedures must be followed.

Small, preliminary fractions of the center cut are checked to ascertain that no more hydroquinone or *p*-methoxyphenol is present. A center cut of approximately 50 ml is retained and handled with the precaution mentioned above. Prior to use the monomers must be deaerated.

The residues are discarded. The clean-up of all equipment should be prompt and thorough to prevent popcorn polymer formation and the coating of the equipment with acidic polymers.

A method of purifying the acrylic acids further consists of the fractional crystallization of the distilled acids [3]. Since this paper by Katchalsky and Eisenberg [3] is widely quoted in the literature on acrylic and methacrylic acid, we presume that this procedure of purification has been used extensively. We cannot recommend it in view of the hazardous nature of the uninhibited monomer. While, superficially, this method of purification appears attractive, the procedure is not sufficiently well described with respect to a number of problems—e.g., does the residuum of inhibitor precipitate with the crystallizing acids or does it stay in solution in the fluid portion? How is moisture kept out of the system during manipulation? How are oligomeric acids separated from the crystal mass, if that is the portion normally retained?

Table II gives data which is of interest for the polymerization of acrylic

TABLE II
Polymerization Data

Property	Polymer of Acrylic acid	Polymer of Methacrylic acid
pK_a	4.75	5.65
Heat of polymerization	18.5 kcal/mole	15.84 ± 0.17 kcal/mole
Solubility in		
Monomer	Insoluble	Insoluble[b]
Water	Increases with increasing temperature	Decreases with increasing temperature
Isopropanol	Soluble	Insoluble
Glass transition temp., °C	~106	~130
Rheological properties of aqueous solution	Thixotropic	Dilatant
Mark–Houwink–Sakurada parameters[a]		
K	85×10^{-5} (dioxane, 30°C)	2.42×10^{-3} (methanol, 26°C)
a	0.5 (30°C)	0.51 (26°C)
Price–Alfrey copolymerization parameters		
Q	1.09	2.37
e	0.98	0.83

	Reactivity ratios			
	Acrylic acid (M_1)		Methacrylic acid (M_1)	
Comonomer (M_2)	r_1	r_2	r_1	r_2
Acrylonitrile	0.40	2.5	5.4	1.8
Butadiene	0.28	1.3	0.53	0.20
Ethyl acrylate	0.37	2.1	0.57	0.17
2-Ethylhexyl acrylate	0.25	2.4	4.0	0.21
Methacrylate ion	—	—	0.66	0.08
Methyl methacrylate	0.14	4.6	2.2	0.37
Styrene	0.35	0.22	0.7	0.15
Vinyl acetate	2.0	0.1	20.0	0.01
Vinyl chloride	2.8	0.23	26.0	0.038
Vinylidene chloride	0.63	1.2	0.93	0.10

[a] Parameters for the molecular weight–intrinsic viscosity relationship: $[\eta] = K\overline{M}^a$. For these acids, these constants are solvent, pH, and dissolved electrolyte concentration dependent.

[b] Miller [4] states that poly(methacrylic acid) is soluble in its monomer. Other sources disagree.

and methacrylic acids. Attention should be called to the fact that poly(acrylic acid) behaves normally on dissolution in water; i.e., as the temperature of the water increases, the solubility of the polymer increases. On the other hand, poly(methacrylic acid) may precipitate from a warm aqueous solution, although it may go back into solution on cooling. This characteristic seems to carry over to some other solvent systems. For example, Katchalsky and Eisenberg [3], purified poly(methacrylic acid) fractions which had been precipitated from methanol with ether by cooling the polymer to redissolve the material and warming it to 25°C to precipitate it.

It should also be noted that the polymeric acids, when thoroughly dried, are difficult to redissolve in water. Customarily these acids, when prepared in an aqueous system, are dried to a point that at least 5% of water is retained [4]. By freeze drying, polymers may be prepared which are readily soluble in water. In the spray drying of polyacid solutions, wetting agents may be added to improve the water solubility of the dry product. In general, the solubility of the polyacids is affected not only by temperature but also by the molecular weight, the pH of the solution, the presence of dissolved electrolytes, and the tacticity of the polymer. Useful reviews are references 1, 4–13.

Glacial acrylic and methacrylic acids are corrosive liquids and should be handled with the usual precautions taken with corrosive acids. In addition, the polymerizability of these acids should be taken into account when these monomers are handled. Reference 13 gives considerable detail about storage and handling of these compounds, particularly on an industrial scale. These compounds are toxic and due precautions should be taken.

2. POLYMER PREPARATIONS

A. Hydrolysis of High Polymers

Particularly before acrylic acid became commercially available, poly(acrylic acid) and copolymers were prepared by the hydrolysis of poly(acrylonitrile), polymeric acrylate esters, or appropriate copolymers [14]. Polymethacrylates have also been hydrolyzed; however, the hydrolysis of methacrylates generally requires more drastic reaction conditions than that of acrylates.

Unless careful control over the hydrolysis process is exercised, incomplete reaction as well as chain-degrading reactions may take place.

The procedure is of some value, since the molecular weights and tacticity of certain esters can be determined with comparative ease. These well-defined poly(acrylic esters), on hydrolysis, are thought to retain the characteristics of the backbone, thus offering information on the molecular weight and tacticity of the resultant poly(acids). If on reesterification of the poly(acid) the original poly(ester) is regenerated, the procedure has a measure of self-consistency. Generally, the atactic and isotactic polymeric esters are significantly more rapidly hydrolyzed than the corresponding syndiotactic resins.

The alkaline hydrolysis, as expected, will frequently result in rigid gels, particular if a dilute alkaline solution is added to the poly(ester) solution to permit good dispersion of the alkaline solution. If the concentration of alkali is too high, the poly(acrylate) salts will tend to precipitate [15].

Procedure 2-1 shows the preparation of a sodium poly(acrylate) gel.

2-1. *Preparation of a Rigid Sodium Poly(acrylate) Gel* [15]

To a solution of 0.5 gm of poly(ethyl acrylate) in 100 gm of benzene is added 0.5 gm of a 1 *N* solution of sodium hydroxide in water. The mixture is stirred for a short period. Within a few hours, the fluid mixture forms a durable solid gel. This gel is said to remain solid even on warming. Even on ignition, a block of the material is said to burn without any benzene flowing away.

A procedure leading to a water-soluble, powdery potassium poly(acrylate) is given in Preparation 2-2 [16].

2-2. *Preparation of Potassium Poly(acrylate)* [16]

In a reflux apparatus fitted with an addition funnel and a mechanical stirrer, a solution of 38.6 gm (0.45 mole) of poly(methyl acrylate) in 59.4 gm of toluene is heated gently to reflux with stirring. Then a solution of 22.5 gm (0.4 mole) of potassium hydroxide in 90 gm of absolute ethanol is added in small portions. The reaction mixture, which gradually becomes viscous, is refluxed for 1 hr. Ultimately, a white powder separates. The white powdered product is filtered off and air dried.

The more drastic requirements for hydrolyzing poly(methacrylates) are illustrated in Procedure 2-3. The tacticity of the polymer seems to have a profound effect on the ease of alkaline hydrolysis. Thus, for example, isotactic poly(methyl methacrylate) even with a $\overline{M}_v$ of 1,250,000 can be hydrolyzed to the extent of 49% within 1.5 hr whereas the conventional polymer of $\overline{M}_v$ 100,000 is hydrolyzed 50% in 140 hr. Syndiotactic and atactic poly(methyl methacrylate) also hydrolyze slowly [17]. In general it was observed that as hydrolysis proceeds, the polymer becomes swellable in water at low degrees of conversion. Only when hydrolysis has reached approximately 15% of the ester groups, does the polymer, which now might be considered a copolymer of methyl methacrylate and sodium methacrylate, become water soluble.

2-3. *Preparation of Sodium Poly(methacrylate)* [17]

To a cooled solution of 4 gm (0.1 mole) of sodium hydroxide in 1.8 gm (0.1 mole) of water and 80 gm of isopropanol is added 10 gm (0.1 mole) of a conventionally polymerized poly(methyl methacrylate), $\overline{M}_v$ 100,000. The mixture is heated at 80°–85°C for 140 hr and cooled. The solid product is filtered off and air dried. Upon filtration, the product is found to have been hydrolyzed to approximately 50% of the available ester groups.

To study the properties of isotactic poly(acrylic acid), it is necessary to prepare an isotactic poly(acrylate ester) first. The common practice is to polymerize isopropyl acrylate in the presence of *n*-butyllithium or of phenylmagnesium bromide at −78°C in toluene [18]. The polymer is then hydrolyzed as completely as possible so as not to disturb the regularity of the conformation, hinder the crystallization of the polymer (if it is capable of crystallizing), alter its solubility, or change the shape of the polymer chains in solution.

It will have been noted already that the degree of hydrolysis of poly(acrylates) and poly(methacrylates) frequently does not appear to be complete. In part this may be due to difficulties with the analytical procedures and with other factors. In part, however, this may also be the result of hydration of the product. In fact, isotactic poly(acrylic acid) may crystallize as a hydrate containing as much as one molecule of water associated with two monomer units [18]. Procedure 2-4, for the preparation of isotactic poly(acrylic acid), is said to be the most effective method of twelve different procedures studied [18].

2-4. *Preparation of Isotactic Poly(acrylic acid)* [18]

To a hot solution of 1 gm (0.009 mole) of isotactic poly(isopropyl acrylate) in 5 ml of toluene is added 10 ml of a solution containing 0.070 gm of potassium hydroxide in 1 ml of isopropanol. The mixture is refluxed for 6 hr at 96°–97°C. Then the reflux condenser is set down for distillation, water is added to the reaction mixture and the water–isopropanol is slowly distilled off while the isopropanol is replaced with water. During this step, which is said to take 10 hr, the polymer goes completely into solution. The solution is cooled, and an excess of methanol is added to precipitate potassium poly(acrylate). The polymer is filtered off, washed with methanol, and dried.

The polymer is dissolved in a small quantity of water, and an excess of concentrated hydrochloric acid is added to precipitate poly(acrylic acid). The polymer is filtered off, washed with concentrated hydrochloric acid, and dried under reduced pressure at room temperature. By titration this polymer appears to be only 84.7–85% hydrolyzed, but X-ray diffraction shows that crystalline species have been isolated. On drying to constant weight at 80°C, titration indicates 89.6% hydrolysis.

Table III summarizes a variety of observations on the alkaline hydrolysis of several acrylic polymers. To be noted is that the difference in the rates of

TABLE III

DEGREE OF HYDROLYSIS OF ACRYLIC POLYMERS IN ALKALINE MEDIA

Polymer	Solvent	Hydrolytic medium	Time (hr)	Percent hydrolysis	Ref.
Poly(ethyl acrylate) (0.5 gm)	Benzene (100 gm)	1 N Alcoholic NaOH (0.5 gm)	Several hours	—	15
	Methylene chloride (100 gm)	1 N Alcoholic NaOH (0.5 gm)	Several hours	—	15
Poly(butyl acrylate) (1 gm)	Low-boiling paraffin (100 gm)	1 N NaOH (?)	Several hours	20	15
Poly(methyl acrylate) (38.6 gm)	Toluene (59.4 gm)	KOH (22.5 gm) in 90 gm of absolute alcohol	1	—	16
Poly(methyl methacrylate), conventional polymer (10 gm)	Isopropanol (80 gm)	NaOH (0.4 gm) in water (1.8 gm)	1.5	13	17
Poly(methyl methacrylate), isotactic (10 gm)	Isopropanol (80 gm)	NaOH (0.4 gm) in water (1.8 gm)	24	39	17
	Isopropanol (80 gm)	NaOH (0.4 gm) in water (1.8 gm)	140	50	17
	Isopropanol (80 gm)	NaOH (0.4 gm) in water (1.8 gm)	0.08	7	17
	Isopropanol (80 gm)	NaOH (0.4 gm) in water (1.8 gm)	0.25	13	17
	Isopropanol (80 gm)	NaOH (0.4 gm) in water (1.8 gm)	1.0	36	17
	Isopropanol (80 gm)	NaOH (0.4 gm) in water (1.8 gm)	1.5	49	17
	Isopropanol (80 gm)	NaOH (0.4 gm) in water (1.8 gm)	24	75	17
	Isopropanol (80 gm)	NaOH (0.4 gm) in water (1.8 gm)	140	75	17
Poly(methyl methacrylate), syndiotactic (10 gm)	Isopropanol (80 gm)	NaOH (0.4 gm) in water (1.8 gm)	1.5	11	17
	Isopropanol (80 gm)	NaOH (0.4 gm) in water (1.8 gm)	24	33	17
	Isopropanol (80 gm)	NaOH (0.4 gm) in water (1.8 gm)	140	45	17
Poly(isopropyl acrylate), isotactic (1 gm)	Toluene (5 gm)	KOH (0.7 gm) in 10 ml of isopropanol	10	89.6	18

hydrolysis of isotactic and syndiotactic poly(methyl methacrylate) is sufficiently great that mixtures of such polymers may be separated on the basis of this behavior [17].

The method lends itself to a demonstration of the fact that certain poly(methyl methacrylate) polymers are block copolymers of isotactic and syndiotactic conformation, not to mention that conventional, i.e., freeradically polymerized, poly(methyl methacrylate) contains a substantial fraction of the isotactic polymer.

It is unfortunate that polymer chemists habitually do not report conversion or yield information. This simple bit of information would go a long way in evaluating many aspects of chemical hypotheses and the suitability of a process for preparative purposes.

The alkaline hydrolysis of acrylic esters requires, as shown in Table III, considerable time. The procedure involves a heterogeneous reaction in its early stages and may actually be accompanied by oxidative degradation of the polymeric chains [19]. The acid hydrolysis, on the other hand, is a homogeneous reaction, and is thought not to lead to chain scission [3]. Again, poly(acrylates) hydrolyze more rapidly than the poly(methacrylates) and the isotactic poly(methacrylates) are more easily converted to poly(methacrylic acids) than the syndiotactic poly(methacrylates) [17].

2-5. *Preparation of Poly(acrylic acid) by Acid Hydrolysis* [3]

A solution of 10 gm of poly(methyl acrylate) in 80 ml of acetic acid and 20 ml of water and containing 2 gm of recrystallized *p*-toluenesulfonic acid is maintained at 120°C for 18 hr under a distillation column. During this time, most of the methyl acetate formed by transesterification distills out.

The residue in the flask is cooled and chloroform is added to the residue to precipitate the crude poly(acrylic acid). The crude poly(acid) is filtered off, dissolved in methanol and reprecipitated either with chloroform or with ether. The process is repeated three times to separate residues of acetic acid and *p*-toluenesulfonic acid. The product is filtered and dried to constant weight under reduced pressure at 110°C. Upon titration of a sample, 85% free carboxylic groups are found to be present.

Table IV outlines the observations of a variety of hydrolyses of acrylic esters in acidic media [3,17,20–22].

On the basis of a study of molecular models, it had been proposed that the polymerization of 2,4,6-triphenylbenzyl methacrylate would only result in the formation of syndiotactic polymer. However, upon hydrolysis of this sterically hindered polyester with phosphonium iodide, the resultant poly(acrylic acid) could be separated into an atactic and a syndiotactic fraction on the basis of differences in solubility. This method of hydrolysis is of particular

TABLE IV

DEGREE OF HYDROLYSIS OF ACRYLIC POLYMERS IN ACIDIC MEDIA

Polymer	Reaction medium	Reaction time (hr)	Temp. (°C)	Percent hydrolysis	Ref.
Poly(methyl acrylate) (10 gm)	Acetic acid (80 ml), water (20 ml), p-toluenesulfonic acid (2 gm)	18	120	85	3
Poly(methyl methacrylate), isotactic, $M_v = 400,000$ (50 gm)	Acetic acid (400 gm), water (100 gm), p-toluenesulfonic acid (30 gm)	336	110	95	17
Poly(methyl methacrylate), conventional, $M_v = 1,000,000$	Acetic acid (400 gm), water (100 gm), p-toluenesulfonic acid (30 gm)	336	110	6	17
Poly(methyl acrylate), isotactic	Conc. sulfuric acid	—	—	"Complete"	20
Poly(tert-butyl acrylate), isotactic (15 gm)	Dioxane (300 ml), conc. hydrochloric acid (25 ml), water (40 ml)	18	Reflux	"Complete"	21
Poly(tert-butyl acrylate), atactic (15 gm)	Dioxane (300 ml), conc. hydrochloric acid (25 ml), water (40 ml)	18	Reflux		21
Poly(methyl methacrylate), isotactic (freeze dried from benzene solution) (300 mg)	97.5% Sulfuric acid (100 ml)	5–12 min	45	—	22
Poly(methyl methacrylate), syndiotactic (freeze dried from benzene solution) (300 mg)	97.5% Sulfuric acid (100 ml)	1–3	45	—	22

interest since it evidently permits relatively rapid saponification of syndiotactic poly(esters).

2-6. *Preparation of Poly(methacrylic acid) from Poly(2,4,6-triphenylbenzyl methacrylate)* [23]

With suitable precautions against the hazards associated with hydrogen and phosphines as well as phosphonium iodide, in a 2-necked, 100-ml flask 0.5 gm (0.0012 mole) of poly(2,4,6-triphenylbenzyl methacrylate) (prepared by free-radical polymerization of the monomer with AIBN in toluene, $[\eta] = 0.013$ liters/gm at 20°) is suspended in 30 ml of glacial acetic acid.

The suspension is warmed in a stream of hydrogen to 60°C. At this temperature, over a 1-hr period, 2.5 gm (0.015 mole) of phosphonium iodide is gradually added, whereupon the polymer almost completely dissolves. Warming at 60°C in the hydrogen stream is continued for 4 hr.

After cooling to room temperature, the crude product is precipitated with approximately 300 ml of petroleum ether (b.p. 40°–60°C) and filtered off. The product is repeatedly reprecipitated from methanol solution with petroleum ether and from acetone with ether. The isolated product, a white powder with softening temperature between 230° and 260°C, weighs 80 mg (75% yield), of which 40 mg is soluble in hot glacial acetic acid.

Upon extraction with hot glacial acetic acid, approximately 40 mg of an acetic acid-soluble fraction is isolated upon precipitation with acetone.

NOTE: Since this procedure makes use of a reaction solvent in which the atactic polymer is said to be soluble and a reprecipitating solvent (acetone) in which the atactic polymer is said to be insoluble, this procedure needs to be reexamined. By the judicious selection of solvents, it may be possible to increase the yield substantially and to obtain an estimate of the ratio of syndiotactic to atactic (or isotactic) polymer.

B. Bulk Polymerization (Thermally or Chemically Initiated)

The heat of polymerization of acrylic acid is 18.5 kcal/mole and that of methacrylic acid 15.7 kcal/mole [24]. Perhaps, when these values are converted to 260 cal/gm for acrylic and 180 cal/gm for methacrylic acid, the hazard involved in bulk polymerizations of these monomers may be appreciated more readily. For this reason, only very small-scale bulk polymerizations should be attempted, with suitable protective measures.

Thermal initiation of acrylic acid polymerization has been reported to be very rapid at elevated temperatures. The resultant product is said to be partially insoluble as well as being partially degraded [25].

Preparation 2-7 is an example of the bulk polymerization using benzoyl peroxide as the initiator. The experiment should probably be carried out only on the scale indicated.

2-7. *Preparation of Poly(acrylic acid) in Bulk* [16]

In a hood, behind suitable shielding, in a modest-sized apparatus fitted with reflux condenser and addition funnel, 5 gm of acrylic acid containing a trace of benzoyl peroxide is cautiously heated to 130°C. When vigorous polymerization starts, as evidenced by refluxing, the source of heat is quickly removed. In several small portions 35 additional grams of acrylic acid are added at such a rate as to maintain the exothermic reaction. The reaction mixture gradually becomes rigid. Heating is continued for an additional 4 hr. The resulting polymer is somewhat rubbery in character and soluble in water. A 3% solution of this polymer in water is said to be readily prepared.

The bulk polymerization of glacial methacrylic acid has been carried out by heating 5 gm of the acid with 0.1 gm of 2,2'-azobis(isobutyronitrile) at 60°C for 0.5 hr. The resultant white mass was soluble in methanol, dioxane, and tetrahydrofuran [23].

C. Suspension Polymerization

The general principle of suspension polymerization has been discussed in Sandler and Karo [25a]. Since both acrylic and methacrylic acids are quite soluble in water, obviously the ordinary suspension polymerization procedures are not applicable for conversion of these monomers to their homopolymers (although they may be applicable to copolymerization systems). If the aqueous phase contains a fairly high proportion of dissolved electrolytes, the monomeric acids are "salted out." Then they may be dispersed by agitation, and suspension polymerization procedures applied. The resulting product is water insoluble but swellable with water. In aqueous solution of the alkaline hydroxides or of ammonia, these polymers are readily soluble.

The method is said to have the advantage over aqueous solution polymerizations in producing products of high molecular weight quite rapidly and in permitting the usual controls associated with suspension polymerization techniques. Thus, particle size may be varied by varying the level of electrolyte and by adding various suspending agents. The molecular weight may be controlled by varying the level of initiator and the polymerization temperature. Copolymers with compatible monomers may also be produced readily. Either monomer-soluble or water-soluble initiators may be used. The method is said to have the advantage over the aqueous solution polymerization of methacrylic acid in that the final product is not in a sticky and gelatinous form.

The technique is said to lend itself not only to the suspension polymerization of pure methacrylic acid added from an external source, but also by generating methacrylic acid *in situ* from its monomeric esters or amides by hydrolysis followed by the addition of electrolytes and the other reagents [14].

2-8. *Preparation of Poly(methacrylic acid) by Suspension Polymerization with a Monomer-Soluble Initiator* [14]

In a flask equipped with an explosion-proof mechanical stirrer with adjustable speed, an addition funnel, a thermometer, and a reflux condenser, 80 gm of anhydrous sodium sulfate is dissolved in 400 gm of water. Then 8 gm of a 10% aqueous solution of sodium poly(acrylate) is added as the suspending agent. After the addition of 100 gm (1.16 moles) of inhibited methacrylic acid containing 0.2 gm (0.0008 mole) of benzoyl peroxide, the reaction system is stirred continuously but not too rapidly and heated at 80°C with a water bath for approximately 2.5 hr. A small quantity of the 10% aqueous solution of sodium poly(acrylate) is maintained in the addition funnel throughout the process so that part of this reagent may be added if there are indications of lump formation.

After the initial heating period, the reaction temperature is raised to 90°–95°C for a short time. The solid product is collected on a nylon chiffon cloth and washed free of the sodium sulfate and the suspending agent with a 5% aqueous solution of hydrochloric acid. Then the product is dried under reduced pressure. The yield is essentially quantitative. The product is said to be a sandy mass, virtually ash free, and soluble in alkali hydroxides to produce highly viscous solutions.

A procedure which uses a water-soluble initiator similar to one to be described below as suitable for solution polymerizations in water is given in Preparation 2-9. Since this procedure also uses gradual addition of monomer, better control of the process is possible, although the product is described as somewhat lumpy.

2-9. *Preparation of Poly(methacrylic acid) by Suspension Polymerization with a Water-Soluble Initiator* [14]

In an apparatus similar to the one described in Procedure 2-8, a mixture of 400 gm of a 22% aqueous sodium chloride, 10 gm of a 10% aqueous solution of sodium poly(acrylate), and 0.25 gm (~0.001 mole) of potassium persulfate is heated with continuous stirring to 75°C. Then, gradually, 100 gm (1.16 moles) of methacrylic acid is added at the rate at which polymerization progresses.

After the addition has been completed, the temperature is raised to 95°C for a short period. The product is collected on a nylon chiffon cloth, washed free of the sodium chloride and the suspending agent with 5% aqueous hydrochloric acid, and dried under reduced pressure. The yield is quantitative. Besides sodium poly(acrylate), other suggested suspending agents are sodium isopropylnaphthalene sulfonate, talc, gelatin, gum tragacanth, water-soluble starches, barium sulfate, and other inorganic colloids or powders.

Monomeric methacrylic acid is soluble in benzene while poly(methacrylic acid) is not. This fact has led to a novel patented continuous slurry polymerization procedure [26]. A 5% benzene solution of methacrylic acid containing 1% the weight of the monomer of benzoyl peroxide is heated at reflux until a haze due to polymer formation is observed. Then further solutions of initiator-containing monomer are added at a modest rate while the temperature is maintained at 82°C. At the same rate, the formed slurry overflows onto a filter. Over a 72-hr period 85–90% conversion of the utilized monomer was observed. The product was a free-flowing, dustless powder.

D. Solution Polymerization

Monomeric acrylic and methacrylic acids are soluble in a large variety of organic solvents. However, the polymers derived from these monomers often exhibit limited solubility in these solvents. Frequently their solubility increases at low temperatures.

In Procedure 2-10 poly(methacrylic acid) is purified by using this phenomenon. In this case, since methanol is a theta solvent for the polymer near 26°C [12], the polymer is repeatedly dissolved in the cold solvent and reprecipitated by gradual warming to a temperature somewhat above 25°C. Poly(acrylic acid) is generally more soluble than poly(methacrylic acid) and may exhibit more normal temperature vs. solubility behavior in some solvents. In aqueous systems such factors as the pH of the medium and the concentration of dissolved electrolytes also influence the solubility characteristics of the polymers.

The polymerizations should only be carried out in solutions no more concentrated than 25–30% in monomer.

2-10. *Polymerization of Methacrylic Acid in Moist Dioxane* [3]

A solution of 30 gm ($\sim$0.35 mole) of methacrylic acid, purified by redistillation at 75°C (16 mm Hg) followed by fractional crystallization, and 0.15 gm (0.0006 mole) of recrystallized benzoyl peroxide in 70 gm of moist 1,4-dioxane is sparged by bubbling purified nitrogen through the solution. During this procedure the polymer forms as a swollen gel.

NOTE: This process may be accelerated by cautious warming on a water bath.

The swollen polymer is dissolved in methanol and precipitated with ether. This procedure is repeated three times to free the polymer from initiator residues and monomer.

A sample of this polymer is fractionated by dissolving 1 gm of polymer in 100 gm of methanol at 25°C and gradually adding ether. As a fraction precipitates, ether addition is stopped and the mixture is *cooled* with stirring to dissolve the polymer. The poly(methacrylic acid) is then reprecipitated by warming the solution slowly to 25°C with gentle stirring. The polymer is filtered off and dried under reduced pressure at 110°C. The filtrate, at 25°C, is treated with more ether to precipitate another fraction of polymer and purified by the same means as the first fraction. The procedure has been repeated until twelve fractions have been obtained.

Procedure 2-11 indicates the vigor with which the polymerization may proceed.

2-11. *Polymerization of Acrylic Acid in Toluene* [16]

With suitable protective devices, a solution of 10 gm ($\sim$0.14 mole) of acrylic acid and 0.1 gm (0.0004 mole) of benzoyl peroxide dissolved in 30 gm of toluene is heated. At the boiling point of the solvent, a very violent reaction takes place. Heating is discontinued. After 15 min the reaction mixture is cooled and the product, a voluminous white powder, is isolated by filtration. After drying at reduced pressure the yield is 9 gm (90% yield). The product is said to be water soluble.

Poly(acrylic acid) has also been prepared by reaction in dioxane which has been dried over sodium. In this case, the polymer is fractionated by precipitation from a 1% solution in 1,4-dioxane with heptane [27].

Acrylic acid has also been polymerized at 100°C in dibutyl ether, a solvent in which the polymer is insoluble [25].

Probably the most widely used method of polymerization of acrylic and methacrylic acids involves aqueous media. Under these circumstances, not only the usual factors such as the nature and concentration of the monomer, type and concentration of initiators, and temperature come into play, but also the pH of the medium and possibly the ionic strength [28]. An additional matter which has been reported upon quite recently and needs to be related to the older literature involves the tendency of acrylic acid to form "popcorn" or "proliferating" polymers under certain conditions.

Briefly summarized, the observations of Breitenbach and Kauffmann [2] are these. In a polymerization system, such as that of acrylic acid which is characterized by the formation of a polymer which is insoluble in its pure monomer, if the monomer cannot penetrate the polymer coil, no proliferating polymerization is possible. If, however, a solvent for the polymer, e.g., water, is present, there are concentration ranges in which the monomer can penetrate the polymer coil and popcorn polymers form. Under the experimental conditions [polymerization of acrylic acid containing 35×10^{-5} moles of 2,2'-azobis(isobutyronitrile) per mole of acrylic acid at 40°C along with some water], compositions containing 96.6 wt.% or more of acrylic acid exhibit no tendency to popcorn polymer formation. Systems containing between approximately 86% and 96% of acrylic acid proliferate only after polymerization times of 15–50 hr. At concentrations of 50–80% of acrylic acid, popcorn polymers form very rapidly (1.5–4 hr). At about 35% of acrylic acid the appearance of popcorn polymers occurs only after approximately 17 hr of polymerization. Unfortunately the behavior of 25–30% solutions of acrylic acid in water was not reported upon. This happens to be the concentration range usually recommended for practical polymer solution preparation.

It will be recalled that the polymerization of acrylic acid is accompanied by the evolution of 260 cal per gram of monomer. From the classical definition of calorie, one gram of monomer is therefore capable of raising the temperature of approximately 3.5 grams of water from room temperature to 100°C if the polymerization is sufficiently rapid and no allowance is made for warming the monomer itself. Therefore, to control the polymerization process, concentrations of 20–30% of acrylic acid in water are reasonable limits. In the case of methacrylic acid, somewhat higher concentrations might be used, but to be on the safe side it is best to limit this concentration to 25% also. If the concentration of monomer is too low, too much heat may be dissipated for efficient polymerization. On the other hand, as polymerization proceeds at the 20–30% concentration level, the viscosity of the solution may build up to such a level that proper agitation may not be possible. It has therefore been suggested [12] that the polymerizing system be diluted with water at approximately the reaction temperature to maintain a suitable viscosity without excessively cooling the mixture below a temperature which will stop the reaction.

The initiators used in aqueous solution polymerizations of the acrylic acids usually are the water-soluble initiators commonly used in emulsion polymerization, such as persulfate, percarbonate, and perphosphate salts. Monomer-soluble initiators have also been used, usually at sufficiently low concentrations or in the presence of water-soluble solvents to form a homogeneous system.

In a study of the polymerization of methacrylic acid in a solvent system consisting of water, dioxane, and ethanol, it was shown that at constant dielectric strength, the rate of polymerization increases with the concentration of water. On the other hand, at a constant concentration of water, the rate of polymerization was not significantly changed as the dielectric constant of the reaction medium was varied [29]. The ionic strength of the medium also seems to have, at best, only a slight influence on the rate of polymerization [28].

The effect of chloride ions on the rate of polymerization of acrylic acid needs further elucidation. Thus Katchalsky and Blauer [28] showed that chloride ions were strong inhibitors of the polymerization of methacrylic acid at a pH below 2.5, an effect not found with sulfate or nitrate ions. Yet their work was carried out in a buffer system consisting of sodium acetate and hydrochloric acid. A more recent patent, on the other hand, claims that the rate of polymerization of acrylic acid in the presence of potassium persulfate, 2-mercaptoethanol, the tetrasodium salt of EDTA, and a sulfonated castor oil was significantly increased when 1–8% of an alkali metal chloride was present [30].

Naturally the influence of the pH on the polymerization of the acrylic acids has received considerable attention. Representative studies are references 28, 31–33. The complexity of the situation can be grasped when we consider that at various pH levels, the reacting system may consist of the undissociated acid, undissociated monomeric radicals, undissociated macromolecular acids, undissociated macromolecular radicals, monomeric anions, monomeric anionic radicals, macromolecular anions, macromolecular anionic radicals, and macromolecular species consisting of undissociated acid anionic moieties —both as radicals and as nonradical materials. There may also be ion pairs, intermolecular and intramolecular hydrogen bonding, hydration of the various species, etc.

Since the polymeric acids are somewhat weaker acids than the monomers, the pH of an aqueous solution of methacrylic acid, for example, increases on polymerization. For this reason, the study of Katchalsky and Blauer [28] recommends the use of buffers to maintain reasonable constancy of pH during polymerization.

Up to a pH of 6, the rate of polymerization of methacrylic acid, in an aqueous system, initiated by hydrogen peroxide, decreases. At a pH of 6, the reaction velocity is virtually zero [28]. Below a pH of 5 the rate of polymerization is reasonably proportional to the fraction of the undissociated monomer [28,31]. At a pH range of 3.5–3.6, the overall reaction is first order with respect to monomer and proportional to the square root of the initiator concentration [28]. It is therefore postulated that below a pH of 5, the growing radical is essentially un-ionized and that the copolymerization between methacrylic acid and methacrylate ions can be neglected [31]. Upon extending this study to a pH of 12 [using 2,2′-azobis(isobutyronitrile) solubilized with a small quantity of ethanol], it was found that the polymerization rate goes through a minimum in the pH range of 6–7 and then increases to a flat maximum at a pH of approximately 11. The rate of polymerization is fairly constant between pH 9 and 12. This effect has been attributed to copolymerization of un-ionized and ionized monomers and radicals [31]. At pH values greater than 6, the propagation rate constant of acrylic acid increases. This is thought to be the result of the formation of ion pairs at the terminals of ionized macro-radicals [32–34].

Preparation 2-12 is an example of the polymerization of methacrylic acid in a buffered aqueous system using the apparatus of Baxendale *et al.* [34]— a flask fitted with a mercury sealed stirrer and an arrangement for bubbling nitrogen through the reaction medium. When samples are to be withdrawn from the flask, closure of a stopcock on a by-pass tube permits residual pressure in the flask to force the sample out through the nitrogen inlet tube. Provisions are also present for injecting the reagents into the reaction flask while the flask is filled with nitrogen.

2-12. *Polymerization of Methacrylic Acid in a Buffered Aqueous Solution* [28]

In a 500-ml reaction flask as described above is placed 250 ml of an aqueous solution 0.25 M in methacrylic acid and 0.20 M in an acetate buffer adjusted to a pH of 4.23 $\pm$ 0.03. To this solution is added enough 30% hydrogen peroxide to afford 2.45 gm of hydrogen peroxide. Nitrogen is passed through the solution with stirring, at room temperature, to remove dissolved oxygen. Then the reaction mixture, under a blanket of nitrogen, is placed in a thermostat at 75.0° $\pm$ 1.0°C. Thermal equilibrium is established in 30–35 min, and samples are withdrawn to study the rate of polymerization. The percentages of polymer formation are found to be approximately as given in Table V. At 75°C, the first-order rate constant is $K = (1.84 \pm 0.05) \times 10^{-4}$ sec^{-1}; the dependence of K on the initiator concentration is given by Eq. (2).

$$K \times 10^4 = 0.37 + 2.90[H_2O_2]^{1/2} \qquad (2)$$

It may be worth noting that just as in the case of acrylic emulsion polymerizations, ferrous sulfate has a catalytic effect on the polymerization of acrylic and methacrylic acids initiated by hydrogen peroxide. To prevent precipitation of ferric oxides, the system must be maintained at a very low pH [24].

A procedure which yields substantial laboratory quantities of poly(meth-

TABLE V

CONVERSION OF METHACRYLIC ACID ON
POLYMERIZATION IN A BUFFERED
AQUEOUS SOLUTION [28]

Time after initiation, t (min)	Percent polymer formed after time t
50	20
100	40
150	50

acrylic acid) of high molecular weight (e.g., 0.84–1.7 × 10^6 as determined by intrinsic viscosity measurements) is given in Procedure 2-13.

2-13. *Polymerization of Methacrylic Acid in an Aqueous Solution* [35,36]

In a suitable reaction vessel, a solution of 180 gm (2.1 moles) of freshly distilled methacrylic acid in 900 ml of distilled water is sparged with oxygen-free nitrogen. Then 5.4 gm (0.05 mole) of a 30% hydrogen peroxide solution is added. The mixture is maintained between 80° and 90°C for 5 hr. The gelled polymer system is allowed to cool to room temperature. The polymer is dissolved in distilled methanol and precipitated with diethyl ether. The resulting polymer is air dried, powdered, and then dried under reduced pressure. Yield: 178 gm (99%) of atactic poly(methacrylic acid).

NOTE: During the preparation and during polymer fractionation, it is important to avoid contact of the poly(methacrylic acid) solution with metallic objects.

The purification of 20% aqueous poly(methacrylic acid) solutions by dialysis against distilled water, followed by concentration of the solutions in an evaporator to a 2% water content and final drying, has been reported. The recovered polymer should be stored in sealed ampoules in the cold since there is said to be a slow degradation at room temperature [37].

The polymerization of methacrylic acid in the presence of hydrogen peroxide is said to give rise to polymers of relatively high molecular weight as measured by the viscosity of 8% aqueous solutions [38]. As indicated in Table VI, at constant reaction temperature the effect of initiator concentration on the viscosity is minor. The effect of reaction temperature at constant initiator concentration, somewhat unexpectedly, is an increase in viscosity with increasing temperature. Benzoyl peroxide exhibits similar effects. If we assume that such single point viscosity measurements are sufficient data to permit a correlation with the intrinsic viscosities of the polymers and, in turn, with their molecular weights, one explanation of this phenomenon would be that the level of initiator free radicals available for initiation is relatively constant at a given temperature in a water–methacrylic acid solution, and this concentration of initiator free radicals decreases as the temperature rises. In view of the thermal instability of these two initiators, this hypothesis is not entirely unreasonable.

Ammonium persulfate and sodium persulfate, on the other hand, seem to behave in a more conventional manner, i.e., as the concentration of initiator increases (at a fixed temperature), the molecular weight of the polymer decreases and as the temperature rises (at a fixed initiator concentration), the molecular weight decreases.

Procedure 2-14 is an example of the polymerization of methacrylic acid initiated by ammonium persulfate in a single batch operation, while Procedure

TABLE VI

COMPARISON OF INITIATOR EFFECTS ON THE VISCOSITY
OF POLY(METHYACRYLIC ACID) SOLUTIONS [38]

Initiator concentration (% based on monomer)	Polymerization temp. (°C; approx. 30 min reaction time)	Viscosity of 8% aqueous solution (poises)
Hydrogen peroxide		
1.0	90	0.08
2.5	90	0.062
5.0	90	0.115
5.0	85	0.075
5.0	90	0.115
5.0	95	0.13
Benzoyl peroxide		
0.25	90	0.09
0.50	90	0.10
1.0	90	0.105
1.5	85	0.10
1.5	95	0.14
Ammonium persulfate		
0.25	90	0.125
1.0	90	0.06
2.0	90	0.04
1.0	85	0.08
1.0	90	0.06
1.0	95	0.048
Sodium persulfate		
1.0	90	0.06
2.0	90	0.04

2-15 makes use of the gradual monomer addition technique. This latter approach potentially permits the preparation of poly(methacrylic acid) solutions of concentration levels greater than the usual 20–25% level.

2-14. *Batch Polymerization of Methacrylic Acid Initiated by Ammonium Persulfate* [38]

A flask fitted with addition funnel, reflux condenser, thermometer, and mechanical stirrer is charged with 90 gm of distilled water. The water is heated to 90°C, 0.2 gm (0.0009 mole) of ammonium persulfate is added, and, after the salt has dissolved, 10 gm (0.12 mole) of methacrylic acid is added with constant stirring. The temperature is maintained, as nearly as possible, at 90°C. After heating for 5 min, the exothermic phase of the reaction starts and may rise to 96°C without additional heating. Since the polymer is less soluble in hot water than in cold, it precipitates from the solution. After the reaction temperature drops, heating is continued at 90°C for a total of 30 min from the time of initiation. Then the product solution is allowed to cool, whereupon the polymer goes back into solution. The resulting solution, diluted to 8% poly(methacrylic acid), has a viscosity of 0.04 poises at 25°C.

2-15. *Polymerization of Methacrylic Acid Initiated by Ammonium Persulfate, Gradual Addition Method* [38]

In a 2-liter flask fitted with a mechanical stirrer, reflux condenser, and an addition funnel, 840 gm of water is heated to 100°C. Then 3.2 gm (0.014 mole) of ammonium persulfate is added with stirring. After 2 min, the addition of 160 gm (1.86 moles) of methacrylic acid is started at such a rate that the addition is completed in 53 min. During this time, the temperature of the reaction system is maintained at 99°–100°C. After the addition has been completed, heating is continued for 10 min. The viscosity of an 8% poly(methacrylic acid) solution prepared from the product solution is 0.025 poises at 25°C.

Similar polymerizations have been carried out at monomer concentrations of 0.2 M/liter using potassium persulfate as the initiator for the polymerization of methacrylic, acrylic, and itaconic acids at 50°C for 5 hr under reduced pressure. In this case, the solid acids were isolated by precipitation from the aqueous solution with ethyl acetate and reprecipitated from a methanol–ethyl acetate system [39].

The polymerization of aqueous acrylic acid in the presence of potassium persulfate is said to be somewhat anomalous [40]. At 100°C, the relationship between the original initiator concentration $[I_0]$, the initial monomer concentration $[M_0]$, and the residual fraction of unpolymerized monomer U in the range of $[M_0] = 0.01$–0.78 M/liter was found to be

$$[I_0] = 0.00185 \{\arctan (60[M_0]) - \arctan (60[M_0]U)\} \tag{3}$$

Equation (3) is said to imply that, at least up to an initial initiator concentration of 29×10^{-4} M/liter, the process is a "limiting polymerization," also called a "dead-end" polymerization. The phenomenon is said to relate to the competition between the termination and propagation steps, possibly at different reaction orders [40].

Another interesting difference in polymerization behavior between methacrylic acid and acrylic acid is the remarkable observation that the reducing agent sodium bisulfite will cause the initiation of the polymerization of methacrylic acid while it does not initiate the polymerization of acrylic acid. It has been suggested that sodium bisulfite and methacrylic acid form a redox couple whereas acrylic acid and sodium bisulfite undergo some sort of addition reaction to form nonradical products [41].

The use of persulfate–bisulfite redox couples to initiate acrylic acid polymerizations has been reported [4,21]. For example, at 60°C, to a solution of 232 gm of water, 0.50 gm (0.0019 mole) of potassium persulfite, and 0.2 gm (0.0017 mole) of potassium metabisulfite, 167 gm (1.16 moles) of a 60% aqueous solution of acrylic acid has been added over a 0.5-hr period, followed by an additional heating period of 0.5 hr to produce a 25% solution of poly(acrylic acid) [4]. While operating in a system which had been freed of dissolved oxygen and while maintaining an inert atmosphere, polymerizations have been carried out at 0°–10°C if a few hundredths of a percent of ferric ions were present in the redox polymerizations system [4].

A potassium permanganate and oxalic acid redox system has been used to initiate the polymerization of acrylic acid in aqueous solutions [42]. However, it is doubtful that this system is of more than theoretical interest in view of the possible formation of manganese dioxide or of the discoloration due to an excess of permanganate ions.

The use of sodium hypophosphite ($NaH_2PO_2 \cdot H_2O$) along with hydrogen peroxide, particularly in the presence of copper salts as chain-transfer agents, results in the formation of poly(acrylic acid) with exceptionally low concentrations of residual monomer and of particularly good storage stability as characterized by a low degree of gel formation [43].

2-16. *Polymerization of Aqueous Acrylic Acid (Hydrogen Peroxide, Sodium Hypophosphite, Cupric Acetate System)* [43]

A solution of 647 gm of distilled water, 0.25 gm (0.1% on monomer) of sodium hypophosphite monohydrate, 0.047 gm (0.19% on monomer) of cupric acetate monohydrate, and 3.75 ml (0.5% on monomer) of 30% hydrogen peroxide is heated with steam. The solution is purged with oxygen-free nitrogen; then 250 gm (3.47 moles) of glacial acrylic acid is added with stirring. When the internal temperature reaches 82.5°C, the polymerization is initiated and the temperature rises rapidly. After the exothermic stage of the reaction subsides, steam heating is continued for 1 hr. The resulting solution is homogeneous, total nonvolatiles are 29%, unreacted monomer is 0.20%, conversion is 99.3%, and the absolute viscosity at 20°C is 62.7 poises.

By varying the concentration of sodium hypophosphite and of the copper salt, the molecular weight distribution of the polymer may be controlled readily. Among the copper salts suggested are the sulfate, chloride, selenide, formate, and lactate, as well as the acetate.

Both acrylic and methacrylic acids have also been polymerized in water solution with small quantities of 2,2'-azobis(isobutyronitrile) at pH levels up to pH 13. If necessary, a few percent of ethanol has been added to stabilize the initiator in the system [31,44,45]. Solutions of benzoyl peroxide in methacrylic acid dissolved in water have been polymerized [46]. The addition of small amounts of sodium hydroxide either prior to or after polymerization, is said to result in a product with enhanced water solubility. Other "conditioning agents" which may be used instead of sodium hydroxide are such bases as the carbonates of the alkali metals or of ammonia, pyridine, or ethanolamine [46]. A surfactant such as 1%, by weight, of a sulfonated lauryl alcohol may be added to a 5% aqueous solution of poly(methacrylic acid) prior to spray drying at 80°C. This is thought to prevent the agglomeration of particles.

E. Charge-Transfer Complex Polymerization

A study of the copolymerization of acrylic acid and acrylamide in the presence of zinc chloride and 2,2'-azobis(isobutyronitrile) (AIBN) in benzene showed that alternating copolymers form. Since copolymers of these two monomers prepared by ordinary free-radical methods normally produce random copolymers, it is presumed that charge-transfer complexes are involved when zinc chloride is present in the system [47].

In benzene solution, the system is heterogeneous. When no AIBN is present, no polymer forms. In the absence of benzene, the acrylamide–acrylic acid–zinc chloride compositions form homogeneous compositions. However, even after 2 months at room temperature, no evidence of polymerization was noted. To form polymers, AIBN and a temperature of 45°C are necessary.

2-17. *Charge-Transfer Complex Polymerization of Acrylic Acid–Acrylamide* [47]

In a flask fitted with a mechanical stirrer, condenser, thermometer, and nitrogen inlet tube, a solution of the monomers given in Table VII are dissolved in 50 ml of benzene and sparged with nitrogen. While the solution is stirred, the requisite quantity of zinc chloride and 300 mg of AIBN is added. Then the reaction flask is placed in a thermostatted water bath at 45° ± 1°C.

After the desired reaction time, the polymerization is stopped by cooling to 0°C. The copolymer is precipitated by addition to a solution of 300 ml methanol, 10 ml of water, and 1 ml of concentrated hydrochloric acid. The copolymers are purified by dissolving them in water and reprecipitating with methanol.

F. Radiation-Induced Polymerization

The polymerization of both acrylic and methacrylic acids has been initiated by ultraviolet, visible light, and by γ-radiation. Some of these preparations

TABLE VII

COPOLYMERIZATION OF ACRYLIC ACID AND ACRYLAMIDE IN THE PRESENCE OF ZINC CHLORIDE [47][a]

Monomer charge			Reaction temp. (°C)	Reaction time (min)	Conversion (%)	Percent acrylic acid in polymer	$\bar{M}_v \times 10^{-5}$ [b]
Acrylic acid (moles)	Acrylamide (moles)	Zinc chloride (moles)					
0.10	0.10	—	45	5	20.6	66.2	1.55
0.10	0.10	0.10	45	5–7	61.9	50.5	1.20
0.10	0.10	0.50	45	5–6	10	47.0	4.79
0.15	0.10	0.50	49	5–6	14.5	53.7	3.89
0.10	0.15	0.50	55	8	19.9	44.2	6.30
0.05	0.45	0.50	—	—	—	47.4	17.4

[a] Solvent: 50 ml of benzene; initiator 300 mg of AIBN.
[b] Viscosities were determined in 1 M sodium nitrate at 30°C; the relationship of intrinsic viscosity and molecular weight, $\bar{M}_v$, used was $[\eta] = 3.73 \times 10^{-4} \bar{M}_v^{0.66}$.

have been carried out with the liquid monomers, with solutions, and with frozen monomers (i.e., in the solid state).

The polymerization in the near ultraviolet (at approximately 300 nm) may involve the decomposition of benzoyl peroxide, present in the monomer, to produce free radicals which in turn initiate the process. In the same wavelength range, acidified ferric ions will also produce free radicals capable of initiating polymerization.

Dyes have been used to extend the effective range of wavelengths into the visible region. In the absence of oxygen, the quantum efficiency was found to be low. However, a system consisting of a dye like Rose Bengal in the presence of ascorbic acid and oxygen enhances the quantum yield considerably [48]. Other dyes which may be used are fluorescein and its halogenated derivatives —eosin, phloxine, and erythrosine. While ascorbic acid or phenylhydrazine hydrochloride are among the more effective reducing agents used in this system, hydroxylamine, hydrazine sulfate, thiourea, alkylthiourea, glutathione, and cysteine may be used. In the absence of oxygen, the dyes are reduced to their *leuco* form without initiating polymerization. When oxygen is present, semiquinones are produced along with hydroxyl free radicals.

Acriflavin in the presence of freshly prepared, acidified stannous chloride and oxygen has been found suitable. The monomer itself may be dissolved in water, acetone, or methanol.

A particularly interesting observation is that if the monomer–dye–reducing agent–oxygen system is exposed to the light of a flash bulb, followed by storage in the dark, polymerization starts after an induction period, as a result of a "dark reaction" between the reduced dye and oxygen [48].

Diazidotetraminecobalt(III) azide—$[Co(NH_3)_4(N_3^-)_2]^{3+}(N_3^-)_3$—has been used as a sensitizer for the polymerization of acrylic acid in aqueous solution upon irradiation by monochromatic light with wavelengths of 385, 405, and 435 nm. The mechanism of polymerization is said to involve:

1. A primary photochemical excitation of the complex followed by a dark reaction of electron transfer within the complex to produce azide radicals.

2. Initiation of polymerization by these azide radicals.

3. Termination of the chain process by the complex molecule [49].

The polymerization of crystalline acrylic acid or methacrylic acid at a temperature only slightly below their respective melting points with ultraviolet radiation gave rise to noncrystalline and disoriented polymers. While under these reaction conditions the solid methacrylic acid contained measurable concentrations of free radicals, acrylic acid contained no detectable free radicals. The formation of free radicals in methacrylic acid was found to be very temperature dependent. As the temperature was lowered, the concentration of free radicals decreased.

As a consequence of the presence of free radicals of finite half-life in methacrylic acid, it was postulated that a postirradiation reaction might be observed during solid state polymerizations. No such observations were made with acrylic acid.

With small mechanical compressive forces, the polymerization could be retarded or stopped altogether. This suggests that the polymerization takes place in association with crystal dislocations which are displaced under mechanical stresses [50].

Other work on the ultraviolet radiation-induced polymerization in the solid state is also discussed in references [51–53].

A brief description of a solution polymerization of acrylic acid is given in Procedure 2-17 [25,54,55].

2-18. *Ultraviolet-Initiated Polymerization of Acrylic Acid in Solution* [25]

With suitable protection of personnel against the effects of ultraviolet radiation, in a flask wrapped in black paper to protect it from light, a quantity of purified acrylic acid dissolved in two volumes of ethanol is sparged with oxygen-free nitrogen for 2 hr to eliminate oxygen. Benzoin (at a level of 13 mg per 1 gm of monomer) is added as a photosensitizer and rapidly dissolved. The solution is cooled at $-78°C$ and irradiated at a distance of 10 cm with a Phillips SP-500 lamp, which is said to be nearly monochromatic at a wavelength of 360 nm.

After 5 min, the monomer is converted to an extent of 50%. The radiation source is turned off; the solution is brought to room temperature. Water is added to reduce the viscosity. The benzoin is filtered off. Concentrated hydrochloric acid is added and the product solution is cooled at 0°C for 24 hr. The polymer is then filtered off and dried at 60°C in the presence of sodium carbonate. For further purification, the polymer may be dissolved in water and lyophilized. The product is syndiotactic poly(acrylic acid).

Acrylic and methacrylic acids have been grafted to polypropylene films [56] and polyester (terylene) fibers [57]. Treatment of the substrate with ferrous ions seems to favor the grafting with acrylic acid while cupric ions favor grafting with methacrylic acid.

The polymerization of methacrylic acid in solution initiated by γ-radiation also produces polymer with considerable syndiotactic character [25,58,59].

2-19. γ-*Radiation-Induced Polymerization of Methacrylic Acid* [58]

In a suitable ampoule, one volume of freshly distilled, inhibitor-free methacrylic acid dissolved in two volumes of anhydrous methanol is degassed by twice freezing in liquid nitrogen and evacuating. The solution is finally sealed off under reduced pressure, warmed briefly to about 0°C and, after complete melting, is cooled to Dry-Ice temperature.

The material is then irradiated for a total dose of 7.1 megarads in a cobalt 60 γ-ray source with a does rate of 0.16 megarads/hr.

The resulting viscous solution, which is slightly hazy, is warmed to room temperature and, after cautious opening of the ampoule, is poured into hexane. The precipitated polymer is collected and dried under reduced pressure. Yield: approximately 35%.

The polymer is methylated with diazomethane in benzene, and the tacticity of the resulting poly(methyl methacrylate) is studied by NMR spectroscopy. The polymer exhibited 85% syndiotactic triads. Poly(methacrylic acid) prepared at 60°C in methyl ethyl ketone solution with AIBN was 57% syndiotactic.

In the liquid state, solvents have a significant effect on the stereoregularity of poly(methacrylic acid) prepared with γ-radiation [59]. In a given solvent (methanol, 1-propanol, and isopropanol were the only solvents considered in this work) the percentage of syndiotactic polymer increases with decreasing polymerization temperature. At a fixed temperature, syndiotacticity increases as the bulkiness of the solvent increases. In 1-propanol, at a given temperature, the fraction of syndiotactic triads increases as the monomer concentration decreases. Polymers with as high as 95% syndiotactic character have been prepared [59].

In a study of the solid-state postpolymerization of methacrylic acid initiated with γ-radiation, distinctly different polymerization characteristics were found when rapidly crystallized (shock crystallized) monomer was compared with slowly crystallized methacrylic acid.

Slowly crystallized methacrylic acid gives rise to a polymer with a narrow molecular weight distribution and a linear dependence of molecular weight on polymer yield. The overall polymerization rate of slowly crystallized monomer is substantially faster than that observed for shock crystallized monomer. These results are attributed to changes in the polymerization mechanism characterized by essentially no termination step and consequently independent chain propagations. In shock crystallized monomers termination is attributed to occlusion of propagating radicals and defects in the crystal structure.

The polymers produced from slowly crystallized monomer are largely atactic, containing a secondary kind of stereosequencing. A gradient of stereosequencing along the chains was considered on the basis that the probability was highest for *meso*-triads forming at the initial end of a chain. The probability of *meso*-triads forming decreases with chain growth. The growing polymer chains introduce defects into the monomer crystals. Consequently, a gradual change in the stereoregularity of the polymer takes place [60].

The polymerization of acrylic acid when initiated by γ-radiation leads to observations which vary significantly from those made on the polymerization of methacrylic acid [61–63]. The hypotheses of Chapiro are such that even polymerizations initiated by the thermal decomposition of free-radical generators should be reexamined in their light.

The tacit assumption in most kinetic treatments of reactions is that the reacting species are randomly distributed throughout the system. This postulate allows the conventional use of reactivities or concentrations of the reactants in the mathematical treatment. The "normal" kinetic relationships do not hold if molecular aggregates are present. The fact that carboxylic acids form aggregates by hydrogen bonding is well known. Acids primarily form dimers although equilibrium concentrations of linear multimolecular structures are also present.

The dimeric structure of acrylic acid may be represented by structure **I**.

I

The multimolecular structure would be structure **II**.

II

Chapiro's structure **II** is probably somewhat oversimplified; e.g., helical aggregates may well be involved. However, the main point is that there is an orientation of vinyl groups. Under favorable conditions this *could* account for the observed fact that poly(acrylic acid) forms stereoregular polymers readily, and that in neat (both in solid phase and in liquid phase) and in certain polar solvents the rate is extremely fast, leading to polymers of molecular weights in the range of $1-3 \times 10^7$.

In the liquid phase there is a significant postpolymerization stage while γ-irradiated acrylic acid in the solid phase exhibits only a short postpolymerization stage. Unlike in previously reported data, there is evidence that free radicals are trapped in the crystalline monomer at those temperatures at which polymerization occurs.

Polar solvents even at temperatures as high as 76°C usually lead to syndiotactic polymers, as does polymerization of the liquid monomer. In this regard, using acetic acid as a solvent is less effective in the production of syndiotactic poly(acrylic acid) than nonacidic polar solvents. This may be caused by acetic acid entering into the multimolecular aggregates and thus interfering with the orientation favorable for such conformation. Carefully dried monomer exhibits a reduced rate of polymerization upon irradiation and forms an atactic polymer. Some spontaneous polymerization does, however, take place when dry acrylic acid is degassed.

Benzoquinone as well as acetone inhibits the radiation-induced polymerization of acrylic acid.

Polymers produced in toluene, hexane, or chloroform, even when the concentration of monomer is 70–80%, are essentially not syndiotactic. The polymers produced in the solid phase consist of an atactic fraction (i.e., a dioxane-soluble fraction), a syndiotactic fraction (i.e., a fraction soluble in dioxane-water mixture), and an insoluble gel.

The behavior of methacrylic acid is quite different. Evidently the aggregates formed either in bulk or in solution do not affect the polymerization kinetics. Perhaps the methyl group causes irregularities in the orientation of the vinyl groups either by steric or by hydrophobic repulsion.

3. POLYMERIZATION OF ITACONIC ACID

The polymerization of itaconic acid seems not to have been studied very extensively although industrial applications in copolymer systems appear to be of considerable interest. In view of the work of A. Katchalsky and co-workers [28] on the effect of pH on the polymerization of acrylic acid and methacrylic acid, analogous research on pH effects on itaconic acid reactions has been carried out to a limited extent [64]. Typically, with a persulfate initiator in an aqueous solution at 50°C, the monomer is converted to an extent of 85–90% to its homopolymer within 35–45 hr. In the pH range of 2.3–3.8, the rate of polymerization is constant. As the pH increases, the rate becomes progressively slower and stops completely at a pH of 9. Generally, the last 5–10% of the monomer seems to be difficult to convert to polymer.

Preparation 3-1 is one of the early examples of the polymerization in a strongly acid system. This polymer evidently seemed to contain only 40% of the expected carboxylic acid groups and its elementary analysis deviated considerably from calculated values [65].

3-1. *Bottle Polymerization of Itaconic Acid* [65]

In a 4-oz. screw-cap bottle, sealable with a Buna-N gasket, are placed 50 ml of 0.5 N hydrochloric acid (prepared by diluting concentrated hydrochloric acid with distilled water that has been deaerated by heating to vigorous boiling and cooling while bubbling oxygen-free nitrogen through it), 20 gm (~0.15 mole) of itaconic acid, and 0.10 gm (0.00037 mole) of potassium persulfate.

The bottle is flushed with oxygen-free nitrogen to remove air, sealed, covered with a protective steel jacket, and placed in a thermostatted bath at 50°C for 68 hr. During this treatment some of the itaconic acid remains undissolved.

The bottle is cooled to room temperature. The solution is slowly added to well-stirred acetone to precipitate the polymer. The product is filtered off, redissolved in water, and reprecipitated with acetone. After drying under reduced pressure, 7 gm of the polymer is isolated (35% yield).

The polymer is described as soluble in water and methanol but insoluble in ethanol and other common solvents. The product, if isolated by freeze drying, contains one molecule of water per repeat unit. On drying at 100°C and 0.1 mm Hg the water is lost but some anhydride formation takes place.

Poly(itaconic acid) has also been prepared in a 0.2 M/liter aqueous solution using potassium persulfate at 50°C over a 5-hr period under reduced pressure. After the polymer is reprecipitated twice into methanol–ethyl acetate, a polymer is isolated with a molecular weight of 1.64×10^5, determined by vapor pressure osmometry of a methanolic solution of the methyl ester prepared from the polymer [39]. Unfortunately Tsuchida and co-workers did not report on the quantitative extent to which poly(methyl itaconate) had been formed from this polymer (presumably by reaction with diazomethane). Consequently, there is little in the literature to confirm or dispute the paper by Braun and Aziz el Sayed [66] which offered evidence that during the free-radical polymerization of itaconic acid, carbon dioxide evolves to a considerable extent. During the process, it seems that hydroxyl and formyl radicals are generated and incorporated in the macromolecule. It is proposed by these authors that the homopolymer of itaconic acid contains virtually no itaconic acid repeat units but rather intramolecular lactone rings and acetal- or hemiacetal-like moieties. Since the polymer remains soluble in the reaction solvent (dioxane), we may presume that no cross-linking by intermolecular acetal or ester formation takes place.

The experimental data for the polymerization of itaconic acid in dioxane with initiators which are soluble in organic solvents are given in Table VIII. The polymers prepared at 50°C were described as being slightly yellow in color and soluble in water, methanol, dioxane, dimethylformamide, dimethyl sulfoxide, formamide, and ethanol. Copolymers prepared at room temperature were colorless and had similar solubility behavior except that they were insoluble in cold dioxane but soluble in dioxane when heated to 70°C. The polymers were insoluble in petroleum ether and in methyl ethyl ketone.

Returning to the polymerization of itaconic acid in an aqueous system, Tate [67] indicates that a 32% solution of itaconic acid in water with 2% of potassium persulfate (calculated on the monomer), upon reaction at 60°C for 47 hr gave 93% conversion (by titration; we presume this to be a double

TABLE VIII
POLYMERIZATION OF ITACONIC ACID IN DIOXANE SOLUTION
IN A NITROGEN ATMOSPHERE [66]

Itaconic acid (moles/liter)	Vol. of dioxane soln. used (ml)	Initiator		Reaction conditions		Yield (%)
		Type[a]	Quantity (%)[b]	Temp. (°C)	Time (hr)	
0.709	50	B	2.15	20	720	54
0.909	550	B	2.00	20	625	54
1.83	21	B	2.15	62	51	85
3.84	20	B	0.537	62	98	38
0.384	200	B	0.537	72	192	14
3.84	20	B	0.268	72	159	44
5.124	15	B	0.537	72	127	69
0.768	200	A	1.58	20	720	48
0.768	200	A	2.38	20	696	50
0.768	200	A	3.16	20	720	52
3.84	200	A	0.174	60	142	65
3.84	20	A	0.398	60	142	47
3.84	20	A	0.833	60	142	84

[a] B = Dibenzoyl peroxide; A = 2,2′-azobis(isobutyronitrile).
[b] In mole% calculated on itaconic acid.

bond determination for unreacted monomer). Partially neutralized itaconic acid (64% acid in water, diluted to 36% during the reaction, and 40% neutralized with sodium hydroxide) on polymerization in the presence of 1% of potassium persulfate (calculated on itaconic acid) at 60°C for 48 hr afforded only a 72% conversion (by titration).

In various solvents, the rate of polymerization increases markedly with increased pressure in the range of 5000 kg/cm^2 (71,100 psi). The polymers are described as white, brittle, somewhat hygroscopic, soluble in methanol, dimethylformamide, and water, but insoluble in other organic solvents. The polymer is stable to 200°C. Elemental analysis indicates that decarboxylation may have taken place since the basic polymer structure seems not to involve itaconic acid repeat units [68].

Reviews of interest in the field of itaconic acid chemistry are references 11, 67, 69.

4. POLYMERIZATION OF SALTS OF THE ACRYLIC ACIDS

It is self-evident that salts of the poly(acrylic acids) may be produced either by the neutralization of the poly(acrylic acids) with bases, i.e., hydroxides, methylates, oxides, carbonates, bicarbonates, basic salts, etc., or by the polymerization of preformed salts of the monomeric acids, either in solution or in the solid state. The neutralization of polymeric acids with polymeric bases such as poly(vinylpyridine) should also be considered.

Particularly in the case of neutralization of polymeric acids and of copolymers containing such acids, a wide range of products may be formed by partial neutralization of the carboxylate function.

Technological applications of these salts are varied and, to some extent subject to temporary fads. For example, in the early 1950's, because of wartime requirements during World War II and the Korean conflict, much effort was expended on soil modification by treatment of swampy areas with calcium and magnesium acrylates followed by polymerization *in situ*. Today, twenty to thirty years later, one hardly hears of this matter. Similarly, from time to time, silver acrylate is considered for photographic applications, but no substantial progress seems to be made on this idea. At the moment, a flurry of excitement has been caused by the recently suggested use of tin methacrylates or acrylates in antifouling paints for ships.

On the other hand, the incorporation of methacrylic or acrylic acid in emulsion copolymers used in floor polishes has a long history. Among the properties leading to the use of coatings containing these components is their increased water solubility upon treatment with caustic. Thus the polish coating may be removed with soap and water.

The acrylates and methacrylates of divalent metals will form insoluble polymeric salts. Similarly, treatment of the polymeric acids with bases derived from divalent elements may result in cross-linked systems. Such cross-links have been termed "temporary" cross-links as distinguished from "permanent" ones, i.e., those containing covalent carbon–carbon bonds [70]. It may very well be that pigments such as titanium dioxide are bonded in an emulsion polymer system by such temporary cross-links.

If we consider all the metallic elements, the amphoteric elements, the various possible valence states of many elements, ammonia and the amines, and other organic bases, both monomeric and polymeric, the possible salts which might be discussed are obviously quite extensive. Consequently we will make mention of a somewhat random selection here.

In Section 2,D on the effect of pH on the polymerization of methacrylic and acrylic acids it was pointed out that the polymerization and copolymerization of the unsaturated anions is substantially slower than that of the un-ionized acids.

In an aqueous solution of sodium acrylate at a pH of 7.2, polymerizations have been carried out with various quantities of ammonium persulfate at $20°–90°C$ for times up to 6 hr. The rate constant of the polymerization was found to be proportional to the concentration of the initiator and to the square of the initial monomer concentration. In the range of 40°–60°C the activation energy was found to be 16.7 kcal/mole [71].

Aqueous solutions of salts of acrylic acid which may contain between 0.5% and 3.0% of a water-soluble initiator have been sprayed into a chamber in which the air is heated between 150° and 485°C. The polymer forms rapidly by this spray drying–polymerization process [72]. Solutions of mixtures of salts of acrylic acid have also been dripped onto oppositely rotating, steam-heated rolls. The dried powder could be scraped from the rolls before the end of a revolution of the roll. In this procedure, 0.5% on the monomer of a persulfate salt was added to the monomer solution. The rolls were 1 ft. in diameter, 4 mils apart, rotating at 5.5 rpm. Table IX gives some observations made on this method of polymerization [73].

By subjecting aqueous solutions of sodium acrylate in concentrations of 25–40 wt.% to high-energy ionizing radiation, polymers were prepared at pH's from 4 to 14. The products were aqueous solutions or gels. The intrinsic viscosities for the product solutions ranged upward from 6 dl/gm at 25°C in 2 N sodium hydroxide [74]. The dosage rate of these polymerizations was less than 200,000 rads/hr for a total of from 1000 rads to that sufficient to convert substantially all of the monomer to polymer.

A study of the solid-state polymerization of the acrylates and methacrylates of several alkali metals was carried out in evacuated sealed tubes. Samples were irradiated at -78°C with a ^{60}Co source (a condition under which no polymerization takes place), stored overnight at Dry-Ice temperature, and

TABLE IX

POLYMERIZATION OF SOLUTIONS OF ACRYLATE SALTS[a]

Conc. of aqueous solution (%)	Monomer composition		Roll temp. (°C)	Yield (%)	Viscosity of 1% aqueous solution (cP)
	Component A (%)	Component B (%)			
38.3	Calcium acrylate (15)	Sodium acrylate (85)	—	97.8	136
38.3	Calcium acrylate (15)	Sodium acrylate (85)	148	97.2	160
40	Calcium acrylate (10)	Potassium methacrylate (90)	127	91	—
40	Calcium acrylate (18)	Ammonium acrylate (82)	127	92	—

[a] Dripped onto steam-heated opposing rolls, 1 ft. in diameter, 4 mils apart, rotating at 5.5 rpm, monomer solution containing 0.5% of ammonium persulfate (based on monomer) [73].

then placed in thermostats at temperatures varying from 0° to 155°C. It was found that oxygen retards the polymerization of potassium acrylates irradiated with 0.2 Mrad but that polymerization at 25.3°C did ultimately take place at approximately half the rate as that observed in evacuated tubes at the same polymer yield.

The polymerization of potassium acrylate proceeded more rapidly than that of any of the monomers studied in this work. At 0°C, potassium acrylate polymerized more rapidly than lithium acrylate at 101°C and sodium acrylate at 120°C.

The reaction rates of all the acrylates decay rapidly at low conversions. The chain length of poly(potassium acrylate) is one order of magnitude greater than that of poly(lithium acrylate).

In the methacrylate series, sodium methacrylate reacts more rapidly than potassium methacrylate. Lithium methacrylate is entirely inert. Initially the polymerization of sodium methacrylate exhibits an accelerated reaction stage.

It is postulated that the difference in polymerization rates of the same unsaturated anions in the presence of different cations must be caused by the geometry of the crystal lattices involved. There is also evidence that the polymers lie in an amorphous phase while the reactive ends of the growing chains are anchored in the monomer lattice [75].

Calcium acrylate dissolves readily in water to give a reasonably stable solution at room temperature. At 24°C, 44 gm of calcium acrylate will dissolve in 100 gm of water. The freezing point of such a solution is approximately −12°C [76].

The interesting phenomenon about this monomer which finds application in soil stabilization, is the insolubility of the polymer in water although water dissolves in the polymer. Under equilibrium conditions, at room temperature, the hydrated polymer contains 25% of water. This polymer is rigid. Higher water concentrations may be achieved. The resultant swollen polymers, depending on the water content, may be mechanically hard, semirigid, soft, or very soft and rubberlike. The swollen polymers are not water permeable.

As expected, the polymeric calcium acrylate may act as an ion-exchange resin. If monovalent cations are exchanged for calcium ions, the ionic cross-linkages are destroyed and water-soluble polymers such as poly(sodium acrylate) may form. Clearly this would interfere with proper soil stabilization [76].

The polymerization of calcium acrylate may be initiated with typical redox systems such as ammonium persulfate–sodium thiosulfate and certain peroxides with reducing agents such as sodium thiosulfate. Hydroxylamine hydrochloride, hydrazine hydrate, and tetramethylenepentamine are also suitable reducing agents with certain oxidizing agents [77].

By the appropriate selection of initiator concentration, gel times may be varied from less than a few minutes to several hours.

The ratio of ammonium persulfate to sodium thiosulfate is not very critical. Approximately equal amounts are usually used.

In the presence of a sandy soil, 5% of initiator based on the quantity of calcium acrylate in the soil will cause gelation in about 80 min; 20–25% of initiator causes gel formation in 6–8 min; and 50% of initiator affords a gel time of about 1 min.

The quantity of water present in the soil influences results. Water levels of up to 25% of the soil weight have been treated satisfactorily with 10% of calcium acrylate.

Strontium and zinc acrylates and calcium methacrylates have also been suggested for soil stabilization [77].

Barium acrylate may be polymerized in solution by dye-sensitized photo-initiation in the presence of a reducing agent. Typical dyes are phenothiazine dye photooxidants such as methylene blue or thionine. The reducing agent may be a sulfinic acid such as *p*-toluenesulfinic acid [78].

4-1. Photopolymerization of Barium Acrylate [78]

(*a*) *Preparation of Sensitizer Solution.* In the dark, to 90 ml of distilled water are added with stirring 2.15 gm of sodium *p*-toluenesulfinate and 0.03 gm of methylene blue (Color Index No. 52015). Stirring in the dark is continued for 24 hr. The solution is then made up to 100 ml with distilled water. The solution is stored in the dark.

(*b*) *Polymerization.* To 4.0 ml of a 40% solution of barium acrylate in distilled water is added, in the dark, 1.0 ml of the sensitizer solution. This mixture is placed in front of a 500-watt projector whose lens is covered with a red filter. The projector is turned on and the course of the polymerization may be followed by the increase in opacity as colloidal polymer particles form.

Hydrogen- and deuterium-atom bombardment have been used to initiate the polymerization of barium methacrylate dihydrate. The process is said to be initiated by the diffusion of hydrogen atoms in solid-state barium methacrylate dihydrate. By this method, large amounts of polymer form very rapidly, unlike in polymerizations initiated by γ-radiation [79]. γ-Radiation has been used in the solid-state polymerization of potassium acrylate, calcium acrylate, barium acrylate, and uranyl acrylate [80].

Poly(methacrylic acid) in aqueous or ethylene glycol solution has been treated with bases such as nicotine, ammonia, *n*-butylamine, isobutylamine, cyclohexylamine, ethanolamine, and piperidine to form polymeric salts. Usually the salts seem to consist of one amine molecule for two carboxylate units. The polymeric salts were isolated by evaporation of the solutions. All materials were amorphous by X-ray diffraction. With the exception of the isobutylaminium and the ammonium salts, the structures have regular shapes and are optically isotropic [36].

Complexes of poly(acrylic acid), poly(methacrylic acid), and poly(itaconic acid) with poly(cations) have been studied. These complexes are said to be more stable than complexes formed between the poly(acids) and the corresponding low molecular weight cations. Poly(ionic complexes) seem to consist of reversible structures involving loops, ladders, and their intermediate structures. The structures are pH dependent; at low pH loops prevail, while at higher pH values ladders are formed [39].

5. MISCELLANEOUS PREPARATIONS

1. Copolymerization of acrylic acid with 1-substituted imidazoles [81].
2. Cyclopolymerization of acrylic anhydride [82].
3. Copolymerization of methacrylic acid and maleic acid in water [83].
4. Preparation of water solutions of maleic acid and acrylic acid [84].
5. Mechanico–chemical initiation of the polymerization of crystalline salts of acrylic acid [85].
6. Copolymers of ethylene with acrylic acid [86].
7. Composite membranes containing poly(acrylic acid) [87].
8. Grafting of acrylic and methacrylic acids to cellulose [88].
9. Graft copolymerization of acrylic acid to Nylon 6 [89,90].
10. Alternating copolymerization of acrylic acid, involving proton transfer of the acid [91,92,93].
11. Solution properties of low molecular weight atactic poly(acrylic acid) and its sodium salt [94].
12. Kinetics and mechanism of free radical polymerization of acrylic and methacrylic acids in aqueous solution [95].

REFERENCES

1. "Glacial Methacrylic Acid, Glacial Acrylic Acid," Bull. CM-41 I/ej. Rohm and Haas Company, Philadelphia, Pennsylvania, Industrial Chemicals Department, 1972.
2. J. W. Breitenbach and H. F. Kauffmann, *Makromol. Chem.* **175**, 2597 (1974).
3. A. Katchalsky (Katzir) and H. Eisenberg, *J. Polym. Sci.* **6**, 145 (1951).
4. M. L. Miller, *Encycl. Polym. Sci. Technol.* **1**, 197 (1964).
5. C. E. Schildknecht, "Vinyl and Related Polymers," pp. 297–306, 308–311, and 322. Wiley, New York, 1952.
6. E. Trommsdorff and C. E. Schildknecht, *in* "Polymer Processes" (C. E. Schildknecht, ed.), p. 102 ff. Wiley (Interscience), New York, 1956; C. E. Schildknecht, *ibid.* p. 193 ff.

7. G. E. Heckler, T. E. Newlin, D. M. Stern, R. A. Stratton, J. R. Witt, and J. D. Ferry, *J. Colloid Sci.* **15**, 294 (1960).

8. K. Weigel, *Fette Seifen, Anstrichm.* **69**, 95 (1967).

9. F. J. Glavis, *in* "Water Soluble Resins" (R. L. Davidson and M. Sittid, eds.), 2nd ed., p. 154 ff. Van Nostrand-Reinhold, Princeton, New Jersey, 1968.

10. L. S. Luskin, *High Polym.* **24**, Part 1, 105 (1970).

11. B. E. Tate, *High Polym.* **24**, Part 1, 205 (1970).

12. "Glacial Methacrylic Acid, Glacial Acrylic Acid," Bull. SP-88 8/60. Rohm and Haas Company, Philadelphia, Pennsylvania, Special Products Department, 1960.

13. "Storage and Handling of Acrylic and Methacrylic Esters and Acids," Bull. CM-17 E/cd. Rohm and Haas Company, Philadelphia, Pennsylvania, 1972.

14. E. Trommsdorff and G. Abel, U.S. Patent 2,326,078 (1943).

15. E. Trommsdorff, U.S. Patent 2,200,709 (1940).

16. G. D. Graves, U.S. Patent 2,205,882 (1940).

17. F. G. Glavis, *J. Polym. Sci.* **36**, 547 (1959).

18. V. A. Kargin, V. A. Kabanov, S. Ya. Mirlana, and A. V. Vlasov, *Vysokomol. Soedin.* **3**, 134 (1961); *Polym. Sci. USSR (Engl. Transl.)* **3**, 28 (1962).

19. H. Staudinger and E. Trommsdorff, *Justus Liebigs Ann. Chem.* **502**, 201 (1933).

20. E. M. Loebl and J. J. O'Neill, *J. Polym. Sci.* **45**, 538 (1960).

21. M. L. Miller, K. O'Donnell, and J. Skogman, *J. Colloid Sci.* **17**, 649 (1962).

22. J. Semen and J. B. Lando, *Polym. Prepr., Am. Chem. Soc., Div. Polym. Chem.* **10**(2), 1281 (1969).

23. G. Greber and G. Egle, *Makromol. Chem.* **40**, 1 (1960).

24. A. G. Evans and E. Tyrrall, *J. Polym. Sci.* **2**, 387 (1947).

25. P. Monjol and G. Champetier, *Bull. Soc. Chim. Fr.* No. 4, p. 1302 (1972).

25a. S. R. Sandler and W. Karo, "Polymer Syntheses," Vol. 1, pp. 282 ff. Academic Press, New York, 1974.

26. G. R. Barrett, U.S. Patent 2,904,541 (1959); *Chem. Abstr.* **54**, 962h (1960).

27. S. Newman, W. R. Krigbaum, C. Langier, and P. J. Flory, *J. Polym. Sci.* **14**, 451 (1954).

28. A. Katchalsky (Katzir) and G. Blauer, *Trans. Faraday Soc.* **47**, 1360 (1951).

29. F. C. De Schryer, J. Smets, and J. Van Thielen, *J. Polym. Sci., Polym. Lett. Ed.* **6**, 547 (1968).

30. D. E. Ballart, U.S. Patent 3,509,114 (1970).

31. G. Blauer, *J. Polym. Sci.* **11**, 189 (1953).

32. T. M. Karaputadze, A. I. Kurilova, D. A. Topchiev, and V. A. Kabanov, *Vysokomol. Soedin., Ser. B* **14**, 323 (1972); *Chem. Abstr.* **77**, 75566n (1972).

33. A. Behzadi and W. Schnabel, *Macromolecules* **6**, 824 (1973).

34. J. H. Baxendale, M. G. Evans, and J. K. Kilham, *Trans. Faraday Soc.* **42**, 668 (1946).

35. J. C. Leyte and M. Mandel, *J. Polym. Sci., Part A* **2**, 1879 (1964).

36. S. Fakirov, D. Simov, R. Baldjieva, and M. Michailov, *Makromol. Chem.* **138**, 27 (1970).

37. Z. Priel and A. Silberberg, *J. Polym. Sci., Polym. Phys. Ed.* **8**, 689, 705, and 713 (1970).

38. H. R. Dittmar and D. E. Strain, U.S. Patent 2,289,540 (1942).

39. E. Tsuchida, Y. Osada, and K. Abe, *Makromol. Chem.* **175**, 583 (1974).

40. C. B. Wooster, *Macromolecules* **1**, 324 (1968).

41. A. R. Mukherjee, P. Gosh, S. C. Chadha, and S. R. Palit, *Makromol. Chem.* **80**, 208 (1964).

42. G. S. Misra and H. Narain, *Makromol. Chem.* **113**, 85 (1968).

43. H. M. Rife and A. H. Walker, U.S. Patent 2,789,099 (1957).

44. C. E. Carraher, Jr. and J. D. Piersma, *J. Appl. Polym. Sci.* **16**, 1851 (1972).
45. C. E. Carraher, Jr. and J. D. Piersma, *Makromol. Chem.* **152**, 49 (1972).
46. D. E. Strain, U.S. Patent 2,566,149 (1951).
47. C. J. Mast and W. R. Cabaness, *J. Polym. Sci., Polym. Lett. Ed.* **11**, 161 (1973).
48. G. Oster, *Nature (London)* **173**, 300 (1954).
49. H. Kothandaraman and M. Santappa, *J. Polym. Sci., Polym. Chem. Ed.* **9**, 1351 (1971).
50. C. H. Bamford, G. C. Eastmond, and G. C. Ward, *Proc. R. Soc. London, Ser. A* **271**, 357 (1963).
51. G. C. Eastmond, E. Haigh, and B. Taylor, *Trans. Faraday Soc.* **65**, 2497 (1969).
52. G. C. Eastmond, *Mol. Cryst. Liq. Cryst.* **9**, 383 (1969).
53. A. Forchioni and C. Chachaty, *J. Polym. Sci., Polym. Chem. Ed.* **10**, 1923 (1972).
54. P. Monjol, *Bull. Soc. Chim. Fr.* No. 4, p. 1308 (1972).
55. P. Monjol, *Bull. Soc. Chim. Fr.* No. 4, p. 1313 (1972).
56. T. O'Neill, *J. Polym. Sci., Polym. Chem. Ed.* **10**, 569 (1972).
57. K. N. Rao, M. H. Rao, P. N. Moorthy, and A. Charlesby, *J. Polym. Sci., Polym. Lett. Ed.* **10**, 893 (1972).
58. E. M. Loebl and J. J. O'Neill, *Polym. Lett.* **1**, 27 (1963).
59. J. B. Lando, J. Semen, and B. Farmer, *Polym. Prepr., Am. Chem. Soc., Div. Polym. Chem.* **10**(2), 586 (1969).
60. J. B. Lando and J. Semen, *J. Polym. Sci., Polym. Chem. Ed.* **10**, 3003 (1972).
61. A. Chapiro, *Eur. Polym. J.* **9**, 417 (1973).
62. A. Chapiro and T. Sommerlatte, *Eur. Polym. J.* **5**, 707 (1969).
63. A. Chapiro and T. Sommerlatte, *Eur. Polym. J.* **5**, 725 (1969).
64. S. Nagai and K. Yoshida, *Kobunshi Kagaku* **17**, 748 (1960); *Chem. Abstr.* **55**, 24086 (1961).
65. C. S. Marvel and T. H. Shepherd, *J. Org. Chem.* **24**, 599 (1959).
66. D. Braun and I. A. Aziz el Sayed, *Makromol. Chem.* **96**, 100 (1966).
67. B. E. Tate, *Adv. Polym. Sci.* **5**, 214 (1967).
68. H. Nakamoto, Y. Ogo, and T. Imoto, *Makromol. Chem.* **111**, 104 (1968).
69. "Physical and Chemical Properties of Itaconic Acid and Esters," Data Sheet No. 577. Pfizer, Inc., New York, New York, Chemical Division, 1970.
70. R. N. Bashaw and B. G. Harper, U.S. Patent 3,090,736 (1963).
71. S. Suzuki, H. Ito, and S. Shimizu, *J. Chem. Soc. Jpn., Ind. Chem. Sect.* **57**, 658 (1954).
72. F. J. Glavis, D. G. Downing, and H. M. Grotta, U.S. Patent 2,956,046 (1960).
73. Rohm and Haas Co., British Patent 869,333 (1961).
74. A. J. Restaino, U.S. Patent 3,764,502 (1973).
75. H. Morawetz and I. D. Rubin, *J. Polym. Sci.* **57**, 669 (1962).
76. "Calcium Acrylate," Bull. SP-42. Rohm and Haas Company, Special Products Department, Philadelphia, Pennsylvania, 1955.
77. V. F. B. de Mello, E. A. Hauser, and T. W. Lambe, U.S. Patent 2,651,619 (1953).
78. J. B. Rust, L. J. Miller, and J. D. Margerum, *Polym. Eng. Sci.* **9**, 40 (1969).
79. H. C. Heller, S. Schlick, H. C. Yao, and T. Cole, *Mol. Cryst. Liq. Cryst.* **9**, 401 (1969).
80. A. J. Restaino, *Nucleonics* **15**, 189 (1957).
81. D. H. Davies, J. D. B. Smith, and D. C. Phillips, *Macromolecules* **6**, 163 (1973).
82. G. B. Butler, A. Crawshaw, and W. L. Miller, *Macromol. Synth.* **1**, 38 (1963).
83. D. A. Topchiev, R. Z. Shakirov, I. K. Chudakova, L. V. Shtyrlina, and V. A. Kabanov, *Vysokomol. Soedin., Ser. B* **13**, 821 (1971); *Chem. Abstr.* **76**, 113623 (1972).
84. D. J. Gale, U.S. Patent 3,635,915 (1972).
85. V. A. Kargin, V. A. Kabanov, and N. Ya. Rapaport-Molodtsova, *Vysokomol. Soedin.* **3**, 787 (1961); *Polym. Sci. USSR* **3**, 657 (1962).

86. D. N. Andreev, A. S. Semenova, M. I. Leitman, L. G. Stafanovich, D. N. Asinov-skaya, and F. I. Duntov, USSR Patent 333,173 (1972); *Chem. Abstr.* **77**, 102431 (1972).
87. S. B. Sachs and H. L. Lonsdale, *J. Appl. Polym. Sci.* **15**, 797 (1971).
88. Y. Ogiwara, H. Kubota, S. Hayashi, and K. Sekine, *J. Appl. Polym. Sci.* **16**, 2197 (1972).
89. M. B. Huglin and B. L. Johnson, *J. Appl. Polym. Sci.* **16**, 921 (1972).
90. K. Matsurzaki, T. Kanai, and N. Morita, *J. Appl. Polym. Sci.* **16**, 15 (1972).
91. T. Saegusa, S. Kobayashi, and Y. Kimura, *Macromol.* **7**, 139 (1974).
92. T. Saegusa, S. Kobayashi, and Y. Kimura, *Macromol.* **7**, 256 (1974).
93. T. Saegusa, *Chemtech.* **5**, 295 (1975).
94. G. J. Welch, *Polymer* **16**, 68 (1975).
95. V. A. Kabanov, D. A. Topchiev, T. M. Karaputadze, and L. A. Mkrtchian, *Eur. Polym. J.* **11**, 153 (1975).

POLY(VINYL CHLORIDE)

I. INTRODUCTION

The production of the monomer vinyl chloride (commonly abbreviated VC or VCM; current *Chemical Abstracts* name: chloroethene, indexed as "Ethene, Chloro-") was estimated at a level of approximately 6.0 billion pounds* per year for the year 1974 [1]. Virtually all of this is converted to poly(vinylchloride) (PVC) or vinyl chloride-containing copolymers. The annual growth rate from 1966 to 1974 for vinyl chloride-based resins was on the order of 9% per annum. It was projected that in the period 1978 to 1980, the annual growth rate would be about 6–7% [1]. These projections probably do not take into account recent restrictions on the use of rigid PVC in food packaging imposed by the United States government and the inroads being made by other polymers in blister packaging, blown containers, and in various non-food-related applications [2]. Poly(vinyl chloride) will undoubtedly remain among the three or four most widely used resins for some time to come.

A. Health and Safety Aspects

As recently as 1969, the monomer vinyl chloride was considered to be a "nontoxic, colorless gas (at atmospheric pressure and room temperature) possessing a faintly sweet odor" [3]. At high concentration it was said to cause anesthesia, and its major hazards were those associated with most gaseous organic chemicals, i.e., fire and explosion. In fact, the monomer was thought to be sufficiently free of ordinary hazards that it was used as a propellant in a variety of household aerosol sprays.

Late in January 1974, this situation changed drastically when it was announced that between 1968 and 1973 three long-time PVC plant operators had died of angiosarcoma, a very rare liver cancer [4,42]. The Occupational Safety and Health Administration (OSHA) issued emergency temporary standards limiting employee exposure in the vinyl chloride and poly(vinyl chloride) industry. These standards have undergone a number of revisions to drastically lower levels of permissible concentrations of VCM where a person might inhale an air–monomer mixture.

The current standards of exposure are still in a state of flux. As of November, 1975, the applicable standards appear to be those given in the Federal Register [5]. The proposed standard for emission into the environment appears in reference [4a]. A brief summary of the standards will be found in reference [6]. A few highlights are given here.

The hazards associated with VCM are not only angiosarcoma of the liver,

---

* Billion = 10^9 is used throughout this chapter.

for recent studies indicate a higher incidence of cancer of the lung, brain, and bone marrow among workers exposed to VCM. The effect of VCM may not be evident until 15 or more years after the beginning of exposure. Particularly severe exposure used to be experienced by workers who had to climb into polymerization reactors to scrape PVC residues from the walls, agitators, etc.

The federal regulations apply to the "manufacture, reaction, packaging, storage, handling or use of VC and PVC" [6]. They evidently do not apply to the handling or use of finished products, i.e., a product which has passed through all processes which involve melting of the resin mass. This includes molding, extrusion, Banbury mixing, and calendering but does not seem to cover thermoforming or blister packaging [6]. The concern about the melting of poly(vinyl chloride) seems to arise from the possibility that residual monomer in the resin may be released to the work environment thus increasing the level of free VCM in the air.

A facility which upon testing its air quality by an approved procedure finds that its vinyl chloride level is less than 0.5 parts per million (ppm) is effectively exempted from most of the rest of the standards. The 0.5 ppm level is called the "action level." If the VCM level in a facility is found to be no greater than 0.5 ppm, no further monitoring of the air quality is required unless there is a change in processing or there are other reasons to believe that the level may have gone up.

The standard sets a concentration limit of one ppm of vinyl chloride on an average over an 8-hour work day and a five ppm allowable peak exposure for 15 min. Above the 1 ppm limit, respirators must be provided. The work area must be restricted to "authorized personnel" with warning signs in the regulated area, and labels on vinyl chloride and unfinished PVC containers must include the words "cancer-suspect agent." A daily roster of persons within the restricted areas must be maintained. There are also regulations about medical examinations and long-term record keeping of exposure for each employee [5,6].

In connection with laboratory work we should like to note that chemists rather generally operate on the mistaken premise that a fume hood represents a universal protection against most hazards such as fire, explosions, and toxic fumes. Naitove [4] mentions that two deaths from angiosarcoma took place a few miles from PVC processing operations. We presume that these deaths could have been caused by vinyl chloride which had been exhausted into the environment. This would also imply that an air quality standard of a maximum of 0.5 ppm of VCM is not sufficiently low [4a].

For additional safety the exhausted air from fume hoods used for VCM should be thoroughly scrubbed before either recirculating it to the hood or exhausting it to the outside. The methods for scrubbing exhaust air need exploration. The recommendations of the monomer supplier and of the hood manufacturers should be considered. Since the safe handling of vinyl chloride rests with the laboratory researcher and his supervisors, proper safety procedures will have to be developed using the up-to-date OSHA Standards as a minimum. We should like to emphasize that in developing these procedures, the matter of disposing of excess vinyl chloride or unfinished poly(vinyl chloride) as well as emergency procedures should not be overlooked.

A few safety suggestions are presented here, based on a Tenneco Chemicals, Inc., Safety Manual [7]. This presentation is not to be considered definitive or complete. It also is not to be construed as an assumption of liability or responsibility either by Tenneco Chemicals, Inc., the authors of this book, or the publisher for any statement or recommendation made here. This material is merely given to indicate some aspects of the scope of the safety considerations which are required when handling vinyl chloride monomer and poly(vinyl chloride).

1. Compliance with the Occupational Safety and Health Administration (OSHA) permanent standard for "Exposure to Vinyl Chloride" is mandatory [5]. Compliance with these, or revised, standards is a matter of law.

2. The permissible exposure limits and air quality monitoring requirements have been discussed above.

3. Liquid vinyl chloride may be used or handled only in an appropriate hood. Supplementary respiratory equipment and protective clothing, including elbow-length rubber gloves, are required when handling VCM, so that there is no direct contact. Storage of vinyl chloride-containing materials must be in the hood and in tight, nonporous containers bearing proper labels —including the words "cancer-suspect agent."

4. A training program for personnel and a medical surveillance program in accordance with OSHA standards must be instituted.

5. Vinyl chloride storage containers should be high pressure (> 800 psig) bombs equipped with high pressure needle valves, a pressure relief at 400 psig, a quick connect fitting for filling, a carrying handle, screw-on caps to protect the valves, and a welded head with clips for grounding. These bombs are to be tested before each filling for leaks under approximately 100 psig nitrogen pressure in a water bath. The bombs shall be labeled "Vinyl Chloride— Extremely Flammable Gas Under Pressure—Cancer-Suspect Agent."

6. The bombs are to be stored in a fume hood which is left on at all times. The bombs are to be grounded. The hood is to bear a sign reading "Danger— Vinyl Chloride—Extremely Flammable Gas Under Pressure—Cancer-Suspect Agent—Leave Hood ON."

7. All monomer must always be used in the fume hood. The fume hood should be fitted something like a dry box so that, when the hood window is lowered to its lowest point, handling inside the hood can be carried out consistently by means of elbow-length rubber gloves. The design of the hood

must be such that safe, flexible movement by the operator is possible, while maximum air flow into the hood is maintained. Respiratory equipment is also used during all operations at the hood.

8. The monomer must be charged to polymerization equipment in the hood. The discharging of materials also must take place in the hood. Excess monomer must be allowed to evaporate completely in the hood with the door closed.

9. OSHA-approved respirators, such as full face, pressure demand fresh air respirators, must be used when charging and discharging reaction equipment, during emergency spills of monomer, in cases of leaking monomer bombs, in case of breakage of bottle reactors or leakage of autoclaves, and when handling any material which is known or suspected of containing unreacted monomers.

10. PVC latexes, slurries, resins, and other PVC compositions may only be stored in areas with adequate ventilation.

11. Samples must be stored and transported only in tightly closed nonporous containers. Latexes and slurries should always be stored in venting fume hoods.

12. PVC samples which are no longer required should be stored in a closed steel drum lined with a polyethylene bag in an outdoor location. The drum is to be labeled "Contaminated with Vinyl Chloride. Cancer-Suspect Agent." When filled, such a drum must be properly discarded.

13. Spills of PVC resin, slurry, and latex are to be treated immediately with a 1% bromine water solution.

CAUTION: Such a solution presents another set of hazards related to the handling of bromine.

After such treatment, the spill is to be cleaned up and disposed of promptly.

Attention on the part of laboratory personnel should be paid to residual monomer which may be trapped (either mechanically or in solid solution) in poly(vinyl chloride).

Naitove [4] states that as of December 1974 a total of over two dozen cases of angiosarcoma were found in the United States and Europe. In the United States alone there are about 365,000 workers involved in the VCM–PVC industry [6]. Even if we assume that only a fraction of these people have been in the field for 20 years and that not all cases of liver cancer have been detected and reported, we do have an indication of the variability of individual resistance to this type of cancer. Of course, this does not mean that extreme precautions should not be taken in preventing further cases of angiosarcoma (which, by the way, cannot be detected until it is terminal).

From the above discussion of the hazards and safety requirements associated with vinyl chloride, it is self-evident that research with this monomer is to be undertaken using great precautions.

B. General Polymer Aspects

While the literature on PVC is voluminous, much of the information necessary to produce a commercially acceptable polymer consists of guarded company secrets. This is particularly true in connection with the all-important morphology of the resin particles. The particle size distribution, bulk density, and porosity of the particles determine the rate of plasticizer up-take and the quantity of plasticizer absorbed. These factors are crucial in the processing of the resin and the final properties of the finished product.

This chapter presents an overview of the methods of polymerization as found mainly in the recent patent literature. It is hoped that by this means some of the principles used to prepare commercially useful PVC will emerge. We do want to state at the outset that since much of the data presented here is gathered from patent disclosures, we cannot make any positive statements as to the reliability of the information. The inclusion of material from patents and the citation of specific patents should not be understood as recommendations to use any of the cited patents.

It should also be noted that many of the procedures described involve partial conversion of monomer to polymer. The elimination of the unreacted monomer must, of course, make use of suitably safe procedures.

The removal of residual vinyl chloride from poly(vinyl chloride) appears to be fairly difficult. For example, in a patented process, 1 kg of PVC containing 34 gm of residual monomer was heated for 0.5 hr at 433 mm Hg at 85°C with steam. Then, as the resinous mass cooled to 40°C, the pressure was reduced to 55 mm Hg. The residual resin (969 gm) still contained about 3 gm of vinyl chloride [8]. While this reduction in residual monomer is impressive, leaving about 0.3% of free monomer in a finished resin may be rather high. Upon injection molding, for example, this monomer may be released to the atmosphere to raise the vinyl chloride concentration to unacceptable levels unless additional ventilation is provided.

The physical properties of monomeric vinyl chloride are given in Table I [9]. Data of interest for polymerization studies are given in Tables II [10], III and IV.

In regard to the glass transition temperature (T_g) given in Table II, considerable variations in this value have been observed. In general, as the polymerization temperature of vinyl chloride is decreased, the glass transition temperatures and melting points of the resulting resins increase. While polymer branching decreases as the polymerization temperature decreases, a study of the reaction temperature effect which goes below the range where changes in the degree of branching may be expected indicates that this is not the cause of the observed effect, nor is it caused by a change in the normal head-to-tail structure of the polymer. It has been shown that poly(vinyl chloride) contains substantial syndiotactic polymer segments. As the polymerization temperature

TABLE I
PHYSICAL PROPERTIES OF VINYL CHLORIDE [9]

Molecular weight	62.5
Boiling point at 760 mm Hg	$-13.8°C$
Vapor pressure (mm Hg)	16,617 at 100°C
	2943 at 25°C
	1293 at 0°C
	470 at $-25°C$
	130 at $-50°C$
	24.4 at $-75°C$
	2.5 at $-100°C$
Melting point	$-153.69°C$
Autoignition temperature	472.22°C
Flammability limits [vol.% in air at room temperature (explosive limits)]	3.6–26.4
Liquid density (gm/ml)	0.746 at 100°C
	0.9013 at 25°C
	0.9834 at $-20°C$
	0.9918 at $-25°C$
	0.9999 at $-30°C$
Refractive index, n_D^{25}	1.3642
Heat capacity (liquid) (cal/mole/°C)	26.25 at 100°C
	20.56 at 25°C
Heat capacity (vapor) (cal/mole/°C)	14.88 at 100°C
	12.82 at 25°C
Latent heat of vaporization (cal/mole)	5250 at $-13.8°C$
Latent heat of fusion (cal/mole)	1172
Heat of polymerization (kcal/mole)	-25.3 ± 0.5
Critical temperature (°C)	147
Critical pressure (atm)	56
Critical volume (cm³/mole)	179
Approximate volume shrinkage upon polymerization (%)	35
Solubility in water (%)	0.11
Inhibitor	Vinyl chloride is considered sufficiently stable to permit shipment and storage in cylinders without inhibitors

is decreased, the number of monomer units experiencing syndiotactic placement in the growing polymer chain increases. Thus the variations in the glass transition temperature may be attributed to changes in the stereospecificity and crystallinity of the resin produced by the polymerization temperature [11]. Table III shows some observations of the effect of polymerization temperature on the glass transition temperature and melting point of poly(vinyl chloride).

TABLE II

PROPERTIES OF POLY(VINYL CHLORIDE)

Glass transition temperature (T_g)	81°C (354°K)
	T_g varies considerably with variations in the method and temperature of polymerization
Solvents[a]	
Theta solvent	Benzyl alcohol at 155.4°C
Solvents for high molecular weight resins	Tetrahydrofuran, acetone–carbon disulfide mixtures, methyl ethyl ketone
Solvents for low molecular weight resins	Toluene, xylene, methylene chloride, ethylene chloride, perchloroethylene–acetone mixtures, 1,2-dichlorobenzene, tetrahydrofurfuryl alcohol, dioxane, acetone–carbon disulfide mixtures, cyclopentanone, diisopropyl ketone, mesityl oxide, isophorone, dimethylformamide, nitrobenzene, hexamethylphosphoramide, tricresyl phosphate
Nonsolvents	Aliphatic and aromatic hydrocarbons, vinyl chloride monomer, alcohols, glycols, aniline, acetone, carboxylic acids, acetic anhydride, esters, nitroparaffins, carbon disulfide, nonoxidizing mineral acids, concentrated alkalies

Mark–Houwink–Sakurada parameters (for viscosity–molecular weight correlation)

Solvent	Temp. (°C)	$K \times 10^5$ (dl/gm)	a
Benzyl alcohol (theta solvent)	155.4	156	0.50
Chlorobenzene	30	71.2	0.59
Cyclohexanone	20	11.6, 13.7, 112.5	0.85, 1, 0.63
	25	24, 12.3, 204, 208, 174	0.77, 0.83, 0.56, 0.56, 0.55
	30	16.3	0.77
Tetrahydrofuran	20	1.63	0.92
	25	49.8, 16.3	0.69, 0.766
	30	219, 83.3	0.54, 0.83

Price–Alfrey copolymerization parameters (See also Table IV)

$e = 0.20$
$Q = 0.044$

[a] Solvents are somewhat dependent on the molecular weight distribution of a polymer sample.

In Table II, the large variations in the parameters for the Mark–Houwink–Sakurada equation [Eq. (1)] relating intrinsic viscosity $[\eta]$ to molecular weight $\bar{M}$ (K and a are parameters which have to be determined for each polymer-solvent system),

$$[\eta] = K\bar{M}^a \qquad (1)$$

are disturbing. Many other parameters are also found in the literature. Among the suggestions made to account for this variation in the average molecular weight range of the samples tested are variations in the degree of branching of the polymer chains, improper fractionation of the polymer, and the presence of aggregates in the polymer solutions [12].

A selected list of reactivity ratios for vinyl chloride with a number of comonomers are given in Table IV. The copolymerization ratios of 1-chloro-1-propene and 2-chloro-1-propene are of interest. These isomers of allyl chloride, strangely enough, seem to be impurities formed in the manufacture of vinyl chloride by some processes. These compounds could find application in the reduction of the cost of poly(vinyl chloride), if they were copolymerized with vinyl chloride. Also to be noted is that the reactivity ratios for the various vinyl chloride–vinyl ester systems do not differ very much from each other. This indicates not only that most vinyl esters copolymerize readily with vinyl chloride, but also that the effect of the carbon chain attached to the acyl group on the copolymerization behavior is minor [13].

Although the original synthesis and polymerization of VCM has been attributed to Regnault, subsequent work indicated that poly(vinylidene chloride) had actually been produced [14]. In 1872, authentic poly(vinyl chloride) was prepared by Baumann [15]. By the early 1930's full-scale commercial production of PVC had started in Germany. The early resins suffered from processing difficulties caused by thermal degradation. The real impetus to the development of the poly(vinyl chloride) industry was the result of the observation of Waldo L. Semon of B. F. Goodrich Co. in 1933 that when the resin was heated with high boiling fluids, leathery or rubbery products formed. This discovery of plasticization revolutionized the industry.

Today the PVC field exhibits considerable complexity. Commercial resins may be classified into general purpose resins, dispersion resins, blending resins, solution resins, and latexes. Within these general categories there are a number of subclasses based on physical properties.

General discussions on the preparation and properties of PVC will be found in such references as 1, 3, and 16–18. The manufacture, processing, stabilization, and applications of PVC are also discussed in references 19–26. A review of the history of ionic and Ziegler catalysis polymerization of vinyl chloride is found in Feldman [27] and Yamazaki [27a]. The recent views of the kinetics and mechanisms of the vinyl chloride polymerization are discussed in Kuchanov and Bort [28] and Ravey et al. [29].

TABLE III
EFFECT OF POLYMERIZATION
TEMPERATURE OF VINYL CHLORIDE ON THE
GLASS TRANSITION AND MELTING POINT OF
POLY(VINYL CHLORIDE) [11]

Polymerization temperature (°C)	Glass transition temperature (°C)	Melting point (°C)
−80	100	> 300
−10	90	265
40	80	220
90	75	—
125	68	155

TABLE IV
SELECTED REACTIVITY RATIOS OF VINYL CHLORIDE (M_1)
WITH COMONOMERS (M_2) [10]

M_2	r_1	r_2
Acrylonitrile	0.04, 0.074, 0.02	2.8, 3.7, 3.28
Allyl acetate	1.2, 1.16	—, 0
Butadiene	0.035	8.8
Butyl acrylate	0.07	4.4
tert-Butylethylene	5	0.0
Butyl methacrylate	0.05	13.5
Butyl vinyl ether	1.0, 1.01	0.5, 0.46
1-Chloro-1-propene	1.13	0.24
2-Chloro-1-propene	0.75	0.58
Dibutyl maleate	1.4	0.0
Diethyl fumarate	0.12	0.47
Di-2-ethylhexyl maleate	0.42	0
Diethyl itaconate	0.06	5.65
Diethyl maleate	0.8, 0.77, 0.9	0.0, 0.009, 0
Diisobutyl maleate	0.65	0.1
Diisopropyl itaconate	0.06	6.0
Dimethyl itaconate	0.053	5.0
Dioctyl itaconate	0.06	7.0
Dioctyl maleate	0.5	0
Ethylene	3.60, 1.85	0.24, 0.20
Ethyl vinyl ether	0.98	0.26
Glycidyl methacrylate	0.04	8.84
Isobutylene	2.05, 1.3, 4.3	0.08, 0.03, 0
Isopropenyl acetate	2.2	0.25
Maleic anhydride	0.296	0.008
Methallyl chloride	0.31	0.0
Methyl acrylate	0.06, 0, 0.12, 0.083	4, 5, 4.4, 9.0

(Continued)

TABLE IV (*Cont.*)

M₂	r_1	r_2
Methyl methacrylate	0, 0.1, 0.02	12.5, 10, 15
Methyl vinyl ketone	0.10	8.3
Octyl acrylate	0.12	4.8
Octyl methacrylate	0.04	14.0
Pentachlorostyrene	0.43	5.3
1-Pentene	0.5	—
Styrene	0.035, 0.02, 0.067, 0.077, 0.05	5.7, 17, 35, 35, 1.0
Vinyl acetate	2.1, 1.8, 1.68, 1.35, 3.74	0.3, 0.6, 0.23, 0.65, 0.033
Vinyl benzoate	1.70, 0.72	0.5, 0.28
Vinyl caproate	1.8, 1.35	0.1, 0.65
Vinylene carbonate	5.2	0.09
Vinyl ethyl diethoxy silane	1.0	0
Vinylidene chloride	0.3, 0.5, 0.23, 0.2, 0.2, 0.5	3.2, 7.5, 3.15, 4.5, 1.8, 0.001
Vinyl isobutyl ether	2.0	0.02
Vinyl laurate	7.4	0.2
Vinyl levulinate	1.40	0.419
Vinyl methyl diethoxysilane	1.0	0
Vinyl pelargonate	1.16	0.282
Vinyl pinonate	1.458	0.446
Vinyl propionate	1.35	0.65
N-Vinylpyrrolidone	0.53	0.38
Vinyl stearate	0.745	0.290
Vinyl undec-10-enoate	1.06	0.358

The current concern with environmental pollution and ecology has led to some interest in the possibility of developing PVC formulations which may be degraded by environmental factors [30].

A cornerstone of the chemistry of polymers is a publication on the structure of poly(vinyl chloride) by Marvel and co-workers [31]. This paper, which is still cited, established the head-to-tail structure of PVC. The approach used involved the dechlorination of uncross-linked poly(vinyl chloride) in dioxane solution with zinc. From statistical considerations, it had been predicted that random attack by zinc on the polymer chains should produce different degrees of dehalogenation depending on whether the polymer were of a head-to-tail or a head-to-head, tail-to-tail structure. Since the experiment resulted in the removal of 84–87% of the available chlorine, as predicted, and the resultant product consisted of polymeric cyclopropane units with occasional, isolated

$$-CH_2CH- \atop \quad | \atop \quad Cl$$

units, the preponderance of the head-to-tail structure was established.

Marvel's work [31] took into consideration the possibility of cross-linking with adjacent polymer chains. This difficulty was overcome by working in dilute solution and ascertaining that the dehalogentated product was still soluble and uncross-linked. From hindsight, it should be pointed out that the effect, if any, of branching and stereoregularity may not have been taken into consideration in the calculations of the degree of dehalogenation to be expected by the random reaction of zinc along the polymeric chains.

The degree of branching found in PVC increases as the polymerization temperature increases. In commercial resins between approximately 1 and 10 branches per 1000 monomer units have been found [1,32]. Experimentally, branching was demonstrated by evaluating variations with polymerization temperature of the infrared absorption bands at 1378 cm^{-1} (for the methyl group, which may occur at a branch) and at 370 cm^{-1} (for the methylene group) [33]. A more recent study involved the complete dehalogenation of a resin by reduction with lithium aluminum hydride [34]. In this work it was found that branching is virtually independent of conversion and partially attributable to the presence of —CH_2Cl side groups.

Talamini *et al.* [34] proposed a mechanism of branching involving chain transfer. Long-chain branches may be the result of transfer involving two polymer chains, short-chain branches were attributed to an intramolecular chain-transfer mechanism. Presumably chain transfer involving monomeric vinyl chloride may also lead to branching.

Natta and Corrandini [35] mention that stretched fibers of PVC produced by a free-radical mechanism have some crystalline character and that short syndiotactic segments exist in these oriented fibers. On theoretical grounds, syndiotactic propagation of the vinyl chloride polymerization is slightly favored over isotactic propagation. For the free-radical polymerization process, the potential energy difference, ΔE, between isotactic and syndiotactic placement was calculated to be of the order of 1.5 and 2 kcal/mole, for poly-(vinyl chloride). The temperature effect on stereospecificity is not particularly large. The activation energy difference between isotactic and syndiotactic propagation was found experimentally to be 0.6 kcal/mole [36]. Commercial PVC's usually are 10–15% crystalline [1]. This aspect of polymer structure manifests itself by an increased processing temperature, tensile strength, softening temperature, and heat distortion temperature, and a greater brittleness over those of the atactic form. The heat stability, however, appears to be independent of the polymer's tacticity [37].

As the polymerization temperature is lowered, Gottesman [1] reports that crystallinity of PVC increases. Thus at a polymerization temperature of −15°C, the resin is 57% crystalline; at −75°C it is 85% crystalline. Talamini and Vidotto [38] give further details on the ratio of the syndiotacticity to isotacticity as a function of the polymerization temperature.

The crystallinity of poly(vinyl chloride) may be substantially reduced when a suspension polymerization is carried out with vinyl chloride in the presence of a seed polymer latex of PVC and a modest quantity of acetic acid [39].

In general, the polymerization of vinyl chloride may be carried out in bulk, solution, suspension, and in emulsion. Free-radical initiators are most commonly used although organometallic initiators and radiation initiation has been considered. Since the monomer is a gas at ordinary temperatures and pressures, suitable equipment is required for VCM polymerization. Sealed tubes and capped bottles have been used for this experimental work. In the use of bottles, safety precautions should be considered both from the standpoint of explosion hazards and the problems of exposure of personnel to VCM.

2. BULK POLYMERIZATION

The bulk polymerization of vinyl chloride offers the potential advantage over the more commonly used procedures (i.e., suspension or emulsion polymerization) that the product is free of protective colloids, suspending agents, surfactants, buffers, water, additives, or solvents [1,40]. Until recently, controlling bulk polymerization on a plant scale was considered difficult and impractical because of the problem of removing the heat generated during polymerization and of controlling the rate of reaction. The Pechiney–Saint Gobain process, which will be outlined below, overcomes these difficulties.

The basic bulk polymerization procedure consists simply of heating the monomer in the presence of a small amount of initiator under a suitable condensing or pressure system until the desired conversion to polymer has been achieved.

At room temperature and atmospheric pressure, vinyl chloride is a gas, albeit a readily condensable one. At ambient temperatures, the solubility of PVC in the monomer (by direct gravimetric determination) is 0.03% [29], although the literature states elsewhere that at about 0.5% conversion of vinyl chloride, precipitation of the polymer is noted [28].

In regard to the solubility of the monomer in its polymer, the diffusion of vinyl chloride into PVC varies with the physical state of the resin. Earlier work frequently dealt with studies involving PVC films; more recently, powdered resins were studied. In the latter case, variations in diffusion rates were found to depend on the method of polymerization (emulsion vs. suspension method) as well as on the physicochemical parameters [41]. The equilibrium solubility of vinyl chloride monomer in poly(vinyl chloride) was found to be a function of polymer type, polymer history, time, temperature, and the VCM partial pressure [42]. Above atmospheric pressure, with the ratio of the partial pressure of VCM to the initial partial pressure of the monomer greater than approximately 0.5, the solubility of vinyl chloride is 0.300 gm per gram of poly(vinyl chloride). At lower pressures, the solubility shows a distinct decrease with temperature. Kuchanov and Bort [28] state that the solubility of VCM in PVC varies between 22.7% and 23.7% between 30° and 60°C.

Berens proposes [42] that his data suggest two processes to be involved in the dissolution of monomer in the polymer: a normal dissolution and a "hole-filling" contribution which approaches saturation with increasing partial pressure of the monomer. In a polymer, above the glass transition temperature (T_g), the interstices in the resin do not persist long enough to be filled, while below the T_g, the pores may be frozen in, and the hole-filling contribution to solubility becomes increasingly significant. In addition, there may be a contribution to the monomer solubility by the known plasticizing effect of the monomer on the polymer.

These factors alone may contribute significantly to variations in experimental results obtained in the polymerization of vinyl chloride. For example, at superatmospheric pressures, variation in the liquid volume of the monomer to the total free space of a closed reactor means a variation in the amount of gaseous monomer which can condense and/or diffuse into the polymer at some stage of the polymerization process. This vapor-phase monomer level may represent a monomer reservoir somewhat analogous to the "monomer droplets" postulated for the emulsion polymerization mechanism by Smith and Ewart.

At high as well as low conversions, the molecular weight of poly(vinyl chloride) appears to be dependent on the degree of conversion [43]. Table V shows that with increasing polymerization temperature, at 10% conversion, the molecular weight of the polymer goes through a maximum. The data presented here was taken at the 10% conversion level to minimize the effect of chain branching on the molecular weight as determined by gel permeation chromatography [44]. On the other hand, when conversion is carried out to a point where the molecular weight becomes independent of conversion, the effect of polymerization temperature on the average molecular weight as measured by the intrinsic viscosity is curvilinear, although the curvature in the range of 38°–71°C, which is of industrial interest, is sufficiently small to be considered linear for practical purposes [43] (cf. Table VI).

As vinyl chloride polymerizes, the reaction mixture undergoes a variety of changes which, in part, may be attributed to the number of phases present as the degree of conversion increases. Both Kuchanov and Bort [28] and Ravey *et al.* [29] divide the process into a number of stages. While the Russian investigators [28] based their kinetic work on bulk polymerizations and the Israeli researchers [29] on suspension polymerizations, the stages described

TABLE V

EFFECT OF POLYMERIZATION TEMPERATURE
ON THE MOLECULAR WEIGHT DISTRIBUTION
OF PVC [44][a,b]

Polymerization temperature (°C)	Weight-average molecular weight $\times 10^{-3}$	Number-average molecular weight $\times 10^{-3}$
-76[c]	40	2
-50	162	26.3
-30	320	50
0	385	73
25	291	57.7

[a] Polymerization conditions: bulk polymerization with 2,2′-azobis(isobutyronitrile) and UV radiation carried to 10% conversion.

[b] Gel permeation chromatography of the PVC produced dissolved in hexamethylphosphoramide (caution: cancer-suspect agent).

[c] Polymerization initiated with γ-radiation.

are similar in their broad outline. This is to be expected since the two procedures, in general, are thought to have identical kinetics. Table VII shows this parallelism.

In the first stage of the polymerization, a homogeneous process takes place until the concentration and molecular weight of the polymer is sufficiently high to bring about precipitation of the polymer. After this, two phases are present simultaneously: a polymer phase swollen with monomer, and a nearly pure monomer phase. The precipitate coalesces to form primary particles of diameter in the range of 0.1–0.6 nm which form aggregates as the rate of polymerization increases up to a conversion of approximately 1% [29]. These aggregates establish a morphology of the resin which seems to persist unchanged throughout the process. A fairly constant number of particles per unit volume is established in the range of 5×10^{10} to 5×10^{11} particles per ml. The rate of polymerization increases from 2 to 200 moles/liter sec [28]. In this early stage of the process, the morphology of the resin may vary from fine particles to large continuous spheres.

In the particle growth stage, the number of particles remains constant. The rate of diffusion of monomer into the polymer is more rapid than the rate of polymerization [29]. Secondary structures form because of the increase in the frequency of collisions between the enlarged particles. In the bulk process, at about 20% conversion, the particles are said to consist of a three-dimensional system of loosely packed spheres which now have lost their fluidity. As the

TABLE VI
EFFECT OF POLYMERIZATION TEMPERATURE ON THE INTRINSIC
VISCOSITY AT INTERMEDIATE CONVERSIONS [43]

Reaction temperature ($\pm 0.1°C$)	Initiator concentration (%)	Reaction time (hr)	Conversion (%)	Intrinsic viscosity ($[\eta]$, dl/gm)
0.0	0.5[a]	168	80	3.90
10.0	0.4[a]	72	40	2.93
20.0	0.3[a]	24	60	2.26
30.0	0.3[a]	6	40	1.66
40.0	0.2[a]	4	50	1.29
50.0	0.05[a]	4	50	1.02
60.0	0.05[a]	6	80	0.797
70.0	0.1[b]	4	50	0.670

[a] Diethyl peroxydicarbonate used as initiator. (This initiator showed no activity for a one-week run at $-10°C$, at 0.5% concentration.)
[b] Lauroyl peroxide used as initiator.

process continues, the reaction mass becomes a single porous block consisting of packed spheres (20–77% conversion). Beyond the 77% conversion range, under certain conditions, nonporous, transparent blocks may form. Since the monomer phase is absent, conventional homogeneous polymerization takes place in the polymer phase with reduction in the free monomer concentration [28]. In the suspension polymerization process, it is thought that in the 1–30% conversion range, polymerization takes place separately in the polymer and the monomer phases, with the newly formed polymer precipitating out on the existing resin almost as soon as it is formed. At about 30% conversion a limited amount of secondary nucleation seems to take place. Once all of the exterior monomer has been used up (in the 70–85% conversion range) there is only absorbed monomer left to polymerize. Since the supply of this monomer is limited, the pressure begins to drop. After the absorbed monomer has been polymerized, this stage of the process is completed. The residual monomer in the vapor phase above the suspension and the monomer dissolved in the suspension medium finally polymerize in the last stage of the process. Since the monomer has to diffuse into a nonswollen, dense resin to the sites of polymeric free radicals, this process is quite slow [29].

It is interesting to note that electron photomicrographs of the particles formed during bulk polymerization show a remarkable uniformity in particle size [45].

In addition to the above features of the polymerization of vinyl chloride, a full description of the process must also take into consideration the following

TABLE VII

STAGES OF THE POLYMERIZATION OF VINYL CHLORIDE

Bulk Polymerization (According to Kuchanov and Bort [28])		Suspension Polymerization (According to Ravey et al. [29])	
Phase	Conversion (%)	Phase	Conversion (%)
Single-phase polymerization	0–0.5	Single-phase polymerization	0–0.1
Heterogeneous polymerization		Heterogeneous polymerization	
Fixed number of particles form	0.5–1.0	Unstable colloidal suspension	0.1–1.0
Particle growth stage	1–10	Particle growth stage, fixed number of particles	1–30
Secondary structures form	10–20		
Reaction mass is single porous block of packed spheres	20–77	Limited amount of secondary nucleation takes place	30–70
Single-phase polymerization	77–100	Single-phase polymerization	70–85
		Slow single-phase polymerization	85–100

aspects mentioned by Kuchanov and Bort in their detailed review of the literature [28].

1. Both in bulk and solution polymerizations, there appears to be a reduction in the rate when dilatometric data is used to plot integral kinetic curves. This discontinuity occurs at the transition from the initial homogeneous stage of the reaction to the heterogeneous process. In Kuchanov and Bort [28] the volume changes associated with the appearance of the insoluble resin phase are thought to be the cause of the apparent change in polymerization rate. When the course of the reaction is followed by other techniques, this reduction in the polymerization rate is not observed.

2. As may be expected, in the presence of a solvent, the precipitation of polymer takes place at a higher degree of conversion. For a given true solvent, there is a critical solvent concentration beyond which no polymer precipitates and beyond which true homogeneous reaction kinetics are observed. As the level of such a solvent increases up to a critical value, this initial rate of polymerization increases. However, when the solvent level is increased after polymer precipitation has already taken place, the rate of polymerization decreases and becomes proportional to the available monomer concentration.

Kuchanov and Bort do not seem to draw any distinction in their discussion between a true solvent for both the monomer and the polymer (such as tetrahydrofuran) and a pseudo-solvent, which is essentially a nonsolvent for the polymer beyond a certain low polymer concentration and/or molecular weight.

3. As has been mentioned above, the rate of diffusion of the monomer into the polymer is generally quite high. Therefore, up to a conversion of approximately 77%, there is a constant equilibrium concentration of monomer present in the polymer particles. At 77% conversion, in effect, all of the available liquid monomer phase is dispersed in the polymer [25]. This diffusion of monomer into the swollen polymer phase does not interfere with the polymerization in the rigid phase. At 50°C, the degree of swelling is 23.1%. About 25% of the monomer, VC, may act as a plasticizer of PVC at 50°C [46].

4. Kinetic polymerization curves show that autoacceleration takes place essentially from the beginning of the process.

5. For the bulk polymerization of vinyl chloride, the relation between the rate of polymerization, R_{pol}, and the conversion, p, is not clearly defined. Some authors state that up to $p = 3.5\%$, the relation between R_{pol} and p is linear; others claim that up to $p = 15\%$, R_{pol} is a function of the $\frac{2}{3}$rd power of p; while still other workers claim that

$$R_{pol} = f(p^n) \tag{2}$$

where n varies between 0.57 and 0.67 (note that 0.67 is essentially $\frac{2}{3}$).

It was also found that at high rates of conversion (i.e., where $n \simeq \frac{2}{3}$), there is a direct proportionality between R_{pol} and the total surface area of the particles, a proportionality which does not hold at low rates (N.B., in Ravey *et al.* [29] no relation between surface area and the rate of polymerization was observed).

6. Over a wide range of conversions

$$R_{pol} \propto [I]^{1/2} \tag{3}$$

where [I] is the initiator concentration, or, in the case of radiation-initiated processes, the dosage rate.

7. When compared to other monomer systems, there is a relatively high chain-transfer-constant-to-monomer in the polymerization of vinyl chloride.

8. Kuchanov and Bort [28] state that the molecular weight of the PVC produced is not greatly dependent on conversion or initiator concentration. Ravey and Waterman [43], on the other hand, indicate that only at intermediate ranges of conversion is the molecular weight reasonably independent of conversion. The effect of polymerization temperature on molecular weight has been discussed above. The degree of branching is said to be fairly constant up to a conversion of 80%. Up to this degree of conversion, the degree of polydispersion is also fairly constant (Table V indicates that there is considerable variation in the degree of polydispersion with changes in the polymerization temperature). The degree of polydispersion, $\overline{M}_w/\overline{M}_n$, increases markedly above 80% conversion [28].

9. Chain-transfer agents, as expected, markedly decrease the molecular weight of the polymer. In systems containing chain-transfer agents, the initial rate of polymerization increases while the autoacceleration of the process proceeds at a lower rate. If the concentration of a chain transfer agent is sufficiently high, the autoacceleration may be completely suppressed, and the reaction kinetics, at low conversion, becomes identical with that of a homogeneous polymerization. Under these conditions, the resultant polymer may separate as lamellar crystals.

10. In radiation-initiated systems, at temperatures above 0°C and for conversions up to 70–80%, the "after-effect" is virtually unobservable. It can be detected only in polymerizations carried out below −20°C [28].

While the above discussion is based primarily on observations involving bulk polymerizations, suitably modified, many of the factors described also apply to other techniques of polymerization. Kuchanov and Bort [28] and Ravey *et al.* [29] discuss the kinetics of the polymerization in considerable detail.

Considering the complexity of the vinyl chloride polymerization, it is remarkable that reasonably reproducible PVC resins are being produced consistently throughout the industrial world.

While industrially poly(vinyl chloride) is primarily produced by suspension polymerization, several bulk polymerization processes have been patented. At least one of these is used to supply considerable tonnage of the resin. Even though the process uses no emulsifiers or suspending agents, the product exhibits desirably high porosity and high bulk density, coupled with good transparency upon plasticization. Various aspects of the so-called Pechiney– Saint Gobain process may be gathered from references 1, 40, and 47–55.

In Table VII, it will be noted that in a bulk polymerization, at a conversion of 10–20%, secondary particles form. That is, swollen polymer particles collide to form larger particles. As the process proceeds, virtually no free liquid monomer is present. Using a special autoclave, in which polymer lumps could be broken up, bulk polymerization of vinyl chloride could be carried out beyond this low conversion range [47]. This concept was improved upon in the basic patent for the Pechiney–Saint Gobain process [48]. (N.B.: Evidently Produits Chimique Pechiney–Saint Gobain is now a subsidiary of Rhone-Poulence S.A.)

In a vertical autoclave of 1000 liter capacity a mixture of 200 kg of vinyl chloride and 32 gm of 2,2'-azobis(isobutyronitrile) (AIBN) is polymerized under an inert atmosphere for 3 hr at 60°–62.5°C (140°–144°F) at a gauge pressure of 0.93 MPa (gauge) (135 psig) with an agitation rate of 100–130 rpm. At the end of this period, conversion is approximately 10%. Then the reacting mass is dropped into a horizontally arranged autoclave where agitation at 30–50 rpm is by a rotating ribbon agitator. This agitator, consisting of two appropriately arranged spiral ribbons, serves also to break up polymer lumps and to transport the product from the entrance port toward the exit port. Since the polymerization is quite exothermic, initially, external cooling is required to maintain the reaction temperature at 60°C. Later in the process, warm water heating is needed to maintain the gauge pressure between 0.91 and 0.93 MPa. The process is completed in the horizontal agitator within 9.5–10 hr. The autoclave is cooled and the excess VCM is passed through a cyclone bag filter (to remove entrained fine particles) to a recovery system. The resin is freed of much of the residual monomer by evacuating the system two times, breaking the vacuum each time with nitrogen. The yield of resin is 125 kg (62.5%), bulk density 0.46 g/ml (38.6 lb/ft^3) [1,49].

If a larger system is used, lower rates of agitation are used. For example, in a 10-m^3 autoclave, the high-speed agitation is 26–40 rpm; the low-speed agitation in the horizontal autoclave is only 5–10 rpm [49].

The two-speed agitation system appears to be important to the process. In a comparable experiment carried out a single speed (500-liter autoclave operating at 30 rpm), after 13 hr the yield was 61% and the bulk density of the product was only 0.30 gm/ml (25.2 lb/ft^3) [49].

According to a more recent report, the temperature should be in the range of 70°C for the first stage of the reaction since at 62°C, the particles formed are such as not to permit controlled growth in the second stage. The product yield is reported to be 80% [50].

A variation, said to lead to a better control of the porosity of the resin, involves polymerizing only half of the monomer in the first stage of the process and feeding the remainder of the monomer to the horizontal autoclave stage [52]. It is likely that this variation would facilitate regulation of the pressure and the temperature in the autoclave.

Another patented process passes gaseous vinyl chloride into a reactor containing solid poly(vinyl chloride) and lauroyl peroxide at 9 kg/cm² and 60°C. As the reactor temperature rises, the rate of vinyl chloride addition is increased to stabilize the temperature. After 5 hr, the product is isolated. Its bulk density is 0.63 gm/ml. It is capable of absorbing 21% of its weight of dioctyl phthalate [56]. In a fluidized bed technique, gaseous vinyl chloride is passed up through a column containing PVC to produce a resin with a bulk density of 0.65 gm/ml [57].

Instead of varying the agitation rate, a recent two-stage process polymerizes the monomer at two successive temperatures. In a frusto-conical autoclave equipped with a special, complex agitator which is said to permit gentle and controlled mixing, 21 kg of vinyl chloride and 69.3 gm of lauroyl peroxide are heated at 93.5°C (200°F) for 3 hr. The reaction is finished by heating at 106–112 psig (0.73 to 0.77 MPa) and for an additional 10 hr at 51.5°C (125°F). Yield: 81%. In this process, the initially formed milky fluid changed, in turn, to a thick slurry, a wet cake, and finally a dry particulate powder, free of impurities [58].

In laboratory preparations of PVC consideration should be given to the purification of the monomer. Three methods mentioned in the recent literature are the following:

1. Passage of vinyl chloride gas in a nitrogen-filled system through a series of wash bottles filled with dilute sulfuric acid, drying by passage through a silica gel drying column, followed by low temperature fractionation. The monomer thus purified still showed some retardation on polymerization [59].

2. After degassing the monomer on a vacuum line, it is partially polymerized under reduced pressure at 0°C in the presence of AIBN with exposure to ultraviolet radiation. The residual VCM is distilled away from the polymer and stored as a gas in darkened 10-liter bulbs. This monomer exhibits virtually no retardation on polymerization [59].

3. What appears to be the most rigorous method of purification is given as Procedure 2-1.

2-1. Purification of Vinyl Chloride Monomer [60]

With suitable safety precautions (see Note), into the reservoir of a high vacuum line system, 200 ml of liquid vinyl chloride (99.9% pure as supplied, for example, by Ethyl Corporation) is trapped by cooling the reservoir with liquid nitrogen.

The reservoir is evacuated to 10^{-5} torr. Then the stopcock to the vinyl chloride reservoir is closed and the monomer is melted at a Dry Ice–methanol temperature. The gases dissolved in the monomer are vaporized into the evacuated space. The vinyl chloride is refrozen with liquid nitrogen, the stopcock is opened to the high vacuum line and the gases in the free spaces are removed at reduced pressure. When the pressure has again been reduced to 10^{-5} torr, the stopcock is again closed, the monomer is melted again at a Dry Ice–methanol temperature and thus the degassing procedure is repeated until the pressure gauge shows no change in pressure when the stopcock is opened after freezing the vinyl chloride in the reservoir. Complete degassing of 200 ml of the monomer may require 18 cycles. The monomer is then transferred to a series of calibrated ampoules, frozen at the liquid nitrogen temperature, and sealed at 10^{-5} torr of pressure.

To facilitate calibration of equipment at various subzero temperatures, the following relation between density and temperature is used [60]:

$$d = 0.9471 - 1.746 \times 10^{-3}T - 3.24 \times 10^{-6}T^2 \tag{4}$$

where d is the density in gm/ml and T is measured in degrees Celsius.

In working with vinyl chloride polymerizations, it is a common practice to create an inert atmosphere in the free space over the reactants by charging a slight excess of liquid vinyl chloride to the reactor and allowing some of the monomer to evaporate, thus displacing the air of lower density. Much experimental work has been carried out with this procedure even though oxygen is not completely eliminated by this method. It must be pointed out that oxygen does have a retarding effect on the initiation. The thermal stability of the resultant polymer is reduced, as is the molecular weight at comparable degrees of conversion [59,61].

NOTE: As indicated in the introductory section, all experimental work with vinyl chloride must be carried out with extreme safety precautions with particular attention being paid to the hazards of inhalation of minute quantities of the monomer and various matters of working in restricted areas with appropriate warning signs. Since many of the cited experimental procedures were published before OSHA regulations went into effect, these may be quite hazardous and are only given in this chapter to illustrate the general principles that have been used.

Much of the early work on vinyl chloride polymerization, like the work on styrene–butadiene copolymers, was carried out in pressure bottles, screw-cap bottles, crown-cap beverage bottles, sealed glass tubes and ampoules, and metal pressure tubes. Such equipment was encased in metal protective tubes and frequently mounted on the periphery of a slowly rotating drum inside a constant temperature bath in such a manner that the tubes would be rotated end-over-end. The rotational speed of the drum was adjusted so that the centrifugal force did not interfere with proper agitation inside the polymerization tubes. In connection with the use of beverage bottles, it should be noted that today these may be manufactured from rather thin glass and covered with a plastic protector. *We believe such bottles to be even more unsafe than the old-fashion crown cap beverage bottles.* The term "pressure bottles" as used by Batzer and Nisch [62] may refer to the old fashioned "citrate of magnesia" type bottle which was closed with a gasketed ceramic plug held on by a heavy wire arrangement. Such bottles are probably difficult to obtain.

Procedure 2-2 is a typical example of a bottle polymerization.

2-2. *Bottle Bulk Copolymerization of Vinyl Pelargonate and Vinyl Chloride* [13]

With appropriate safety precautions (see Note), to a tared 4 oz. (approximately 125 ml) screw-cap bottle fitted with a GR-N rubber gasket, protected against atmospheric moisture, and cooled with a Dry Ice–acetone bath, are added 8.96 gm of vinyl pelargonate, 15 mg of benzoyl peroxide, and slightly more than 1.36 gm of vinyl chloride. The bottle is gradually warmed to allow displacement of air by evaporating vinyl chloride, and, when the content of vinyl chloride is 1.36 gm, the bottle is capped and placed in a protective metal sleeve. This assemblage is placed in a thermostat at 60°C for 6 hr. The bottle is then cooled in a Dry Ice–acetone bath, opened (in an appropriate hood), and allowed to come to room temperature. The content of the bottle is poured into a nonsolvent, and the precipitated polymer is collected on a filter and dried under reduced pressure for a prolonged period of time. Conversion is only approximately 2% under these conditions. The low conversion in this particular preparation may be a result of the presence of the vinyl pelargonate as a comonomer.

NOTE: Only pressure rated bottles or metal pressure cylinders should be used.

Table VIII summarizes the effect of polymerization temperature, initiator type, and activation with ultraviolet radiation on the conversion of VCM. Since this is based on research dealing with determining the degree of syndiotacticity produced on polymerization, these data only serve to indicate that the rate of PVC formation, in bulk, is usually substantially greater than indicated in the copolymerization experiment described above. In the preparation used for the data in Table VIII, the polymer was purified by dissolving the resin in tetrahydrofuran (or cyclohexanone) and precipitating the product by adding the polymer solution to methanol, followed by filtration and drying under reduced pressure [36].

TABLE VIII

TEMPERATURE EFFECT ON THE BULK
POLYMERIZATION OF VINYL CHLORIDE [36][a]

Polymerization temperature (°C)	Initiator[b]	Reaction time (hr)	Conversion (%)
50	AIBN	1.5	8.3
30	AIBN	2.8	8.5
0	AIBN[c]	2.8	2.8
−15	TBB	2.5	1.6
−30	AIBN[c]	5.7	2.5
−45	TBB	2.5	3.5
−50	AIBN[c]	8.5	0.5
−70	AIBN[c]	7.7	0.1
−70	TBB	6.3	0.6

[a] Polymerization conditions: 1 mmole of initiator per mole of vinyl chloride; reaction temperature as indicated; polymerization by bulk technique under nitrogen of high purity.

[b] AIBN = 2,2'-azobis(isobutyronitrile); TBB = tri-*n*-butylboron.

[c] Activated with ultraviolet radiation.

Bulk polymerizations in sealed tubes are described in references 34 and 63–65. The description in Sorenson and Campbell's book [63,63a] is reasonably detailed. In this procedure a Pyrex tube is cooled with Dry Ice–acetone, flushed with nitrogen, charged with initiator and monomer, again flushed with nitrogen and carefully sealed, keeping in mind that considerable pressure may develop during the process. The tube is placed in a protective jacket and then heated. Table IX outlines some of the polymerizations carried out by similar procedures.

In their study of the effect of oxygen on the polymerization, Garton and George [61] carried out bulk polymerizations in a 4.5-liter steel reactor with an efficient stirrer under reduced pressure. By use of compressible lead gaskets and seals, a relatively tight system was obtained. A typical reaction charge consisted of 1 kg of vinyl chloride and a mixture of 1.5 gm of bis(*tert*-butylcyclohexyl) peroxydicarbonate and 1.8 gm of capryloyl peroxide. The polymerizations were carried out at 54°C for a suitable time to isolate resins at 10% conversion.

Other initiators which have been used in polymerizing vinyl chloride are diisopropyl peroxydicarbonate, di-*sec*-butyl peroxydicarbonate, *tert*-butyl peroxypivalate, as well as lauroyl peroxide and benzoyl peroxide. All of these initiators, after allowances are made for variations in their half-lives at a given temperature, behave similarly as far as the two-phase polymerization of

TABLE IX

BULK POLYMERIZATION OF VINYL CHLORIDE IN SEALED GLASS TUBES

Quantity of VCM charged	Initiator	Polymerization time (hr)	Polymerization temperature (°C)	Remarks	Refs.
50 ml	Benzoyl peroxide (0.15 gm)	24	50	Polymerization under nitrogen. Mol. wt. of polymer 50–75000	63,63a
100 gm	a	4	0	55% conversion; DP of polymer 1950	64
—	Lauroyl peroxide	—	50 ± 0.1		65
—	Benzoyl peroxide	—	50 ± 0.1		65
—	AIBN	—	50 ± 0.1		65
—	Lauroyl peroxide (0.5% of VCM charged)	—	50	Polymerization under greatly reduced pressure	34
—	^{60}Co γ-radiation (4.14 rad/sec)	—	50, 90, 110	Polymerization under greatly reduced pressure	34

^{a} Cobalt naphthenate (0.9 gm, previously treated with H_2O_2), 0.11 gm of benzenesulfinic acid, and 5 ml of benzene, charged at −30°C.

vinyl chloride is concerned [66]. There is considerable advantage, from the economic standpoint, in using an initiator such as diisopropyl peroxydicarbonate, which is active at a relatively low temperature.

Experimental confirmation that the polymerization of VCM is subject to chain transfer to the monomer and to existing polymer is given in Cotman *et al.* [45] and Abdel-Alim and Hamielec [67]. Disproportionation seems to be the dominant mode of chain termination [67].

Aside from conventional chain-transfer agents for vinyl chloride polymerizations, such as chlorinated hydrocarbons or mercaptans, the molecular weight distribution may also be reduced with 2-iodopropane [68] and aldehydes such as propionaldehyde [69].

One example of polymerization in the gas phase is found in the work of Jones and Melville [70]. These investigators initiated the polymerization of vinyl chloride in the gaseous state with methyl radicals from the photolysis of acetone. The monomer vapor was also polymerized by ultraviolet radiation generated from a zinc-spark source. This spark source emits radiation in the range of 2100 Å. Over-all the quantum yield of polymer was found to be low.

3. SUSPENSION POLYMERIZATION

Industrially, poly(vinyl chloride) resins are produced primarily by suspension polymerization procedures. This is particularly true of the so-called "general purpose resins." Unlike resins isolated from emulsion polymerizations (see Section 4 below), a relatively small concentration of foreign materials (suspending agents and surfactants) will be left on the polymer particles produced by suspension and efforts are constantly being made to reduce the level of suspending agents used further [71]. Simply from geometric considerations, suspension polymer particles, with diameters in the millimeter range, need only be covered by small quantities of suspending agent when compared to emulsion particles whose diameters are in the micron range. In addition, the large suspension polymers may represent less of a dusting problem than emulsion particles. While this may seem trivial from the scientific point of view, it is important in plant operations.

Since most applications of poly(vinyl chloride) involve plasticization of the resin, the ease with which a plasticizer is taken up by the resin is one of the important properties of the material.

For this reason, rather irregular, porous beads are desired. The shapes of the resin "spheres" have been variously described as "shrunken orange, popcorn, or modified popcorn" with dull, spongy surfaces [72].

The kinetic equivalence of bulk and suspension polymerizations of vinyl chloride has been demonstrated [65]. Thus, the suspension polymerization process may be considered as the polymerization of individual monomer droplets in an inert solvent phase. Naturally, some characteristics peculiar to the suspension process may be superimposed on those of the bulk process. This matter has been nicely summarized by Eliassaf [73].

In the first volume of this series [73a], we have tried to draw a fairly sharp distinction between suspension and emulsion polymerization. Basic characteristics of the suspension process are, as mentioned above, the kinetic equivalence between bulk and suspension polymerization, the use of monomer-soluble initiators (usually), the use of a suspending agent whose function is to protect individual monomer droplets from adhering to each other and to control the dimensions of the droplets, and the need for agitation. In PVC technology, the suspending agent is probably the most important single factor in determining the final properties of the resin [73,74]. The precise compositions of the commercially used suspending agents are closely guarded secrets. However, a few guidelines as to the nature of these compounds can be found in the published literature and patents. Among them are various types of poly(vinyl alcohols) [particularly a fairly high molecular weight variety derived from about 80% hydrolyzed poly(vinyl acetate), i.e., a grade which is soluble in water with some difficulty], ethoxylated or propoxylated cellulosic materials, other ethers of cellulose, vinyl acetate–maleic anhydride copolymers, salts of acrylic acid copolymers with vinyl esters or acrylic esters, hydrolyzed vinyl acetate–styrene copolymers, metallic stearates (lithium stearate may be of particular interest), vinyl ether–maleic anhydride copolymers, poly(vinyl pyrrolidone), and combinations of two or more of these sometimes with the addition of small quantities of surfactants, etc. Some efforts have been made to correlate the hydrophilic–lyophilic balance (HLB) of the suspending agent on the final properties of PVC [73,75]. By the proper selection of the HLB of the suspending agent, the interfacial tension between VCM and the aqueous phase is controlled. The larger the degree of dispersion of the monomer, the smaller the initial droplets formed, and, ultimately, the larger the porosity and plasticizer absorptive capacity of the resin.

From the discussion on bulk polymerization (see Section 2), it will be recalled that shortly after initiation of polymerization, a fixed number of particles of approximately 1 μm diameter precipitate. In the case of suspension polymerization, this would be within the suspended droplet (whose diameter is in the 100–200 μm range). The particles increase to a diameter of about 2 μm. At low conversion these are opaque in nature but at high conversions, particularly at elevated temperatures or high rates of polymerization, they tend to become transparent and glassy (i.e., nonporous).

Time-lapse photography has been used to follow the course of the suspension polymerization of vinyl chloride. This study showed that the spherical droplets of monomer burst under the action of the shearing force of the agitator to produce deformed and irregular resin particles. Without agitation, on the other hand, opaque spots appear in the spherical monomer droplets which increase in size as the polymerization proceeds [76].

Careful control of the agitation rate is thus necessary to develop reproducible suspension polymerization procedures. With experience, it appears to be possible empirically to correlate laboratory procedures to plant practices.

The most commonly used initiators in this system are AIBN, benzoyl peroxide, lauroyl peroxide (LPO), and diisopropyl peroxydicarbonate (IPP). The last two are probably of most interest industrially. In aromatic solvents, LPO has a half-life of 10 hr at 62°C while that of IPP is 10 hr at only 35°C [76a]. Therefore one may expect that at the same temperature, polymerizations initiated with IPP proceed significantly more rapidly than LPO-initiated ones; this, indeed, has been found.

Mention has already been made of the observation that the polymerization of vinyl chloride is autoaccelerating virtually from the start of initiation. In addition, toward the end of the reaction there is a "heat kick," i.e., an unusually steep increase in the rate of heat evolution [74]. This heat may be difficult to dissipate in large reactors. It is the practice, in the industry, to adjust the initiator level to such a concentration that the polymerization is essentially complete within a time interval of 10–20 hr at 50°–55°C (120°–130°F). One may estimate that, in view of the fact that the half-life in an aromatic solvent of LPO is of the order of 10 hr at this temperature, at the end of the process nearly half of the initiator is still present. This may contribute to the instability of the final resin. Also during the polymerization itself, there is a build-up of initiator in the monomeric phase caused by its exclusion from the polymer phase. Therefore, toward the end of the process, when the monomer concentration is low, the ratio of initiator to free monomer may be inordinately high [29].

The rate of polymerization during a cycle is not uniform. Typically, after an induction period, there is a stage at which the rate of polymerization may be 50% per hour to convert about half of the available monomer. This may take about 10 hr in the typical temperature and initiator concentration range. The last 50% of the conversion may take 3 hr, and, in effect, fully 25% of the initial monomer concentration may be polymerized in the last hour [74].

The effect of variation in the concentration of an initiator on the conversion is shown in Table X.

While most suspension polymerizations are carried out with monomer-soluble initiators, there are cases where totally water-soluble initiators have been used; an interesting intermediate situation is a redox initiation system consisting of lauroyl peroxide and ferrous caproate or of *tert*-butyl hydroperoxide and tributyl borate. These initiators seem to be useful at temperatures as low as −15°C (presumably methanol is added to such systems as an anti-

TABLE X
APPROXIMATE EFFECT OF CHANGES
OF INITIATOR CONCENTRATION
ON THE CONVERSION OF
VINYL CHLORIDE [77][a]

Percent AIBN on monomer	Approximate percent conversion
0.05	10
0.1	20
0.2	40[b]
0.4	70

[a] Polymerization conditions: suspension process; AIBN initiation at 65°C; measured after 1.5 hr.
[b] 70% Conversion in approximately 2 hr.

freeze). While these syntheses are said to produce polymers of increased molecular weight, there is also an increase in polydispersity. This has been attributed to a sharp decrease in the redox initiation as the process proceeds [78].

The effect of pressures up to about 2500 atm (252 MPa) on the rate constants, conversion, molecular weight, and monomer transfer constants for the polymerization of vinyl chloride has been studied [79].

From a computerized study, optimized reactor designs have been proposed. One point which may profitably be considered even in laboratory autoclave experiments is the proposal that the most convenient method of maintaining a constant polymerization temperature consists of heating the reaction system rapidly to initiate the polymerization, then applying cooling as soon as the reaction has been initiated. Presumably toward the end of the process additional heating may be needed to complete the conversion [80].

While the suspension polymerization of vinyl chloride is the most widely used method industrially, there are a few additional problems which need to be mentioned [81].

1. During the polymerization there is a build-up of PVC on the interior surfaces of the reactor. This coat is quite hard and difficult to remove. Obviously it will interfere with the critical heat-transfer problems associated with the process. Particles which may be torn from these reactor scales may contribute to fish-eye formation. The general procedure to overcome this problem is to scrape reactor walls and agitators periodically. It will be recalled

from our discusion of the health problems associated with VCM that the personnel engaged in this activity had the highest incidence of cancer among persons associated with the industry. Attempts have been made to reduce scale formation by modifying the reactor charge, by treating the reactor surfaces, by modification of the reactor design, etc. [82–84].

The proposal has also been made that PVC encrustation which fouls equipment may be removed by treatment with such solvents as tetrahydrofuran or *N*-methylpyrrolidone [1]. This approach may have merit but raises the problem of recovering the solvent from a solution containing a film-forming solute. Also, slightly cross-linked PVC may only be swollen by the solvent and some sort of scraping operation may still be required to clean the equipment.

2. According to Garud *et al.* [81], "PVC wool" forms to some extent during both the polymerization and the drying process. While modifications in the suspending agent system may reduce this problem to some extent, it cannot be completely eliminated. Current U.S. processes do not seem to produce any "PVC wool."

A stirred autoclave usually is the equipment of choice when the concern is with the nature of a completed polymerization product. For precise data, bottle polymerizations are not recommended. The reader is urged to devise suitably safe procedures for himself in the light of applicable OSHA regulations.

3-1. Bottle Suspension Polymerization of Vinyl Chloride [65]

With suitable safety precautions, a 12-oz. (see Note a) beverage bottle (internal capacity approximately 375 ml) is fitted with a rubber stopper equipped with both a gas inlet tube which can be closed off with a suitable valve and an exhaust tube which leads to a drying tube. The gas inlet tube should extend approximately half-way into the bottle. This equipment is tared. The stopper and its tubes are removed and 170 gm of doubly deionized and deaerated water is placed in the bottle followed by 0.2 gm of the suspending agent [vinyl acetate–maleic anhydride copolymer or poly(vinyl alcohol) such as Elvanol 50-42] and 0.2 gm of the initiator (lauroyl peroxide, benzoyl peroxide, or AIBN). The rubber stopper with its inlet and outlet tubes is attached to the bottle, the valve is closed, and the assembly is again tared. Then it is thoroughly cooled in a Dry Ice–acetone bath. While the bottle is in this bath, the gas inlet tube is attached to a cylinder of uninhibited vinyl chloride and the valve on the gas inlet tube is opened. Then the cylinder valve is opened slowly and vinyl chloride is passed into the bottle.

From time to time, the valves are closed, the gas inlet tube is disconnected from the cylinder, and the weight of the bottle is checked. Slightly more than 100 gm of vinyl chloride is condensed into the bottle. The cylinder valve is closed, the cylinder is disconnected (see Note b), and the bottle is allowed to warm to room temperature in a suitable fume hood while standing upright on the pan of a beam balance. Condensate is periodically removed from the bottle and pan. When enough vinyl chloride has boiled away to leave exactly 100 gm of monomer in the bottle, the rubber stopper is removed and the bottle is rapidly sealed with an appropriate crown cap. If other bottles have to be prepared, the capped bottles are stored in a Dry Ice–acetone bath until all bottles have been prepared.

The bottles are then placed in individual protective metal devices. These assemblies are placed in a constant temperature bath maintained at 50° $\pm$ 0.1°C. It is customary to rotate the protected bottles end-over-end by attaching them to the outside of a rather large drum arrangement which rotates slowly in the bath. Reciprocating shaking devices have also been used.

After established times have elapsed, some bottles are removed from the bath and cooled in an ice bath. The bottles are cautiously opened with proper safety precautions. The excess monomer is allowed to vent off. The polymer is recovered by filtration, washed with distilled water repeatedly, and dried to constant weight under reduced pressure at 50°C (see Note c). A high degree of conversion to polymer will be found within 18 hr [85].

Exhaustive extractions with boiling methanol may be used to remove the suspending agent [85].

NOTES: (a) Same precautions as in Preparation 2-2.

(b) Particular care for the safe disposal of residual monomer in the connecting tubing and valves must be taken.

(c) The exhaust vapors from the vacuum oven must, at all times, be treated with the same precautions as other vinyl chloride-bearing gases.

Procedure 3-2 is a somewhat unusual procedure for a suspension polymerization. It makes use of a modest concentration of a surfactant as the only dispersing agent along with a typical totally water-soluble redox initiation system usually associated with emulsion polymerizations. Since the procedure calls for premixing the oxidizing and reducing agents without the presence of any monomer and then adjusting the pH to 2.4, it is not clear that a reasonable level of initiating radicals is present by the time the polymerization tubes are ready for warming. The procedure is given here mainly to illustrate the technique. The reaction charge may, of course, be varied.

3-2. Suspension Polymerization of Vinyl Chloride in Sealed Tubes [63,85a]

(a) *Preparation of aqueous phase.* To 1600 ml of doubly deionized and deaerated water are added 30 gm of a sodium salt of a sulfonated paraffin oil (dispersing agent), 4.3 gm of ammonium persulfate, and 1.6 gm of sodium bisulfite. The pH of this solution is adjusted to 2.4 with dilute sulfuric acid.

(*b*) *Suspension polymerization.* With suitable safety precautions, to a 250-ml heavy-walled pressure tube bearing a tube suitably constricted for subsequent sealing, 100 ml of the aqueous phase prepared above is added. The tube is cooled in a Dry Ice–acetone bath. The free space is swept with dry nitrogen. Then (possibly adopting the techniques indicated in Procedure 3-1) 50 gm of vinyl chloride is added to the tube. The vapor space is again displaced with dry nitrogen, and the tube is rapidly sealed.

After the tube is placed in a suitable protective device, it is agitated at 40°C for approximately 2 hr. The tube is then cooled in a Dry Ice–acetone bath again, then cautiously opened, and allowed, with suitable precautions, to warm to room temperature. The product consists of coarse particles which are filtered off, washed with distilled water, and dried. Yield: approximately 49 gm (98% of theoretical yield).

When it comes to the design of a stirred autoclave, a somewhat costly investment is usually involved. Therefore, consideration must be given to the many operations one may want to carry out in such equipment before it is built. Since such apparatus may have to be used for homopolymerizations and copolymerizations; for suspension as well as emulsion polymerizations, features which may have to be included are thermocouple or thermometer wells, jackets for heating and cooling, an appropriate agitator configuration (anchor, half-moon, or rectangular blades are probably the most common types), baffles, provisions for removing samples and discharging the product, provisions for maintaining and measuring the pressure up to approximately 200 psi (1.4 MPa), blow-out discs, capability to evacuate the vessel, piping for venting, provisions for charging the monomer and other reagents, provisions for adding one or more monomers during the reaction, provisions for adding measured quantities of additional surfactant and two or more components of a redox initiation system, an explosion-proof motor, etc. The autoclave shell may be of stainless steel, glass-lined steel, or, possibly, glass. While stainless steel is widely used in PVC polymerization equipment, it should be kept in mind that the polymer has a tendency to decompose with the evolution of hydrogen chloride, which tends to act on some stainless steels.

Meeks [86] mentions the use of a glass apparatus designed by Sutherland and McKenzie [86a]. This equipment consists of an explosion-resistant light globe (Crouse-Hinds, EV-530) fitted with a stainless steel plate which carries a stainless steel cooling coil, serum cap addition ports, an agitator with a hydraulic motor, thermoregulators, and assorted piping for additions and venting. It is claimed that this equipment has a capacity of approximately 7.5 liters (2 gal.) and can withstand an internal pressure of 200 psi (1.4 MPa) from −55° to 150°C. Unfortunately the description of this apparatus does not give the details of how rubber serum caps withstand a potential pressure of 200 psi.

Among commercial suppliers of laboratory-scale autoclaves are Parr Instrument Co., Moline, Ill., and Ingenieurbüro S.F.S., Zürich, Switzerland.

The addition of vinyl chloride to an autoclave is accomplished conveniently by volume from a calibrated shot tank fashioned from a stainless steel pipe and a pressure sight glass (Jergusen Gauge and Valve Co., Burlington Massachusetts). It is common to introduce the monomer under pressure into an evacuated autoclave using nitrogen as the pressurizing gas.

3-3. Suspension Polymerization of Vinyl Chloride in a Stirred Reactor [87]

With autoclave safety precautions, in a 6-liter stirred autoclave are placed 3000 gm of deionized and deaerated water, 3 gm of ammonium carbonate, 0.9 gm of methyl hydroxypropylcellulose (Methocel 65HG/50 cP from Dow Chemical Co.), and 3 gm of AIBN. The reactor is purged with nitrogen under a pressure of 10 atm. Then the reactor is evacuated and 1500 gm of vinyl chloride is sucked into the autoclave. The autoclave is closed and the agitator is started. The reaction mixture is heated to 65°C and maintained at 65°C by passing either hot or cold water through an external jacket as required. After approximately 2.2 hr, about 70% of the monomer is converted.

The autoclave is cooled to room temperature and, with appropriate safety precautions, is cautiously vented. The resin is separated by filtration, washed with distilled water, and dried under reduced pressure.

Since the final porosity of PVC is so dependent on the nature of the suspending agent, many variations of the suspending agents have been studied. One common practice is the use of a suspending agent such as the cellulosic materials along with a small amount of surfactant. Usually only 10–20% of the suspending agent is surfactant.

Fortunately some preparative details were published in the paper by Ravey *et al.* [29], which deals with the stages through which the monomer passes as the conversion during suspension polymerization is increased. Preparation 3-4 outlines their method. From the discussion by Ravey *et al.* [29] we may be reasonably certain that resins of controlled porosity will result from this preparation.

3-4. Suspension Polymerization of Vinyl Chloride with a Mixture of a Suspending Agent and a Surfactant [29]

With suitable safety precautions and by techniques similar to those of Preparation 3-3, a 1.5-liter stirred glass autoclave (Ingenieurbüro S.F.S., Zürich) is charged with 750 ml of deionized and deaerated water, 0.525 gm of Methocel 90 HG/100 cP (Dow Chemical Co.), 0.1125 gm of Tensaktol A (B.A.S.F.), and 0.6 gm of lauroyl peroxide. The reactor is sparged with nitrogen, evacuated, then charged with 375 gm of vinyl chloride and closed. With agitation at 500 rpm by an impeller with a rectangular blade, the polymerization is carried out at 60°C. Samples are taken at various degrees of conversion. At about 70% conversion, the pressure in the reactor begins to drop from its constant value. The limit of conversion by this procedure is about 85%. Beyond this point, polymerization involves residual monomer which can diffuse only with great difficulty in the nonswollen, dense polymer. Porous resins are evidently formed by this procedure at conversions up to approximately 70%.

To be noted in Preparation 3-4 is that the suspending agent is 0.14% of the monomer and the surfactant is 0.03% of the monomer. By way of comparison, when only hydroxypropylstarch was used in a very similar procedure, 2.22% of this suspending agent was used to prepare a resin with a median particle size of 0.140 mm [88].

The kinetics of VCM polymerization with two initiators (acetyl cyclohexanesulfonyl peroxide and diethyl peroxydicarbonate), which decompose at distinctly different rates at a given temperature, has been studied in an isothermal pressure calorimeter. In this work, the ratio of water to monomer was 1.7:1, and the level of the unspecified suspending agent was 0.3% of the monomer [89]. The use of two such initiators may reduce the problems caused by the "heat kick."

A recent patent describes the suspension polymerization of vinyl chloride in laboratory equipment in which no suspending agent as such has been added. In this procedure, between 10% and 14% by weight of dioctyl phthalate is dissolved in the monomer. The initiator used is 0.25% by weight on the monomer of ammonium persulfate. The resulting product presumably is a plasticized PVC [90]. This procedure is not as strange as it would seem at first. There are examples in the literature of emulsion polymerizations of vinyl acetate in which no external surfactant has been added. These too use persulfate or persulfate–bisulfite redox initiation systems. It may be postulated that fragments of the initiating species form a sulfate or sulfonate end group on a growing polymer chain. This, in effect, would constitute an *in situ* synthesis of surfactant-type molecules. A low level of a surfactant may, of course, act as a dispersing agent in a vinyl chloride polymerization. Thus, in this patent our assumption is that a certain amount of a sulfated (or sulfonated) poly(vinyl chloride) forms to control the size and shape of the final resin particle. The idea of forming a plasticized polymer in a single operation, as carried out in this patent, is one that should be explored further. Unfortunately, the ultimate uses of the resin have such a diversity of plasticizer requirements that this approach may present commercial problems.

The use of poly(vinyl alcohol) or gelatin as the suspending agent is said to lead to resins with rather low absorption capacity for plasticizer [74]. However, particularly on an industrial scale, these reagents are still used. Their use goes back to World War II in Germany [91].

Preparation 3-5 is a recently published variation of this procedure. The ratio of water to monomer is slightly different, but is in the usual range of 1.7:1– 2:1. The suspending agent is, for once, reasonably defined, and a mixture of peroxides is used as the initiator. The odd size of the reactor probably results from a conversion of imperial gallons to metric units.

3-5. Suspension Polymerization of Vinyl Chloride with Mixed Peroxides [61]

With suitable safety precautions, to a 87 ± 0.5-liter stainless steel autoclave whose agitator is arranged off-center, are added 39.6 kg of deionized and deaerated water and 36 gm of poly(vinyl alcohol) (Elvanol 50-42, a du Pont product). Under oxygen-free nitrogen, the suspending agent is vigorously stirred in the water for 30 min. The vessel is sealed and pressure tested to 150 psi (1.05 MPa) with oxygen-free nitrogen. The vessel is evacuated and a solution of 9.0 gm capryloyl peroxide and 11.2 gm of bis(*tert*-butylcyclohexyl) peroxydicarbonate (Perkadox 16 from Novadel, Ltd.) in 22.5 kg of vinyl chloride is introduced into the autoclave. The reaction vessel is maintained at 54°C until the internal pressure drops to 60 psi (0.4 MPa). The autoclave is cooled to room temperature and cautiously vented. The polymer is filtered off, washed, and air dried at 60°C for 3 hr.

Since the ability of PVC resins to absorb plasticizers is an important property of the material, control of the particle size distribution of the resin is a major concern. Obviously the smallest particles are obtained by emulsion polymerization. Efforts are also being made to produce moderately fine resin by suspension techniques.

With a constant polymerization recipe, the rate of agitation is said to influence the particle size distribution of the product. At slower stirring rates, finer particles are said to form than at faster rates. For example, in one preparation, when the reaction mixture was stirred at 150 rpm, 75% of the polymer had a particle size greater than 0.125 mm. On the other hand, when the stirring rate was 100 rpm, only 27% of the resin had a particle size greater than 0.125 mm [92]. This concept was logically extended to a procedure in which stirring was interrupted 10–20 min after polymerization had proceeded for from 15 min to 1.5 hr. This procedure also is said to produce smaller particles than when continuous stirring is used [93,94]. One may assume that when stirring is stopped in an industrial autoclave of considerable size, the monomer droplets will remain in motion for quite some time, although at a reduced rate. Whether this effect can be observed in equipment holding only up to about 10 liters needs to be determined.

Preparation 3-6 illustrates the effect of a change in stirring rate on the average particle size distribution, as measured by the bulk density. If we assume that small particles pack more closely than larger particles in a bulk density determination, then a large bulk density may be considered to be evidence of the presence of small particles. To be noted in this procedure also is the use of a small amount of buffering agent. This procedure is patented and therefore is given here only for reference.

3-6. The Effect of Stirring Rate on Particle Size in the Suspension Polymerization of Vinyl Chloride [95]

With suitable safety precautions, using charging procedures similar to those previously indicated, an autoclave with an internal diameter of 165 cm, fitted with a 3-bladed Pfaudler stirrer whose blade span is 110 cm, is charged with 75 liters of a 2% aqueous solution of Methocel 65 HG/50 cP (methyl hydroxypropylcellulose from Dow), 1.0 kg of dipotassium phosphate, 225 gm of diisopropyl peroxydicarbonate (as a 50% solution), and 2500 kg of de-ionized and deaerated water. The mixture is agitated at a "tip speed" of 4.6 m/sec (equivalent to 79.9 rpm). Then 1500 kg of vinyl chloride is added and the polymerization is carried out at 57°C at a stirring rate of 79.9 rpm. The product yield is 87% with a bulk density of 0.66 gm/ml. When a comparable polymerization was carried out with the agitation at a "tip speed" of 8.0 m/sec (equivalent to 140 rpm), the resultant resin had a bulk density of 0.53 gm/ml. Thus, at the higher rate of agitation, a larger particle size polymer forms than at the slower rate.

Mention has been made above that a serious industrial problem in the suspension polymerization of VCM arises from the accumulation of resin on the reactor walls and on the agitator (see Section 1 and 3). One recent patent claims that by running the polymerization in an acidic medium and in the presence of oxidized polyethylene having a minimum number-average molecular weight of about 1000, this deposition of PVC is substantially reduced [84]. This patented process is outlined in Procedure 3-7. It is given here for reference only and not as a recommendation to practice this patent.

3-7. Suspension Polymerization of Vinyl Chloride with Reduced Tendency to Form Deposits on Reactor Walls [84]

With suitable safety precautions, to a 30-gallon (113.6-liter) glass-lined autoclave equipped with an agitator are added 44 kg of deionized and deaerated water, 50 gm of poly(vinyl alcohol), 8 gm of disodium phosphate, 137 gm of dioctyl phthalate, 14 gm of a silicone antifoaming agent, 6 gm of lauroyl peroxide, and 25 ml of a 30% solution of diisopropyl peroxydicarbonate in toluene. Then enough phosphoric acid is added to bring the aqueous medium to a pH of 2.5 (approximately 25 ml of a 75% aqueous solution of phosphoric acid). This is followed by the addition of 33 gm of finely divided oxidized low density polyethylene (such as AC-629 from Allied Chemical Corporation). Thereupon, using techniques similar to those indicated previously, 33.1 kg of vinyl chloride is added. The polymerization is carried out

TABLE XI

SUSPENSION POLYMERIZATION OF VINYL CHLORIDE

Water:VCM Ratio	Suspending agents (% on monomer)	Initiators (% on monomer)	Other additives (% on monomer)	Reaction time (hr) (initial pressure)	Reaction temp. (°C)	Remarks	Ref.
2:1	Gelatin (0.3) Sorbitan sesquioleate (0.2) NH₄ lauryl sulfate (0.05)	Lauroyl peroxide (0.25)		20	50		97
1.7:1	Hydroxypropylcellulose: low mol. wt. (0.06) high mol. wt. (0.01)	tert-Butyl perpivalate		(180 psi, 1.27 MPa)	69	90% conversion K value 58 bulk density 0.46 gm/ml dioctyl phthalate absorption 25%/100 gm PVC	98
2:1	1% aqueous poly(vinyl alcohol)	Di(tert-butyl) peroxy-oxalate (0.001–5)		6	20	41.5% conversion exhibits improved color stability on heating	99
2:1	Poly(methacrylic acid) (0.8)	$K_2S_2O_8$ (0.02)	After 8 hr pH (adjusted to 8 with Na_2CO_3 solution. Dibutyltin S-(monobutyl-β-mercaptopropionate)	8	55	Resin with improved heat stability formed	100
See Remarks	None	tert-Butyl perpivalate 1.28 gm $K_2S_2O_5$ gradually	$CaCl_2$ (see Remarks)		50	To 1 liter of water containing 40 gm $CaCl_2$ and 400 gm PVC (particle size 0.25 mm), 4020 gm VCM is added during polymerization while pressure is maintained at 94.3 psi 85% conversion bulk density 0.51 gm/ml av. particle size 0.39 mm porosity 19 ml/gm	101
3.6:1 (initial) 1.8:1 (final)	Methyl hydroxypropyl-cellulose (50cP) (0.022) Poly(vinyl alcohol) (21.4% acetyl groups, 50 cP)(0.1)	tert-Butyl perpivalate (0.036)		After 1 hr a second charge of VCM is added over a 5 hr period. Polymerization continued for 6 hr after completion of addition	55	94% conversion bulk density 0.48 gm/ml dioctyl phthalate absorption. 46 gm/100 gm PVC	102
2:1	Poly(vinyl alcohol) (19% acetyl groups) (0.08) Cetyl alcohol (0.5) Sorbitol monooleate (0.2)	Diisopropyl peroxy-dicarbonate (0.015)		10	57	After completion of polymerization, 0.01% (based on monomer) of Bisphenol A is added. Dioctyl phthalate absorption 40% PVC has good thermal dimensional stability and color stability	103

Ratio	Suspending agent	Initiator (phr)	Additives		Temp	Remarks	Ref.
2:1	Sodium stearate (1.5) Cetyl alcohol (0.75)	Azobis(dimethylvalero-nitrile) (0.05)	After completion of polymerization, 0.3% calcium chloride in water added	14.5	47.5	This PVC is suitable as a resin	104
1.5:1	Poly(vinyl alcohol) (20% acetyl groups) (0.1)	Lauroyl peroxide (0.05)	α-Monostearin (1.68) Dioctyl phthalate (0.16)	(21.3–149.3 psi, 0.15–1.1 MPa)	64	Transparent, antistatic PVC 80% conversion DP = 800	105
2:1	Poly(vinyl alcohol) (unspecified % acetyl groups) (0.08) Sodium lauroyl sulfate (0.08)	Diisopropyl peroxydi-carbonate (0.015)	Calcium chloride (0.005) as a 5% aqueous solution when conversion reaches 26%	— (to completion)	57	High porosity uniform particle size PVC dioctyl phthalate absorption 38%	106
2:1	Poly(vinyl alcohol) (0.1)	Azobis(dimethyl-valeronitrile) (0.03)		4–5	56	No sudden exothermic stage toward end of process Heat stable, fish-eye-free PVC	107
2:1	Hydroxyethyl hydroxy-butylcellulose (0.1)	Acetyl cyclohexyl-sulfonyl peroxide (0.012) Lauroyl peroxide (0.15)		18	55	PVC with more rapid plasticizer absorption rate than one prepared with poly(vinyl alcohol)	108
2:1	Hydroxypropylcellulose (0.05)	Isopropyl hydroperoxide percarbonate (0.25)	Mineral oil (0.50)	10	50	90% conversion	109
2:1	Hydroxypropylcellulose [molar ratio of propylene oxide to cellulose = 105:305 viscosity (2% aqueous solution) less than 50 Cp] (0.1) Methyl hydroxypropyl-cellulose (0.34)	Lauroyl peroxide (0.1)		12	58	Product said to have high bulk density, uniform particle size, good plasticizer absorption	110
2:1		Diisopropyl peroxydi-carbonate (0.03)	n-Decylmalonic acid (0.075) pH adjusted to 4	Until pressure drops	60	PVC with narrow particle size distribution average particle size 0.1 mm	111
~1.33:1	Poly(vinyl alcohol) (26% acetyl groups, $\overline{DP}$ 750) (0.05 initially used)	Diisopropyl peroxydi-carbonate (0.15)	At 5% conversion, added poly(vinyl alcohol) (20% acetyl groups, DP 1700) (0.05 in aqueous solution)	10	57	85% conversion $\overline{DP}$ of PVC 1100 bulk density 0.54 gm/ml particle size, 99.5% less than 0.354 mm	112
2.33:1 (monomer consisted of 1.5 gm of 2-ethylhexyl acrylate and 28.5 gm of VCM)		tert-Butyl peroxy-pivalate (0.2) NaHSO$_3$ (1.67) CuCl$_2\cdot$2H$_2$O (0.00039)			25		113
2:1	Ultrawet K (0.05) Ethylenediamine hydrochloride (0.2)	Lauroyl peroxide (0.2)		16	50	93% conversion exceptionally large particles, 99% greater than 40 mesh	113a

at 56°C for 5–6 hr, i.e., until the internal pressure in the reactor has dropped from 120 psig [0.9 MPa (gauge)] to 90 psig [0.6 MPa (gauge)]. The reactor is then cautiously vented, and the polymer is filtered off and worked up.

After four consecutive runs, only 52 gm of PVC had adhered to the reactor walls and this coating was soft and easily removable. The product had improved drying characteristics and was generally hydrophobic.

While the above procedure uses an aqueous medium with an acidic pH, diisopropyl peroxydicarbonate may also be used at a pH greater than 9 [96]. Whether the pH of the medium has an effect on the deposition of PVC on the walls of the reactor will require further objective evaluations.

In Table XI [97–113a], a selection of recently patented polymerization recipes are outlined. The original patents must be consulted in evaluating the details of the procedures used.

In connection with reducing the particle size of PVC suspension polymers, a series of patents claim that incorporation of 0.5–15% of the weight of the monomer of alkyl phthalates (alkyl groups being C_1–C_{10}), glycol esters of fatty acids in which the glycols consist of up to five polyethylene oxide units and the acids are C_1–C_6 aliphatic acids, and phosphate esters of phenolic compounds such as C_6H_5OH, C_7H_7OH, or C_8H_9OH reduces the particle size [114].

Table XII gives the half-lives of several initiators which are currently of interest in suspension polymerizations of vinyl chloride. While these data are a rough guide to the temperature–time relationship to be expected, many other factors need to be considered. For example, lauroyl peroxide brings about a polymerization process of vinyl chloride in which the maximum rate takes place between 60% and 80% conversion whereas the rate of polymerization with diisopropyl peroxydicarbonate is much more uniform throughout the process. Also, particle structure varies with the initiator used [115]. Recently, the development of initiators has been in the direction of molecules with unsymmetrical substitution about the peroxide bridge. Data on these compounds are also included in Table XII [115,116].

4. EMULSION POLYMERIZATION

While most of the commercial production of poly(vinyl chloride) in the United States makes use of suspension procedures, some emulsion polymers are being produced. The method is particularly popular in Europe. Emulsion polymers find applications in latex form and as spray-dried powders.

In the latex, the resin may be present as particles as small as 0.2 μm (0.0002 mm) in diameter. On spray drying, some form of agglomeration takes

TABLE XII

HALF-LIFE DATA FOR SELECTED INITIATORS OF INTEREST IN SUSPENSION POLYMERIZATIONS OF PVC

Initiator	Half-life (hr)							Ref.
	40°C	50°C	60°C	70°C	80°C	100°C		
Acetyl 2-chlorooctanoyl peroxide	5.6	1.55	0.4	—	—	—		116
Acetyl cyclohexylsulfonyl peroxide	2	0.4	—	0.05	—	—		115
Acetyl isobutyryl peroxide	4.9	1.2	0.3	—	—	—		116
Acetyl isobutyl peroxycarbonate	—	15.8	3.8	1.0	—	—		116
Azobis(dimethylvaleronitrile)	—	13.8	3.1	—	0.2	—		116
Azobis(isobutyronitrile)	—	—	18.6	5.0	1.3	—		116
tert-Butyl peroxyneodecanoate	25.4	—	1.7	0.5	—	—		116
tert-Butyl peroxypivalate	84	20	—	1.5	—	0.05		115
		18.5	5.3	1.5	—	—		116
Diisobutyl peroxydicarbonate	9.5	2.6	0.7	—	—	—		116
Diisopropyl peroxydicarbonate	40	7	—	0.5	—	0.01		115
	4.0	1.4	0.3	—	—	—		116
Lauroyl peroxide	123	54	—	3.4	—	0.1		115
	—	54.2	12.8	3.4	—	—		116

place to yield resins with particle diameters in the 1–2 μm (0.001–0.002 mm) range. Thus the dry emulsion-produced resins are approximately one-hundredth the size of suspension polymers. On the other hand, this method of manufacture frequently requires the use of relatively large quantities of surfactants, which remain adsorbed on the particle surface.

The types of surface active agents used include a range of anionic reagents (such as sodium lauryl sulfate and the "Aerosols"). To stabilize the latex, enhance its mechanical stability, reduce agglomeration during polymerization, etc., nonionic surfactants as well as "protective colloids" may also be added to a formulation. In the final resin, the water absorption characteristics, electrical resistivity, clarity, weathering, and compatibility with plasticizers and other compounding ingredients is affected by the nature and concentration of the surfactants used [117].

Table XIII compares the bulk, suspension, and emulsion processes and some of the properties of resulting resins. This comparison may assist in the selection of a process for specific applications.

Classically, emulsion polymerizations involve monomers which are solvents for the homopolymer. Strictly speaking only these systems follow the theory of Harkins as quantified by Smith and Ewart. According to this theory, the propagation rate is a function of the number of polymer particles which are present in the reacting system. In the emulsion polymerization of vinyl chloride, however, the reaction rate increases up to a conversion of about 20–25% since the number of growing polymeric radicals increases. Thereafter the number of particles and the rate of polymerization remains constant until the conversion is in the range of 70–80%. Ultimately the rate decreases as the concentration of monomer declines [1,118].

Talamini and Peggion [119] visualize the process as a modified hetero-geneous solution polymerization. The monomer has an appreciable solubility in the aqueous phase. These authors estimate the solubility to be on the order of 0.5 moles per liter. (Presumably this is under the pressure conditions of a typical reactor. Our Table I gives the solubility as 0.11%, or only approximate-ly 0.02 moles per liter at standard temperature and pressure.) Polymerization starts in the aqueous solution. The polymer which forms separates. The emulsifier in the solution protects the particle from coagulation. By imbibing monomer on the surface of the polymer particle, growth takes place until latex-sized particles form. When the surfactant is consumed by adsorption on these particles, radicals precipitate from solution onto existing particles. Then the number of particles remains constant, very much as in a conventional emulsion polymerization. The total surface area of the polymer particle appears to be involved in the polymer process.

The fact that most PVC emulsion polymerization procedures involve the use of a water-soluble surfactant along with a monomer-soluble emulsifier

TABLE XIII

COMPARISON OF BULK, SUSPENSION, AND EMULSION POLYMERIZATION PROCESSES [81]

Bulk polymerization	Suspension polymerization	Emulsion polymerization
Monomer is polymerized as such without any other medium.	Monomer is suspended in another liquid phase (usually water), aided by suspending agents and strong agitation.	Monomer is emulsified in water by addition of surfactants and moderate agitation.
Monomer-soluble initiators are used.	Monomer-soluble initiators are usually used.	Water-soluble initiators are usually used.
Heat removal is difficult and involves risks. Sometimes controlled by evaporation of monomer. Product does not need to be dried.	Heat is easily controlled by dissipation to the suspending medium. Safety hazards are less than in bulk processes. Product is readily separated from slurry and can be dried easily.	Heat is easily controlled by the aqueous system which acts as heat transfer agent. Safety hazards are low. PVC-latex is formed. To isolate resin, a coagulation step or spray drying is required. Drying costs are high.
The improved two-step process permits control of particle size within a narrow angle.	Particle size, shape, and porosity can be controlled easily.	Particle size can be controlled easily.
Molecular weight distribution is less readily controlled than by other methods.	Good control of molecular weight distribution.	Good control of molecular weight distribution.
Polymer is free of extraneous materials. Heat stability, dynamic stability, and clarity are better than in polymers produced by other methods.	Polymer has lower heat stability and clarity than bulk polymer.	Poor heat stability because of excessive emulsifying agents.
Good processability, suitable for rigid polymer applications because of good heat and dynamic stability and good transparency. Used in packaging articles.	Good processability. Used for general purpose applications, e.g., in extrusion, calendering, molding, and in electrical grade PVC compounds. This process can also be used for specialty-type resins.	The extremely fine resin particles give specific properties such as excellent fusion power to form rigid solids and its ability to form plastisols. This latter property is an important factor in favor of this process.

[16] may very well be related to the fact that the reaction actually starts in solution.

There are polymerization procedures in which the monomer and aqueous phases are mechanically homogenized. The products are dispersions with particles of nearly colloidal dimension [74]. However, kinetically this may be a variation of the suspension polymerization process since a monomer-soluble initiator was indicated. This does not mean that mechanical homogenization of monomer in an aqueous phase might not offer certain advantages in a "true" emulsion process also.

Bik [120] and Goryusko and Vilesov [121] are cited here because the former is a fairly recent review of PVC emulsion polymerization while the latter deals with optimization of the emulsion polymerization of vinyl chloride.

In regard to surfactants used in emulsion polymerizations, the presence of small quantities of iron salts or complexes in this reagent is usually desirable particularly in redox-initiated systems. At least one manufacturer of surfactants furnishes a grade specifically designed for use in emulsion polymerizations. This material probably contains controlled amounts of iron compounds.

Preparation 4-1 is an adaptation of a German procedure used in World War II [91] to laboratory equipment. The level of surfactant used in this process is rather high.

4-1. Emulsion Polymerization of Vinyl Chloride (Hydrogen Peroxide Initiated, Laboratory Scale) [63]

With appropriate safety precautions, in a 3-liter stirrable autoclave are placed 1 liter of distilled and deaerated water, approximately 50 gm of a soap or an anionic surfactant, and 5.5 gm of 30% aqueous hydrogen peroxide. (CAUTION: strong oxidizing agent.) The vessel is cooled in a Dry Ice–acetone bath. The air is displaced with oxygen-free nitrogen, and approximately 500 gm of uninhibited vinyl chloride is condensed from a cylinder into the reactor. The autoclave is closed and warmed to room temperature. With agitation, the reaction mixture is maintained between 40° and 50°C for approximately 20 hr. The polymerization is essentially complete when a notable drop in internal pressure takes place.

The autoclave is cooled to room temperature and cautiously vented. Depending on the nature of the soap or surfactant used, the polymer may be coagulated by the additon of aqueous sodium chloride, aqueous calcium chloride, acidification, or by drying the latex directly. Coagulated resins are washed with distilled water repeatedly and dried under reduced pressure in a warm oven.

Another technique of precipitating poly(vinyl chloride) from a latex consists of adding approximately 10% on a dry solids basis of xylene or toluene to the latex along with more than an equal volume of water. After vigorous agitation to permit the solvent to swell the polymer particles, a 10% aqueous solution of aluminum sulfate is added with agitation and steam is led into the mixture to drive off the solvent [122].

It appears to be the common practice today to carry out emulsion polymerizations in the presence of "seed polymers." The term "seed polymer" refers to a small amount of a latex which has been prepared separately and is added to the aqueous phase of a fresh polymerization prior to the addition of the monomer and emulsifier required by the process. Two obvious advantages arise from the use of seed polymers:

1. If the amount of fresh emulsifier is judiciously limited, virtually no new polymer particles form as the fresh monomer is polymerized. Since the monomer must migrate to existing particles, i.e., the seed latex particles in this case, a well-known phenomenon takes place. The absorption of monomer is related to the surface area of the existing particles. Since the surface area of small particles is disproportionately larger than that of larger particles, more monomer is taken up by the smaller particles. Thus the smaller particles tend to grow at the expense of the larger ones. The effect is to produce a latex with particles of uniform size, or a so-called "monodispersed" latex.

If an excess of surfactant is present, and new particles can form, the rates of polymerization on the seed polymer and on the new monomer–polymer particles is independent of the emulsifier concentration. It appears that polymerization on the seed polymer increases markedly as conversion increases [123].

2. Use of seed polymers introduces a new degree of flexibility into the kind of product that may be produced. Within wide limitations, the seed polymer can be a latex based on any monomer—not only one based on vinyl chloride. For example, the seed may be a hexyl acrylate-based polymer or copolymer which may confer internal, permanent plasticization to the vinyl chloride polymer ultimately associated with it. Furthermore, as in many of these processes, the monomer added to the seed may consist of a mixture of several monomers to yield a large variety of copolymers which have significantly different properties from copolymers prepared without the use of a seed (co)polymer. Whether the products of such procedures are graft copolymers, intertwined chains within the latex particle, mixtures of latex particles of different chemical composition, or combinations of these probably varies with each system. Investigation of the fine structure of such latex systems is difficult. Therefore the technique itself is widely used. The physical properties of the system are related to the operations involved in the preparation rather than with the overall composition and conformation of the polymer chains.

Since the seeding technique generally leads to monodispersed systems, there is also no reason why two or more monodispersed latices with widely different particle sizes cannot be prepared and combined. This would lead to latices with a controlled bi-(or higher) modal distribution of particle sizes.

A further aspect of the emulsion polymerization technique involves the gradual addition of the monomer, of surfactants, and in some cases, of additional initiator solutions to a seeded, reacting system. While this may seem like a technique involving virtually insurmountable complications (particularly if one also intends to add several monomers at separate rates), the fact is that this approach is quite commonly used. With a modest investment in equipment to control and measure flow rates and a little experience, the necessary skills are readily acquired.

Major advantages of this approach are (1) the control which it offers over the particle size distribution of the product, since the amount of surfactant is limited so that no new particles form during the process; (2) the reduction of the amount of surfactant actually required, since only enough is used to cover perhaps 30% of the available surface area of the latex particles; and (3), since essentially all of the surface active agent is adsorbed on the resin, the surface tension of the latex may be controlled as desired.

The patent by Powers [124] is recommended for its discussion of the gradual addition technique to control particle size. The calculations which may be used to determine the amount of emulsifying agent necessary as monomer is added and polymer forms are discussed by Powers in considerable detail. Procedure 4-2 is an example of this technique. Since the point of the procedure is the method of handling the surfactant and the monomer, we do not give details of some of the other aspects of the polymerization procedure which may be worked out by referring to other descriptions given in this chapter. The procedure is outlined here for reference purposes only.

4-2. Seeded Thermal Emulsion Polymerization of Vinyl Chloride with Gradual Addition of Monomer and Surfactant [124]

(*a*) *Preparation of seed latex.* With suitable safety precautions, in suitable pressure equipment, the following latex is prepared. Since only 0.8 lb of this latex will be required in the second step of this procedure, the polymerization should be carried out on a suitably small scale.

To a solution of 804 parts of deionized, deaerated water, 8.25 parts of sodium stearate, 0.84 parts of 28% aqueous ammonia, and 0.84 parts of potassium persulfate is added, with agitation, 187 parts of vinyl chloride. The polymerization is carried out at 50°C for 25 hr. After venting, the resulting latex has an average particle size of 0.0342 μm.

(*b*) *Emulsion polymerization with gradual addition.* With suitable safety precautions, a 15-gallon autoclave, fitted with an off-center, T-shaped agitator rotating at 200 rpm, is charged with 36.7 lb of deionized and deaerated water, 0.27 lb of potassium persulfate, 0.27 lb of 28% aqueous ammonia, 0.80 lb of the above seed latex, and 0.0066 lb of sodium stearate. The autoclave is evacuated and 9.0 lb of vinyl chloride is pumped in. The

TABLE XIV

EMULSION POLYMERIZATION OF VINYL CHLORIDE WITH GRADUAL
ADDITION OF MONOMER AND SURFACTANT [124][a]

Time (hr)	Vinyl chloride added (cumulatively) (lb)	Polymer present in latex (lb)	Surfactant present (lb; dry solids basis)	Percent of calculated surfactant requirement
0	9.0	0.15[b]	0.013	35
1	11.4	4.2	0.070	22
2.1	14.4	8.2	0.164	30
4	18.9	14.1	0.20	26
6	23.2	18.0	0.21	23
8	27.8	22.8	0.37	35
10	33.3	26.4	0.37	32
16	52.4	39.6	0.53	35
20	60.0	49.1	0.60	34
23.2	60.0	57.0[c]	0.60	31

[a] Reaction conditions: 50°C; surfactant added as a 5.25% aqueous solution.
[b] From seed latex.
[c] After venting.

reaction mixture is maintained at 50°C. After the polymerization has been initiated, additional monomer is pumped into the autoclave at a rate of about 2.5 lb/hr and a 5.25% aqueous solution of sodium stearate (which must be maintained between 40° and 60°C to prevent gelling) is pumped in at a rate of approximately 1.2 lb/hr. Table XIV gives more precise details of the addition schedule. The addition of monomer and surfactant is completed in 20 hr. Heating is continued for an additional 3.2 hr after completion of the additions.

The latex is then cooled and vented. The final latex contains 57.0 lb of PVC (95% conversion) with an average particle diameter of 0.253 μm. The latex is quite stable, of low viscosity, and of lower surface tension than if it had been prepared without the controlled addition of surfactant.

A more recent patent not only involves the continuous addition of surfactants, but also of the initiator. The whole process appears to be quite rapid. In part this is thought to be associated with the presence of a dialkylsulfosuccinate surfactant [125]. The procedure is outlined here for reference only.

4-3. Seeded Thermal Emulsion Polymerization of Vinyl Chloride with Gradual Addition of Initiator and Emulsifying Agent [125]

With suitable safety precautions, a 15-m³ glass-lined stirred autoclave is charged with 7.5 parts of a PVC seed latex (35% solid content, average particle diameter about 0.30 μm), 130 parts of deionized and deaerated water and 0.1 part of sodium bicarbonate. The autoclave is evacuated and 20 parts of vinyl chloride is pumped in. The mixture is heated at 50°C with agitation.

Meanwhile two aqueous solutions are prepared at convenient concentrations to permit addition over a 6–7-hr period. Solution A contains 0.02 parts of potassium persulfate; solution B contains 0.25 parts of sodium lauryl sulfate and 0.5 parts of sodium bistridecylsulfosuccinate. The polymerization is initiated by adding about 20% of the potassium persulfate solution A.

After about one hour, the gradual addition of 80 parts of vinyl chloride and the remainder of solutions A and B is started. The total addition time, while the reaction temperature is maintained at 50°C, is 6–7 hr. At the end of the reaction period, the latex is cooled, 0.2 parts sodium bistridecylsulfo-succinate is added quickly to stabilize the latex. The excess monomer is vented off and the latex is spray dried.

An alternative procedure to the continuous, gradual addition of reagents to the polymerizing latex which may be convenient under certain circumstances involves the introduction of small fixed quantities of the materials quite rapidly at fixed time intervals ("slug-wise" addition). The additions may also be tied to the percent conversion of the monomer in the system, which is said to permit the preparation of a latex with 48% solids content even though only 0.7%, based on monomer, of an emulsifier is used [126]. Another patent claims the advantages of adding a sodium lauryl sulfate solution to a vinyl chloride emulsion polymerization linearly in relation to the heat of polymerization as it evolves during the process [127].

The effect of the pH on the rate of polymerization has been investigated to some extent. Thus, Powers [124] carried out the polymerization in the presence of aqueous ammonia while Chatelain buffered the system at pH 7 [128]. Liegeois [129] observed that the initiation of the emulsion polymerization of vinyl chloride parallels the rate of decomposition of the persulfate ion (at various reactor pressures below the vapor pressure of the system). In the pH range from 3 to 9, as the rate of decomposition of the persulfate ion increases, the rate of initiation of the polymerization also increases. Thus, it would seem that the pH of the system does have an effect on the rate of conversion. Yet since the process can take place in acidic, neutral, or basic media, the choice of the pH to be used does not appear to be very critical.

The cation associated with the surfactant is said to have some influence on the properties of the final resin. Thus, for example, lithium, potassium, ammonium, or zinc sulfonate salt emulsifiers are said to produce polymers with improved transparency as compared to those formed when the analogous sodium salts are used [130]. The addition of magnesium, calcium, or zinc soaps to the aqueous phase during polymerization is claimed to improve the stabilization of the resin by organotin, barium–cadmium, and lead stabilizers [131].

Emulsion polymerizations of vinyl chloride have not only been initiated by the thermal decomposition of the persulfate ions, but also by combinations of oxidizing and reducing agents (the so-called "redox systems"). These redox systems permit initiation at substantially lower temperatures than thermally initiated emulsion polymerizations. For example, a vinyl chloride–vinylidene chloride copolymer was prepared in an acidic medium (nitric acid) using potassium persulfate and sodium bisulfite at $30° \pm 0.5°C$ in 67% conversion within 7 hr [132]. The two components of the redox system may be added gradually as the polymerization proceeds [133]. (N.B.: The polymerization described in Fischer and Lambling [133] is carried out at 0°C.) To prevent freezing of the system during polymerization, a quantity of methanol is added to the continuous phase.

Some typical redox systems mentioned in the recent literature are potassium persulfate–potassium metabisulfite [133]; the monomer-soluble oxidizing agent *tert*-butyl perpivalate and potassium metabisulfite [134]; hydrogen peroxide–sodium formaldehyde sulfoxylate [135]; hydrogen peroxide–sodium formaldehyde sulfoxylate in the presence of cupric sulfate and EDTA [136]; hydrogen peroxide–oxalic acid in the presence of ferrous sulfate [137]; and persulfates and sodium formaldehyde sulfoxylate [138].

In Procedure 4-4, a redox system is used as the initiator for a nonseeded system. Since this procedure is based on a study which was carried out for polymerizations at temperatures ranging from $-30°C$ to $+20°C$, one third of the continuous phase is methanol to prevent freezing. In this study, the PVC formed has the highest yield, highest glass transition temperature, and the greatest degree of syndiotacticity if the polymerization is carried out at 0°C and the ratio of ferrous sulfate to hydrogen peroxide is 0.09:1. A satisfactory glass transition temperature and level of syndiotacticity is also obtained at 15°C. The procedure is given in generalized form.

4-4. *Redox Emulsion Polymerization of Vinyl Chloride* [137]

With suitable safety precautions, a stainless steel autoclave fitted with a magnetic stirrer and a coil for heating and cooling is charged with a solution of 150 ml of methanol, 300 ml of water, 0.0315–0.538 wt.% (based on vinyl chloride) of ferrous sulfate, 0.250 gm of oxalic acid, and 0.975 wt.% (based on the methanolic water phase) of the emulsifier (a sodium alkylsulfonate in the C_{12}–C_{18} range of alkyl groups).

The autoclave is purged with pure nitrogen and evacuated three times for 15 min at 20°C to 10 mm Hg. Between evacuations, the system is purged with nitrogen. The autoclave is then cooled to $-30°C$ and charged with 186 gm of vinyl chloride, followed by 0.247–0.860 wt.% (based on vinyl chloride) of hydrogen peroxide. The polymerization is carried out with stirring at a designated constant temperature between $-30°C$ and $+20°C$. The reaction time is on the order of 7 hr for conversions of 45–48%.

Upon completion of the reaction, the autoclave is cautiously vented and coagulum is filtered off. The latex may be coagulated by the addition of a 5% aqueous solution of potassium aluminum sulfate. The coagulum is filtered off, washed with distilled water, and dried at reduced pressure at 60°C. The syndiotactic index of this latex is determined from the infrared absorptions at 635 and 692 cm^{-1}. When prepared at 0°C, the resin has a syndiotacticity index of 2.1 and a glass transition temperature of 105°C.

An example of a complex seeded polymerization is given in Procedure 4-5. In this process the pH of the seed latex should be maintained between 1 and 3.5, possibly by gradually adding carboxylated monomers or by the use of appropriate acids or bases. The system is interesting since it is a patented method for the production of a vinyl chloride–ethylene copolymer with a 2-ethylhexyl acrylate seed polymer and additions of acrylic acid. The procedure is given here for reference only. Note the use of preemulsified monomers in this procedure.

4-5. Seeded Emulsion Copolymerization of Vinyl Chloride and Ethylene with Acrylic Acid [138]

(*a*) *Preparation of seed latex.* In an autoclave are placed 140 gm of a 25% aqueous solution of sodium lauryl sulfate, 4200 ml of distilled water, and a solution of 46.5 gm of ammonium persulfate in 500 ml of water. The mixture is warmed to 80°C while, over a period of 1 hr, a monomer emulsion obtained by dispersing 3500 gm of 2-ethylhexyl acrylate and 70 gm of a 25% aqueous solution of sodium lauryl sulfate in 1-liter of distilled water is added. By the end of the addition period, the polymerization is substantially complete. The latex has a solid content of 37%, a surface tension of 56 dynes/cm, a pH of 2.0, and an average particle size between 0.03 and 0.06 μm.

(*b*) *Seeded polymerization of vinyl chloride and ethylene with acrylic acid.* With suitable safety precautions, in a stirred, high pressure autoclave, with provisions of introducing a number of separate streams, is placed 3840 gm of distilled water, 160 gm of the seed latex described above and 1200 gm of vinyl chloride. (The pH of this dispersion is 1.9.) With stirring, this dispersion is heated to 60°C and maintained at this temperature. The autoclave is then charged with ethylene until the pressure reaches 1200 psi (8.3 MPa). While this ethylene pressure is maintained, 200 gm of a 3% aqueous solution of sodium persulfate is charged to the autoclave. Then, simultaneously and continuously three streams are added to the apparatus over a 13-hr period. These consist of (1) 4800 gm of vinyl chloride monomer, (2) 900 gm of a 3% aqueous solution of sodium persulfate, and (3) 1200 gm of a 10% aqueous solution of acrylic acid.

After the addition has been completed, heating and stirring are continued for an additional 3 hr at 60°C. The latex is cooled to room temperature and vented.

The product latex has a solid content of 48.7%, a surface tension of 59.7 dynes/cm, and a pH of 1.65. The ethylene content of the dry resin is 18 wt.%, acrylic acid content 1.9 wt.%, and vinyl chloride content 80.1 wt.%. The latex itself is said to be stable to ten twelve-hour freeze–thaw cycles between −18°C and +25°C.

5. SOLUTION POLYMERIZATION

In discussing solution polymerizations of vinyl chloride we must recall that poly(vinyl chloride) is insoluble in its monomer as well as in many common solvents. Therefore we have to distinguish between *true solution polymerizations*, i.e., systems in which the monomer, the added solvent, and the polymer are truly in solution; and *pseudo-solution polymerizations*, i.e., systems in which the monomer is in true solution but from which the polymer separates as a swollen phase. Among the true solvents are tetrahydrofuran (THF), chlorobenzene, 1,2-dichloroethane, diethyl oxalate, 2,4,6-trichloroheptane, and many plasticizers. Examples of pseudo-solvents are methanol, aliphatic hydrocarbons, and cyclohexane [16].

In a pseudo-solvent such as cyclohexane, the polymerization process is similar to that of a bulk polymerization except that the rate of conversion is lower. In methanol the rate is enhanced.

In true solvents, the situation is more complex since there may very well be ratios of monomer to solvent where the insolubility of PVC in the monomer overcomes the solvating power of the solvent. In such situations, autoacceleration is observed, while in dilute true solutions, in which the polymer remains in solution, the polymerization rates are linear with time, i.e., no autoacceleration is observed [16]. Unfortunately, the literature is not always very precise in defining the type of solution polymerization under consideration.

Commercially, only a very small proportion of the total production of PVC is carried out in solution. The cost of the solvent and of solvent recovery makes the process unattractive. Its primary application is in certain processes such as the copolymerization of vinyl chloride with vinyl acetate (10–25% vinyl acetate) or the terpolymerization of vinyl chloride–vinyl acetate and certain maleates. The resultant polymer solutions are valuable coating materials [1]. The solution processes are said to produce highly uniform copolymers with narrow molecular weight distributions [1,139].

5-1. *Solution Polymerization of Vinyl Chloride in a Pseudo-Solvent* [63,139a]

With suitable safety precautions, 47 gm of cyclohexane and 0.8 gm of 2,2′-azobis(isobutyronitrile) (AIBN) are placed in a 250-ml pressure vessel.

The vessel is flushed with pure nitrogen. It then is cooled in a Dry Ice–acetone bath and 40 gm of vinyl chloride is condensed into the reactor. The free space is again swept with nitrogen. Then the reactor is sealed and, with suitable protection, it is warmed for 20 hr at 40°C.

The pressure vessel is cooled and the excess monomer is vented off. The precipitated polymer is filtered off. Yield: approximately 34 gm (85%). The resin is washed with ethanol, then with water and is dried under reduced pressure.

In a recent patent, a polymer prepared in a pseudo-solvent was treated with a plasticizer such as epoxidized linseed oil or di-2-ethylhexyl phthalate before the solvent was removed. Then, after the solvent had been evaporated under reduced pressure, a PVC paste containing up to 50% of plasticizer was readily produced [140].

While the product yield is low, probably because it is carried out near atmospheric pressure, a polymerization procedure [141] has been reported which uses a typical redox system in a methanolic system.

In a recent publication, it was demonstrated that in dimethyl formamide polymeric PVC radicals can be terminated by reaction with ferric chloride at 40°–70°C [142]. In a redox system the metal ions may be present both in the ferrous and the ferric state, hence there is the possibility that a sufficiently high level of this activator will interfere with the polymerization.

Another initiating system which has been used in the solution polymerization of vinyl chloride is a combination of silver nitrate and tetraethyltin [143].

It is believed that true solution polymerization leads to PVC with minimal branching in its chains, particularly at low conversions. Such polymerizations have been carried out in di-2-ethylhexyl phthalate [144].

To determine whether the solvent itself can bring about crystallinity or stereoregularity in the free-radical polymerization of vinyl chloride, a large number of polymerizations in a variety of solvents was carried out. Among the solvents studied were nitro compounds, amines, amides, nitriles, ethers, esters, ketones, aldehydes, carboxylic acids, anhydrides, and heterocyclic compounds. In all, 38 solvents were evaluated. The experiments were carried out in sealed glass ampoules with 0.25–2 moles of solvent per mole of VCM, under nitrogen, at temperatures between 0° and 60°C for 18 hr with 0.0008 mole of AIBN per mole of monomer.

In general, it was found that the majority of these solvents did not lead to stereoregular PVC. The resins produced in some aliphatic acids did exhibit a somewhat enhanced stereoregularity. Only in aliphatic aldehydes was crystalline PVC formed. Considering that aldehydes are chain-transfer agents for VCM polymerizations, it was not entirely surprising that yields and molecular weights were low [145].

6. RADIATION-INITIATED POLYMERIZATION

The polymerization of vinyl chloride has been initiated with various radiation sources. Since this technique has little practical value and usually requires equipment not readily available in the ordinary laboratory, this field will be reviewed here only briefly.

An early experiment with the bulk polymerization of vinyl chloride in sealed quartz tubes at 40°C with ultraviolet radiation indicated that while conversion is quite high, partial degradation of the polymer accompanied the process [146]. Since the degradation of PVC by ultraviolet radiation is quite well known, this observation is not particularly startling. The average molecular weight was about 50,000. The polymer exhibited a high degree of polydispersion.

With uranyl nitrate as a photosensitizer, vinyl chloride was polymerized at low temperatures using a tungsten–iodine lamp as the radiation source. Since this lamp produces primarily visible light, degradation of the polymer was minimized. The degree of crystallinity of the resulting polymer was said to be inversely related to the polymerization temperature [147].

The solid-state polymerization of vinyl chloride at $-196°C$ with ultraviolet radiation using a hydrocarbon glass as a medium has been studied [148].

The solution polymerization of vinyl chloride in N,N-dimethylacetamide at $0°–4°C$, irradiated for 5 hr with light with wavelengths in the range of 300–400 nm gave a 67% yield of PVC [149]. The kinetics of the photoinitiated polymerization of VCM at $-20°C$ in tetrahydrofuran, 1,2-dichloroethane, and diethyl oxalate has been studied dilatometrically [150].

While one would expect that polymerizations of a monomer in a heterogeneous system with ultraviolet radiation would be difficult because of the blocking effect of the continuous phase, emulsion polymerizations of vinyl chloride have been carried out. The design of a small-scale pilot plant has even been reported upon [151]. With anionic emulsifiers, the process is actually mechanistically similar to chemically initiated latex polymerization of VCM [152]. At high conversions, inversion of the latex and the formation of a free-flowing powder containing 35–70% PVC may take place [153]. Stable latexes were also formed in the presence of cationic surfactants [154].

The gas-phase polymerization in the presence of radiation doses in the range of 10–60 Mrad have been studied. The tacticity of the polymer was found to be independent of the polymerization temperature [155]. Upon irradiation of PVC with an electron beam, the polydispersity of the resin increases due to simultaneous chain scission and cross-linking [156].

The polymerization of vinyl chloride has also been initiated by γ-radiation in bulk [34,157], in solution [158,159], and in emulsion [160].

Plasma-initiated polymerizations of vinyl chloride are mentioned in Thompson and Mayhan [161] and Liepins and Sakaoku [162]. Polymerization in an electrodeless glow discharge is discussed in Yasuda and Lamaze [163], and the effect of electric field voltage between two metal plates on the polymerization of vinyl chloride is discussed in Volkov et al. [164].

7. POLYMERIZATION WITH ORGANOMETALLIC INITIATORS

It will be recalled that vinyl chloride polymerized with conventional free-radical initiators exhibits a certain degree of tacticity and crystallinity. In an effort to enhance the properties of PVC, attempts have been made to initiate the polymerization with Grignard reagents, Ziegler–Natta catalysts, and related reagents. It had been hoped that by the use of ionic or coordinated polymerization mechanisms, greater degrees of crystallinity might be achieved. References 165–167 are reviews dealing with this topic.

The nature of the polymerization initiated by *tert*-butylmagnesium chloride is still subject to some controversy. Initially Guyot and Tho [168], stated that while there was no direct evidence for an anionic polymerization mechanism, such a process was indicated by the nature of the end groups of the polymer chain, the high degree of crystallinity of the resin formed, and its low molecular weight.

A few years later Guyot and Mordini [169] proposed a different process. This proposal visualized that an anionic polymerization leading to a low molecular weight, oily polymer was accompanied by a simultaneous free-radical reaction producing solid polymers.

The radicals were thought to be generated from the Grignard reagent in such a manner that *tert*-butyl groups were attached to each chain. The high polymer was not obtained when the ionic mechanism predominated, as, for example, when a solvating agent such as dimethoxyethane had been introduced. The free-radical mechanism was also supported by the fact that the reaction could be inhibited with nitric oxide, that copolymerization with vinylidene chloride was possible, and that the reaction had temperature and solvent characteristics similar to well-established free-radical polymerizations.

Procedure 7-1 briefly outlines the experimental procedure used in this work. The reaction is said to exhibit no induction period. The propagation rate constant at 25°C, k_p, is 26×10^{-4} sec^{-1} and the termination rate constant at 25°C, k_t is 4.7×10^{-5} sec^{-1} [168].

7-1. *Polymerization of Vinyl Chloride Initiated by a Grignard Reagent* [168]

With suitable safety precautions, a 1-liter autoclave fitted with a magnetic stirrer and an internal cooling coil is charged with 371 gm of tetrahydrofuran,

131 gm of vinyl chloride, 11 gm of butane, and 0.090 moles of *tert*-butyl-magnesium chloride. The system is maintained at 25°C for 25 hr. The reaction is terminated by injecting concentrated hydrochloric acid into the reactor. The autoclave is vented. From the reaction mixture 66.3 gm of poly(vinyl chloride) is isolated.

Polymerizations initiated by Ziegler–Natta catalysts may also proceed by a radical rather than a coordinated anionic mechanism [170]. In the polymerization of vinyl chloride with a catalyst system comprising tetrabutoxytitanium, triethylaluminum, and epichlorohydrin, the crystallinity of the resulting PVC was similar to that obtained by radical polymerization. In heptane, the process proceeded at one-half to one-third the rate as when the solvent was absent [170].

Procedure 7-2 is an outline of a patented procedure illustrating the proportions of reagents which may be used.

7-2. Polymerization of Vinyl Chloride with a Ziegler–Natta Type Catalyst [171]

With suitable safety precautions, a pressure vessel filled with nitrogen is charged, in turn, with 36 gm of tetrahydrofuran, 11.8 gm of vinyl chloride, and 0.40 gm of triisobutylaluminum. The mixture is warmed to 30°C, and a suspension of 0.16 gm of vanadium trichloride in 5 gm of heptane is added. The mixture is warmed at 30°C for 19 hr. Then the reaction is stopped by the injection of 4 gm of anhydrous ethanol and the polymer is separated by centrifugation. The product is said to be crystalline.

When a conventional Ziegler–Natta catalyst such as titanium tetrachloride-triethylaluminum in tetrahydrofuran is mixed with vinyl chloride, the rate of conversion is quite low (0.014%/hr at 30°C). On the other hand, when a mixture of titanium tetrachloride and vinyl chloride is mixed with a composition consisting of vinyl chloride, triethylaluminum, and tetrahydrofuran, a substantially faster rate (1.52%/hr at 30°C) is observed in the formation of isotactic poly(vinyl chloride) [172].

When the catalyst system consists of vanadium oxychloride, triisobutyl-aluminum, and tetrahydrofuran, the activation energy of the process is found to be 16.1 kcal/mole, similar to that found for other Ziegler–Natta systems. The PVC prepared by this system has a decomposition temperature of 275°–335°C, compared to 250°–295°C for PVC prepared by free-radical initiation [173,174].

While Haszeldine's group imply that their processes are true coordinated polymerizations [173,174], Mitani and co-workers [175] presented evidence that in a system initiated by tetrabutoxytitanium–ethylaluminum sesqui-chloride, propagation proceeds by a free-radical mechanism according to Bernoulian statistics with syndiotactic placement favored by 0.550 kcal/mole.

The enhanced regularity of the polymer conformation is responsible for increased crystallinity.

In the initiation of the polymerization of vinyl chloride with triethylaluminum–benzoyl peroxide in the presence of allyl acetate, dibutyl ether, and pyridine, the process was thought to be a free-radical one [176,177].

The novel complex triethylaluminum–hexamethylphosphoramide acts as an anionic initiator for the polymerization of methyl methacrylate or acrylonitrile at 80°C. However, with vinyl chloride, at 30°C, it imitates a free-radical process [178].

8. MISCELLANEOUS

A. Cross-Linking of Poly(vinyl chloride)

A variety of chemical cross-linking agents has been suggested for the curing of poly(vinyl chloride) systems to improve properties such as impact resistance. Listed here are a number of these cross-linking agents with representative references to the literature:

Allyl esters [179–183].
Triallyl cyanurate and isocyanurate [184–188].
Ethylene dimethacrylate [185,189–191].
Triethylene glycol dimethacrylate [192,193].
Polyethylene glycol dimethacrylate [185,194,195].
1,3-Butylene dimethacrylate [191].
Trimethylolpropane trimethacrylate [194,196,197].
Pentaerythritol tetramethacrylate [198,199].

B. General

1. Preparation of syndiotactic PVC with cumyl hydroperoxide, sulfur dioxide, and sodium methyl mercaptide as initiator [200].

2. Emulsion polymerization of VCM in presence of chain-transfer agents [201].

3. Emulsion polymerization with ammonium-containing emulsifiers to produce PVC with improved electrical resistivity, with preemulsification of the monomer [202].

4. Emulsion polymerization with sodium dialkyl phosphite as emulsifier [203].

5. Improvement of emulsions by polymerization in the presence of higher alcohols [204–207].

6. Continuous emulsion polymerization of vinyl chloride [208].

7. Conversion of PVC latex to a suspension polymer [209,210].

8. Emulsion polymerization initiated by sulfur dioxide and sodium ferripyrophosphate [211].

9. Formation of stable PVC latex by bubbling VCM through a persulfate solution at atmospheric pressure [212].

10. Use of butyl thioglycolate or 2-ethylhexyl thioglycolate as chain-transfer agent in suspension polymerizations [213].

11. Suspension polymerizations with *in situ* preparation of diethyl peroxydicarbonate [214].

12. Improvement of the porosity of PVC by adding barium chloride or magnesium chloride during suspension polymerization [215].

13. Preparation of a vinyl chloride copolymer powder by copolymerizing a PVC latex under suspension polymerization conditions with a comonomer [216].

14. Reduction of porosity by adding vinyl chloride during suspension polymerization at reduced pressure [217].

15. Use of triethylboron and methyl ethyl ketone peroxide, gelatin, and a nonionic suspending agent in the suspension polymerization of VCM [218].

16. Copolymerization of vinyl chloride, a higher alkyl acrylate, and acrylonitrile by suspension [219].

17. Uniaxially oriented poly(vinyl chloride) [220].

18. Graft copolymerization of vinyl chloride onto poly(vinylpyrrolidone) [221].

19. Suspension polymerization of vinyl chloride with preemulsified monomer and interruption of agitation [222].

20. Polymerization of vinyl fluoride in an aqueous medium with azobis-(isobutyramidine) hydrochloride and ammonium iodide [223].

21. Preparation and applications of vinyl chloride–carbon tetrachloride telomers [224,225].

22. Preparation of a photodegradable copolymer of vinyl chloride and carbon monoxide [226].

23. Separation of PVC into monodispersed fractions [227].

24. Kinetics of emulsion polymerization [228].

REFERENCES

1. R. T. Gottesman, *Coat. Plast. Prepr., Am. Chem. Soc., Org. Coat. Plast. Chem.* **34**, No. 2, 16 (1974). Complete preprint kindly supplied to us by the author.
2. R. L. Rawls, *Chem. Eng. News* **53**, September 15, p. 11 (1975); Anonymous, *ibid.*, November 10, p. 13.

3. H. A. Sarvetnick, "Polyvinyl Chloride," p. 14. Van Nostrand-Reinhold, Princeton, New Jersey, 1969.
4. M. H. Naitove, *Plast. Technol.* December, p. 37 (1974).
4a. While this chapter was in preparation, the Environmental Protection Agency issued proposed emission standards for vinyl chloride in the *Federal Register* **40**, December 24, No. 248, Part II. Vinyl Chloride. Title 40CFR, Part 61, p. 59531 (1975).
5. *Federal Register* **39**, October 4, No. 194, Part II. Title 29—Labor. Chapter XVII—Occupational Safety and Health Administration, Department of Labor. Part 1910—Occupational Safety and Health Standards. Standards for Exposure to Vinyl Chloride, p. 35890 (1974).
6. U.S. Department of Labor, Occupational Safety and Health Administration, "Vinyl Chloride," Job Health Hazard Ser., OSHA 2225. U.S. Dept. of Labor, Washington, D.C., June 1975.
7. Tenneco Chemicals, Inc., Intermediates Division, Safety Manual, "Handling and Use of Vinyl Chloride Monomer (VCM) and Products Containing VCM in the Piscataway Research and Development Laboratories," Sect. RPD-3. Tenneco Chem., Inc., Piscataway, N.J., 1975.
8. J. Golstein, G. Coppens, and J. Devoine, German Patent 2,331,895 (1974); *Chem. Abstr.* **80**, 146608o (1974).
9. Dow Chemical U.S.A., "Physical Properties and Constants of Pure Vinyl Chloride." Dow Chemical, Midland, Michigan, 1975.
10. J. Bandrup and E. H. Immergut, "Polymer Handbook." Wiley (Interscience), New York, 1966.
11. F. P. Reding, E. R. Walter, and F. J. Welch, *J. Polym. Sci.* **56**, 225 (1962).
12. W. R. Moore and R. J. Hutchinson, *Nature (London)* **200**, 1095 (1963).
13. C. S. Marvel and W. G. DePierri, *J. Polym. Sci.* **27**, 39 (1958).
14. V. Regnault, *Ann. Pharm. (Lemgo, Ger.)* **15**, 63 (1835); *Ann. Chim. Phys.* [2] **69**, 151 (1838).
15. E. Baumann, *Ann. Chem. Pharm.* **163**, 308 (1872).
16. J. L. Benton and C. A. Brighton, *Encycl. Polym. Sci. Technol.* **14**, 320 (1971).
17. L. G. Shelton, D. E. Hamilton, and R. H. Fisackerly, *in* "Vinyl and Diene Monomers" (E. C. Leonard, ed.), Part 3, p. 1205. Wiley (Interscience), New York, 1971.
18. B. Sedlátek, *J. Polym. Sci., Part C* **33**, (1971).
19. J. T. Barr, Jr., *in* "Manufacture of Plastics" (W. M. Smith, ed.), Vol. 1, p. 303. Van Nostrand-Reinhold, Princeton, New Jersey, 1964.
20. F. Chevassus and R. de Bronbelles, "The Stabilization of Polyvinyl Chloride." St Martin's Press, New York, 1963.
21. E. Iida, *Raba Daijesuto* **23**, 73 (1971); *Chem. Abstr.* **76**, 87226u (1972).
22. R. J. Fox, R. W. Gould, and G. F. Grant, *Rep. Prog. Appl. Chem.* **55**, 77 (1971).
23. W. S. Penn, *in* "PVC Technology" (W. V. Titov and B. J. Lanham, eds.). Wiley (Interscience), New York, 1972.
24. R. Krüger, K. Hoffmann, and W. Praetorius, *Kunststoffe* **62**, 602 (1972).
25. W. D. Davis, *Tech. Pap., Reg. Tech. Conf., Soc. Plast. Eng., Cleveland Sect.* p. 1 (1972).
26. I. B. Kotlyar and E. N. Zil'berman, *Usp. Khim. Fiz. Polim.* p. 258 (1973); *Chem. Abstr.* **81**, 4278s (1974).
27. D. Feldman, *Mater. Plast. (Bucharest)* **10**, 640 (1973); *Chem. Abstr.* **81**, 78241w (1974).
27a. N. Yamazaki, *Prog. Polym. Sci., Jpn.* **2**, 171 (1971); *Chem. Abstr.* **78**, 30230c (1973).

28. S. I. Kuchanov and D. N. Bort, *Vysokomol. Soedin., Ser. A* **15**, 2393 (1973); *Polym. Sci. USSR (Engl. Transl.)* **15**, 2712 (1973).
29. M. Ravey (Rogozinski), J. A. Waterman, L. M. Shorr, and M. Kramer, *J. Polym. Sci., Polym. Chem. Ed.* **12**, 2821 (1974).
30. R. P. Westhoff, F. H. Otey, C. C. Mehltretter, and C. R. Russell, *Ind. Eng. Chem., Prod. Res. Dev.* **13**, 123 (1974).
31. C. S. Marvell, J. H. Sample, and M. F. Roy, *J. Am. Chem. Soc.* **61**, 3241 (1939).
32. G. Bier and H. Kramer, *Kunststoffe* **46**, 498 (1956).
33. M. H. George, R. J. Grisenthwaite, and R. F. Hunter, *Chem. Ind. (London)* p. 1114 (1958).
34. A. Rigo, G. Palma, and J. Talamini, *Makromol. Chem.* **153**, 219 (1972).
35. G. Natta and P. Corradini, *J. Polym. Sci.* **20**, 251 (1956).
36. J. W. L. Fordham, P. H. Burleigh, and C. L. Sturm, *J. Polym Sci.* **41**, 73 (1959).
37. O. C. Bockman, *Br. Plast.* June, p. 364 (1965).
38. G. Talamini and G. Vidotto, *Makromol. Chem.* **100**, 48 (1967).
39. A. C. Sturt, German Patent 2,219,402 (1972); *Chem. Abstr.* **78**, 30517b (1973).
40. H. A. Sarvetnick, "Polyvinyl Chloride," pp. 38 and 45–47. Van Nostrand-Reinhold, Princeton, New Jersey, 1969.
41. A. R. Berens, *Polym. Prepr., Am. Chem. Soc., Div. Polym. Chem.* **15**, 203 (1974).
42. A. R. Berens, *Polym. Prepr., Am. Chem. Soc., Div. Polym. Chem.* **15**, No. 2, 197 (1974).
43. M. Ravey and J. A. Waterman, *J. Polym. Sci., Polym. Chem. Ed.* **13**, 1475 (1975).
44. Q.-T. Pham, J-L. Millan, and E. L. Madruga, *Makromol. Chem.* **175**, 945 (1974).
45. J. D. Cotman, M. F. Gonzalez, and G. C. Claver, *J. Polym. Sci., Part A-1* **15**, 1137 (1967).
46. I. Ya. Ibragimov, D. N. Bort, and V. N. Efremova, *Vysokomol. Soedin., Ser. B* **16**, 376 (1974); *Chem. Abstr.* **81**, 92026m (1974).
47. Compagnie de Saint Gobain, French Patent 1,261,921 (1961).
48. M. Thomas (Produits Chimiques Pechiney-Saint-Gobain), French Patent 1,357,736 (1963); *Chem. Abstr.* **61**, 3231 (1964).
49. A. Krause, *Chem. Eng.* **72**(12), 72 (1965).
50. J. Chatelain, *Nuova Chim.* **48**(8), 85 (1972); *Chem. Abstr.* **77**, 140583g (1972).
51. L. Jourdan, *Chim. & Ind., Genie Chim.* **106**, 475 (1973); *Chem. Abstr.* **79**, 5669r (1973).
52. D. Feldman and M. Macoveanu, *Stud. Cercet. Chim.* **20**, 863 (1972); *Chem. Abstr.* **78**, 59097a (1973).
53. J. Chatelain, *Br. Polym. J.* **5**, 457 (1973).
54. J. Marzec and J. Groborz, *Polimery (Warsaw)* **16**, 426 (1971); *Chem. Abstr.* **76**, 100452m (1972).
55. H. Huber and W. D. Mitterberger, *Kunststoffe* **63**, 762 (1973).
56. Solvay et Cie., Belgian Patent 762,552 (1971); *Chem. Abstr.* **77**, 6083r (1972).
57. G. Coppens and J. C. Dovaine, German Patent 2,301,859 (1973); *Chem. Abstr.* **80**, 37616h (1974).
58. P. P. Rathke, U.S. Patent 3,799,917 (1974); *Chem. Abstr.* **81**, 64355z (1974).
59. A. Garton and M. H. George, *J. Polym. Sci., Polym. Chem. Ed.* **11**, 2153 (1973).
60. K. Arita and V. T. Stannett, *J. Polym. Sci., Polym. Chem. Ed.* **11**, 2127 (1973).
61. A. Garton and M. H. George, *J. Polym. Sci., Polym. Chem. Ed.* **12**, 2779 (1974).
62. H. Batzer and A. Nisch, *Makromol. Chem.* **22**, 131 (1957).
63. W. R. Sorenson and T. W. Campbell, "Preparative Methods of Polymer Chemistry." Wiley (Interscience), New York, 1961.

63a. E. Jenkel, H. Eckmanns-Mettegang, and B. Rumbach, *Makromol. Chem.* **4**, 15 (1949).

64. H. Okao and K. Kamio (Nippon Carbide Industries, Ltd.), Japanese Patent 60/18,774, (1960); *Chem. Abstr.* **55**, 21670a (1961).

65. A. Crosato-Arnaldi, P. Gasparini, and G. Talamini, *Makromol. Chem.* **117**, 140 (1968).

66. A. H. Abdel-Alim and A. E. Hamielec, *J. Appl. Polym. Sci.* **18**, 1603 (1974).

67. A. H. Abdel-Alim and A. E. Hamielec, *J. Appl. Polym. Sci.* **16**, 783 (1972).

68. H. Klinkenberg and K. Schrage, German Patent 2,045,491 (1972); *Chem. Abstr.* **77**, 20393y (1972).

69. M. Ryska, M. Kolinsky, and D. Lim, *J. Polym. Sci., Part C* **33**, 357 (1971).

70. T. T. Jones and H. W. Melville, *Proc. R. Soc. London, Ser. A* **187**, 37 (1946).

71. R. S. Holdsworth, W. M. Smith, and J. T. Barr, Jr., *Mod. Plast.* **35**(10), June, 131 (1958).

72. H. A. Sarvetnick, "Polyvinyl Chloride," p. 40. Van Nostrand-Reinhold, Princeton, New Jersey, 1969.

73. J. Eliassaf, *J. Macromol. Sci., Chem.* **8**, 459 (1974).

73a. S. R. Sandler and W. Karo, "Polymer Syntheses," Vol. 1, pp. 282 ff. and 290 ff. Academic Press, New York, 1974.

74. J. T. Barr, Jr., *in* "Advances in Petroleum Chemistry and Refining" (J. J. McKetta, ed.), Vol. 7, pp. 364–404. Wiley (Interscience), New York, 1963.

75. J. Eliassaf, *Polim. Vehomarim Plast.* **3**(2), 9 (1973); *Chem. Abstr.* **80**, 121619u (1974).

76. T. Ueda, K. Takeuchi, and M. Kato, *J. Polym. Sci., Polym. Chem. Ed.* **10**, 2841 (1972).

76a. S. R. Sandler and W. Karo, "Polymer Syntheses," Vol. 1. Academic Press, New York, 1974.

77. A. F. Hauss, *J. Polym. Sci., Part C* **33**, 1 (1971).

78. J. Ulbricht and V. N. Thank, *Plaste Kautsch.* **21**(3), 186 and 190 (1974).

79. A. Crosato-Arnaldi, G. B. Guarise, and G. Talamini, *Polymer* **10**(5), 385 (1969).

80. A. L. Pierru and C. Alexandre, *Hydrocarbon Process* **52**(6), 97 (1973).

81. B. S. Garud, A. V. Kothari, and C. L. Rastogi, *Pet. & Hydrocarbons* **6**(3), 156 (1971), published in *Chem. Age India* **22**(11), (1971).

82. S. Koyanagi, S. Arai, S. Tajima, and K. Kurimoto, German Patent 2,117,084 (1971); *Chem. Abstr.* **76**, 73280v (1972).

83. S. Koyanagi, S. Tajima, T. Shimizu, and K. Kurimoto, German Patent 2,044,259 (1972); *Chem. Abstr.* **77**, 20374t (1972).

84. W. M. Reiter and A. A. Reventas, U.S. Patent 3,757,001 (1973).

85. E. F. Jordan, Jr., B. Artymyshyn, G. R. Riser, J. Nidock, and A. N. Wrigley, *J. Appl. Polym. Sci.* **17**, 1545 (1973).

85a. L. Plambeck, U.S. Patent 2,462,422 (1949).

86. M. R. Meeks, *Macromol. Synth.* **4**, 137 (1972).

86a. J. D. Sutherland and J. P. McKenzie, *Ind. Eng. Chem.* **48**, 17 (1956).

87. G. Mücke in Hauss [77].

88. F. Wolf and I. Schuessler, German (East) Patent 88,399 (1972); *Chem. Abstr.* **77**, 115087r (1972).

89. C. Hoheisel, *Angew. Makromol. Chem.* **34**, 19 (1973).

90. M. J. Frederickson and A. C. Sturt, German Patent 2,337,341 (1974); *Chem. Abstr.* **81**, 50471 (1974).

91. J. M. DeBell, W. C. Goggin, and W. E. Gloor, "German Plastics Practice." DeBell and Richardson, Springfield, Massachusetts, 1946.

92. A. Czekay, B. Kraemer, and K. Kaiser, German Patent 2,014,016 (1971); *Chem. Abstr.* **76**, 46700a (1972).

93. A. Czekay, B. Kraemer, and K. Kaiser, German Patent 2,014,015 (1971); *Chem. Abstr.* **76**, 46701b (1972).

94. Knapsack, A.-G., French Patent 2,085,048 (1972); *Chem. Abstr.* **77**, 75797p (1972).

95. I. Ito, T. Sekihara, and T. Emura, German Patent 2,252,340 (1973); *Chem. Abstr.* **79**, 19407r (1973).

96. S. E. Porrvik and F. Kolacny, Swedish Patent 315,404 (1969); *Chem. Abstr.* **78**, 16785f (1973).

97. H. Kawakami, Japanese Patent 59/2196 (1959); *Chem. Abstr.* **53**, 19460b (1959).

98. E. Gulbins, A. Hauss, W. Oschmann, and J. Kovacs, German Patent 2,041,372 (1970); *Chem. Abstr.* **76**, 154443c (1972).

99. K. Nakanishi and T. Tatsuo, Japanese Patent 70/36,514 (1970); *Chem. Abstr.* **74**, 64645z (1971).

100. M. Onozuka and K. Iida, Japanese Patent 70/40,057 (1970); *Chem. Abstr.* **74**, 126786g (1971).

101. C. Lambling and J. Boissel, German Patent 2,125,015 (1971); *Chem. Abstr.* **76**, 100367n (1972).

102. O. Schott, F. Kieferle, W. Deuschel, and D. Lansberg, German Patent 2,037,043 (1972); *Chem. Abstr.* **76**, 154440z (1972).

103. S. Koyanagi, H. Kitamura, and S. Tajima, German Patent 2,121,393 (1972); *Chem. Abstr.* **77**, 49182q (1972).

104. S. Tajima and K. Kurimoto, Japan Kokai 73/64,182 (1973); *Chem. Abstr.* **80**, 60537s (1974).

105. I. Ito, T. Sakihara, T. Emura, and T. Ueda, Japan Kokai 73/89,992 (1973); *Chem. Abstr.* **81**, 26451v (1974).

106. S. Koyanagi, H. Kitamura, and T. Shimizu, U.S. Patent 3,706,705 (1972); *Chem. Abstr.* **78**, 73103f (1973).

107. S. Koyanagi, H. Kitamura, and K. Ogawa, British Patent 1,312,102 (1973); *Chem. Abstr.* **79**, 43043t (1973).

108. S. Koyanagi, K. Ogawa, Y. Onda, and A. Yamamoto, Japan Kokai 73/56,773 (1973); *Chem. Abstr.* **80**, 71347f (1974).

109. Borden, Inc., French Patent 2,086,861 (1972); *Chem. Abstr.* **77**, 102614k (1972).

110. S. Sekigawa, K. Nishizawa, and K. Ohashi, Japan Kokai 72/43084 (1972); *Chem. Abstr.* **78**, 125395z (1973).

111. F. Wolf, I. Schuessler, and H. G. Kramer, German (East) Patent 103,448 (1974); *Chem. Abstr.* **81**, 106,291h (1974).

112. I. Ito, T. Sekihara and T. Emura, German Patent 2,358,099 (1974); *Chem. Abstr.* **81**, 153,093e (1974).

113. R. K. Shen, U.S. Patent 3,668,194 (1972); *Chem. Abstr.* **77**, 89108h (1972).

113a. R. S. Holdsworth and W. M. Smith, U.S. Patent 3,017,399 (1962); *Chem. Abstr.* **56**, 1185a (1962).

114. M. Baer, U.S. Patents 2,470,908, 2,470,909 and 2,470,910 (1949); *Chem. Abstr.* **43**, 6002b–c (1949).

115. G. Bier, *Kunststoffe* **55**, 694 (1965).

116. R. Lewis and R. Friedman, *Mod. Plast.* **50**, March, 88 (1973).

117. H. A. Sarvetnick, "Polyvinyl Chloride," pp. 37 ff and 43 ff. Van Nostrand-Reinhold, Princeton, New Jersey, 1969.

118. S. B. Rath, *Pet. & Hydrocarbons* **6**(3), 158 (1971); published in *Chem. Age India* **22**(11) (1971).

119. G. Talamini and E. Peggion, *in* "Vinyl Polymerization" (G. Edlee Ham, ed.), Vol. 1, Part 1, pp. 332 ff. Dekker, New York, 1967.
120. K. Bik, *Polimery* **16**, 421 (1971); *Chem. Abstr.* **76**, 100451k (1972).
121. V. E. Goryushko and N. G. Vilesov, *Model Khim. Protsessov Reakt.*, *Dokl. Vses. Konf. Khim. Reakt.-KHIMEREAKTOR-71, 4th 1970* Vol. 5, pp. 143–160 (1972); *Chem. Abstr.* **79**, 66868b (1973).
122. L. Plambeck, Jr., U.S. Patent 2,458,636 (1949); *Chem. Abstr.* **43**, 2469d (1949).
123. T. Ohishi, *Kobunshi Kagaku* **29**, 835 (1972); *Chem. Abstr.* **78**, 72721u (1973).
124. J. R. Powers, U.S. Patent 2,520,959 (1950); *Chem. Abstr.* **46**, 3798f (1952).
125. G. Benetta, V. Bresquar, J. Gatta, and F. Testa, British Patent 984,487 (1965).
126. H. Winter, German Patent 1,964,029 (1971); *Chem. Abstr.* **76**, 114230z (1972).
127. D. E. Evans and B. N. Hendy, German Patent 2,322,886 (1973); *Chem. Abstr.* **80**, 121687g (1974).
128. J. Chatelain and S. Soussan, French Patent 2,070,251 (1971); *Chem. Abstr.* **77**, 20358r (1972).
129. J. M. Liegeois, *J. Polym. Sci., Part C* **33**, 147 (1971).
130. Farbwerke Hoechst, A.-G., British Patent 277,289 (1972); *Chem. Abstr.* **77**, 89106z (1972).
131. H. Jadamus, German Patent 2,054,103 (1972); *Chem. Abstr.* **77**, 62508v (1972).
132. P. K. Isaacs and A. Trofimov, U.S. Patent 3,033,812 (1962); *Chem. Abstr.* **57**, 2430a (1962).
133. N. Fischer and C. Lambling, French Patent 2,086,634 (1972); *Chem. Abstr.* **77**, 89394s (1972).
134. N. Fischer and C. Lambling, French Patent 2,086,635 (1972); *Chem. Abstr.* **77**, 89149r (1972).
135. J. K. Pierce, Jr., U.S. Patent 3,642,740 (1972); *Chem. Abstr.* **77**, 6365j (1972).
136. C. Sueling, German Patent 2,050,723 (1972); *Chem. Abstr.* **77**, 62498s (1972).
137. K. Dimov and L. Slavcheva, *Polymer* **14**, 234 (1973).
138. S. J. Makower, P. A. Cautilli, and J. Dickstein, U.S. Patent 3,721,636 (1973).
139. H. A. Sarvetnick, "Polyvinyl Chloride," p. 38. Van Nostrand-Reinhold, Princeton, New Jersey, 1969.
139a. M. Hunt, U.S. Patent 2,471,959 (1949).
140. W. P. May, U.S. Patent 3,795,649 (1974); *Chem. Abstr.* **81**, 14333v (1974).
141. D. Braun, H. Cherdron, and W. Kern, "Techniques of Polymer Syntheses and Characterization," p. 135. Wiley (Interscience), New York, 1971.
142. W. I. Bengough and N. M. Chawdry, *J. Chem. Soc., Faraday Trans 1* **68**, 1807 (1972).
143. S. Nakano and T. Yamawaki, Japanese Patent 65/7064 (1965); *Chem. Abstr.* **63**, 703c (1965).
144. R. K. S. Chan and C. Worman, *Polym. Eng. Sci.* **12**, 437 (1972).
145. K. S. Minsker, A. G. Kronman, B. F. Teplov, E. E. Rylov, and D. N. Bort, *SPE Trans.* **2**, No. 4, 294 (1962).
146. M. T. Bryk and A. S. Shevlyakov, *Plast. Massy* No. 7, p. 45 (1966).
147. J. A. Manson, S. A. Iobst, and R. Acosta, *J. Polym. Sci., Part A-1* **10**, 179 (1972).
148. G. N. Gerasimov, A. D. Smirnov, E. B. Kotin, T. M. Sabirova, and A. D. Abkin, *Proc. Tihany Symp. Radiat. Chem., 3rd 1971* Vol. 1, p. 463 (1972); *Chem. Abstr.* **78**, 124939t (1973).
149. N. Harumiya and H. Miyama, Japan Kokai 72/14,287 (1972); *Chem. Abstr.* **78**, 59031 (1973).
150. J. Ulbricht and W. Seidel, *J. Prakt. Chem.* [4] **315**, 668 and 675 (1973).

151. V. T. Stannett, E. P. Stahel, J. Barriac, E. Oda, S. Omi, T. O'Neill, S. Russo, N. B. Harawala, and G. S. Prehoda, Final Report ORO-3687-1. Available Dept. NTIS, 1971; *Chem. Abstr.* **77**, 62403g (1972).

152. A. M. Smirnov, V. I. Lukhovitskii, R. M. Pozdeeva, and V. L. Karpov, *Vysokomol. Soedin., Ser. B* **13**, 747 (1971); *Chem. Abstr.* **76**, 60138e (1972).

153. A. M. Smirnov, V. I. Lukhovitskii, and V. L. Karpov, *Vysokomol. Soedin., Ser. B* **15**, 726 (1973); *Chem. Abstr.* **80**, 133947a (1974).

154. A. M. Smirnov, V. I. Lukhovitskii, and V. L. Karpov, *Vysokomol. Soedin., Ser. B* **14**, 6 (1972); *Chem. Abstr.* **76**, 127539g (1972).

155. V. S. Tikhomirov, V. I. Serenekov, G. I. Zalkind, R. G. Gumen, and A. V. Pavlov, *Plast. Massy* No. 11, p. 37 (1971); *Chem. Abstr.* **76**, 72837v (1972).

156. R. Salovey and R. C. Gebauer, *J. Polym. Sci., Part A-1* **10**, 1533 (1972).

157. M. Tavan, G. Palma, M. Carenza, and S. Brugnaro, *J. Polym. Sci., Polym. Chem. Ed.* **12**, 411 (1974).

158. G. N. Gerasimov, T. A. Bespyatkina, T. M. Sabirova, and A. D. Abkin, *Dokl. Akad. Nauk SSSR* **209**, 628 (1973); *Chem. Abstr.* **79**, 53860t (1973).

159. G. N. Gerasimov, A. D. Smirnov, T. M. Sabirova, and A. D. Abkin, *Vysokomol. Soedin., Ser. A* **16**, 1530 (1974); *Chem. Abstr.* **81**, 169977q (1974).

160. V. I. Lukhovitskii, A. M. Smirnov, V. V. Polikarpov, A. M. Lebedeva, R. M. Lagucheva, and V. L. Karpov, U.S. Patent 3,709,804 (1973).

161. L. F. Thompson and K. G. Mayhan, *J. Appl. Polym. Sci.* **16**, 2291 (1972).

162. R. Liepins and K. Sakaoku, *J. Appl. Polym. Sci.* **16**, 2633 (1972).

163. H. Yasuda and C. E. Lamaze, *J. Appl. Polym. Sci.* **17**, 1519 (1973).

164. N. G. Volkov, G. M. Gorbachenko, Yu. S. Deev, and V. K. Lyapidevskii, *Khim. Vys. Energ.* **7**, 256 (1973); *Chem. Abstr.* **79**, 79281e (1973).

165. N. Yamazaki, *Prog. Polym. Sci., Jpn.* **2**, 171 (1971); *Chem. Abstr.* **78**, 30230c (1973).

166. A. D. Pomogailo, P. E. Matkovskii, and G. A. Beikhol'd. *Tr. Inst. Khim. Nefti Prir. Solei, Akad. Nauk. Kaz. SSR* **3**, 194 (1971); *Chem. Abstr.* **78**, 4532h (1973).

167. I. Karwat, *Polimery* **18**, 333 (1973); *Chem. Abstr.* **80**, 15222z (1974).

168. A. Guyot and P. Q. Tho, *J. Polym. Sci., Part C* **4**, 299 (1964).

169. A. Guyot and J. Mordini, *J. Polym. Sci., Part C* **33**, 65 (1971).

170. Y. Suzuki and M. Saito, *J. Polym. Sci., Part A-1* **9**, 3639 (1971).

171. Hercules Powder Co., British Patent 834,937 (1960); *Chem. Abstr.* **54**, 26009e (1960).

172. R. W. Cochran, G. E. Helfrich, L. D. Hoblit, and G. Y. T. Liu, U.S. Patent 3,632,799; *Chem. Abstr.* **72**, 127807t (1972).

173. R. N. Haszeldine, T. G. Hyde, and P. J. T. Tait, *Polymer* **14**, 215 (1973).

174. A. G. Cherworth, R. N. Haszeldine, and P. J. T. Tait, *J. Polym. Sci., Polym. Chem. Ed.* **12**, 1703 (1974).

175. K. Mitani, T. Ogata, and H. Awaya, *J. Polym. Sci., Polym. Chem. Ed.* **11**, 2653 (1973).

176. E. B. Milovskaya, E. L. Kopp, O. S. Mikhailicheva, V. M. Denisov, and A. I. Koltsov, *Polymer* **13**, 288 (1972).

177. E. L. Kopp, O. S. Mikhailicheva, and E. B. Milovskaya, *Vysokomol. Soedin., Ser. A* **14**, 2653 (1972); *Chem. Abstr.* **78**, 72714u (1973).

178. A. Akimoto and T. Okada, *J. Macromol. Sci., Chem.* **7**, 1555 (1973).

179. S. Matsuoka and T. Kaku, Japanese Patent 58/297 (1958).

180. R. H. Martin, Jr., U.S. Patent 2,898,244 (1959).

181. F. Suzuki, Japanese Patent 70/11,828 (1970).

182. J. Barton, M. Pegorare, L. Szilagy, and G. Pagani, *Makromol. Chem.* **144**, 245 (1971).

183. J. W. Hirzy, U.S. Patent 3,389,168 (1968).
184. S. H. Pinner, *Nature (London)* **183**, 1108 (1959).
185. G. G. Odian, B. S. Bernstein, J. Schaeffer, L. J. Friedman, and J. Kelly, *U.S. At. E. C.* **NYU–2481**, 1–28 (1961).
186. W. W. Howerton, U.S. Patent 3,291,857 (1966).
187. Raychem Corp., British Patent 1,047,053 (1966).
188. S. Ando and H. Higuchi, Japanese Patent 68/09,376 (1968).
189. Dow Chemical Co., British Patent 829,512 (1960).
190. S. Koyanagi, A. Maruyama, and H. Kitamura, German Patent 1,938,911 (1970).
191. J. W. Calantine, F. J. Maurer, and W. J. Van Essen, French Patent 1,431,547 (1966).
192. J. A. Cornell, U.S. Patent 3,066,110 (1962).
193. S. H. Pinner, British Patent 1,050,781 (1966).
194. International Standard Electric Corp., Netherland Patent Appl. 6,613,716 (1967).
195. S. L. Burt, U.S. Patent 2,618,621 (1952).
196. W. Brenner, German Patent 2,021,218 (1970).
197. I. F. Muskat, German Patent 2,027,070 (1970).
198. A. Ya. Drinberg, Sy. N. Golant, and L. I. Gol'dfarb, *Zh. Prikl. Khim.* **24**, 1181 (1951).
199. E. G. Galen, German Patent 1,808,905 (1970).
200. H. Fassy, P. Lalet, and A. Miletto, French Patent 2,159,558 (1973); *Chem. Abstr.* **80**, 15491 (1974).
201. V. I. Lukhovitskii, A. M. Smirnov, and V. L. Karpov, *Vysokomol. Soedin., Ser. A* **14**, 202 (1972); *Chem. Abstr.* **76**, 127526a (1972).
202. F. E. Condo and H. A. Newey, U.S. Patent 2,674,585 (1954).
203. T. Uno and K. Yoshida, *Kobunshi Kagaku* **14**, 345 (1957); *Chem. Abstr.* **52**, 5027g (1958).
204. Chemische Werke, Huels A.-G., French Patent 1,603,794 (1971); *Chem. Abstr.* **76**, 100366m (1972).
205. S. Tajima and K. Kurimoto, German Patent 2,259,997 (1973); *Chem. Abstr.* **79**, 67261k (1973).
206. S. Tajima, K. Kurimoto, and K. Miyoshi, German Patent 2,260,957 (1973); *Chem. Abstr.* **79**, 92818e (1973).
207. S. Ueno and S. Takeuchi, Japanese Patent 74/02,337 (1974); *Chem. Abstr.* **81**, 106631u (1974).
208. A. R. Berens, *J. Appl. Polym. Sci.* **18**, 2379 (1974).
209. A. Sturt, British Patent 1,341,386 (1973); *Chem. Abstr.* **81**, 4353n (1974).
210. A. Sturt, German Patent 2,111,152 (1972); *Chem. Abstr.* **78**, 85061d (1973).
211. V. G. Fryling, U.S. Patent 2,356,925 (1944).
212. J. B. O'Hara and C. F. Prutton, *J. Polym. Sci.* **5**, 673 (1950).
213. I. Ito, S. Ito, and T. Emura, German Patent 2,214,591 (1972); *Chem. Abstr.* **78**, 16756x (1973).
214. C. Todereanu, P. Gluck, N. Boicescu, I. Szabo, and R. Negretu, Rumanian Patent 55,821 (1973); *Chem. Abstr.* **81**, 26,483g (1974).
215. N. I. Sokolova and A. Panfilov, *Plast. Massy* No. 6, p. 7 (1973); *Chem. Abstr.* **79**, 79236u (1973).
216. A. Sturt, British Patent 1,341,385 (1973); *Chem. Abstr.* **81**, 4354p (1974).
217. D. E. Moore, U.S. Patent 3,661,881 (1972); *Chem. Abstr.* **77**, 49411p (1972).
218. R. Kato and I. Soematsu, Japanese Patent 72/21,576 (1972); *Chem. Abstr.* **77**, 127437t (1972).
219. R. J. Wolf and A. A. Nicolay, U.S. Patent 2,608,552 (1952); *Chem. Abstr.* **47**, 1431b (1953).

220. R. A. Isaksen and E. H. Merz, U.S. Patent 2,984,593 (1961); *Chem. Abstr.* **55**, 21670i (1961).
221. R. A. Isaksen and E. H. Merz, U.S. Patent 2,984,593 (1961); *Chem. Abstr.* **55**, 21670i (1961).
222. Shin-Etsu Chemical Industry Co., Ltd., British Patent 1,093,866 (1967); *Chem. Abstr.* **68**, 30412g (1968).
223. Dynamit Nobel A.-G., French Patent 2,004,758 (1969); *Chem. Abstr.* **72**, 112058t (1970).
224. H. Rosin, S. L. J. Daren, M. Asscher, and D. Vofsi, *J. Appl. Polym. Sci.* **16**, 1687 (1972).
225. E. Trebillon and G. Wetroff, German Patent 2,215,185 (1972); *Chem. Abstr.* **78**, 16759a (1973).
226. W. Kawai and T. Ichibashi, Japan Kokai 73/30,789 (1973); *Chem. Abstr.* **79**, 67072z (1973).
227. Z. Roszkowski, *Polimery (Warsaw)* **16**, 445 (1971).
228. J. Ugelstad, P. C. Mörk, and J. O. Aasen, *J. Polym. Sci., Part A-1* **5**, 2281 (1967); see also Beren's [208] for further references to Ugelstad's contributions.

AUTHOR INDEX

Numbers in parentheses are reference numbers and indicate that an author's work is referred to although his name is not cited in the text. Numbers in *italic* show the page on which the complete reference is listed.

371

SUBJECT INDEX

A

Acetaldehyde–urea, resin, 12–13
Acetal interchange reaction, 173
Acetic acid in polymerization of vinyl
    chloride, 318
Acetyl cyclohexanesulfonyl chloride, 339
Acids, 143
Acriflavin, 291
Acrylamide, 289
Acrylate salts, polymerization of, 265
Acrylic acid
  dimer (acryloxypropionic acid), 268
  dimeric, hydrogen bonded, 294
  inhibitor free, 269
  multimolecular, hydrogen bonded, 294
  polymerization of, 265
    differences in behavior with methacrylic
      acid, 287
  syndiotactic, 294
Acryloxypropionic acid, (acrylic acid di-
    mer), 268
Adhesion improvement, 265
Aerosols, 346
Agitation rate, effect on particle size dis-
    tribution in suspension polymeriza-
    tion, 340
Aldehydes as polymerization solvents, 356
Alkoxy silanes, reactions of, 121
Alkyd paint, 145
Alkyd resin(s), 140–167
  based on aromatic polycarboxylic acids,
    160
    on phthalic anhydride or phthalic acid,
      146–158
  benzoic acid to control molecular weight,
    154–155
  infrared spectra, 157

manufacture of, 143–146
  medium oil, 154
  modified with other thermosetting resins,
    166–167
    with vinyl monomers, 165–166
  oil-modified, 150–154
  processing equipment, 146
  quality control, 144
  reactions of, 145
    of glycerol with various acids, 156
  styrene-modified linseed trimethylol-
    ethane, 165–166
  trimellitic anhydride based, 161
Alkylhalosilanes, condensation polymeriza-
    tion to alkylpolysiloxanes (silicones),
    124–125
Amines, effect on vinylpyrrolone polymeri-
    zation, 243, 244
p-Aminophenol, 84
  polymerization of glycidyl derivative of,
    83–84
Aminoplasts, 1–39
Amino resins, 1–39
  pattern of consumption, 56
  production of, 3–6
  uses, 5–6
Ammonia, effect on vinylpyrrolidone
    polymerization, 243, 244
Amphoteric copolymers, 265
n-Amyltriethoxysilane, 134–135
  condensation of, 134–135
Anhydrides, 143
Antimony trichloride as initiator of
    vinylpyrrolidone polymerization, 240
Arrhenius equation for polymerization of
    vinylpyrrolidone, 239
Ascorbic acid, 291

ORGANIC CHEMISTRY
A SERIES OF MONOGRAPHS

EDITORS

ALFRED T. BLOMQUIST*
Department of Chemistry
Cornell University
Ithaca, New York

HARRY H. WASSERMAN
Department of Chemistry
Yale University
New Haven, Connecticut

1. Wolfgang Kirmse. CARBENE CHEMISTRY, 1964; 2nd Edition, 1971

2. Brandes H. Smith. BRIDGED AROMATIC COMPOUNDS, 1964

3. Michael Hanack. CONFORMATION THEORY, 1965

4. Donald J. Cram. FUNDAMENTALS OF CARBANION CHEMISTRY, 1965

5. Kenneth B. Wiberg (Editor). OXIDATION IN ORGANIC CHEMISTRY, PART A, 1965; Walter S. Trahanovsky (Editor). OXIDATION IN ORGANIC CHEMISTRY, PART B, 1973

6. R. F. Hudson. STRUCTURE AND MECHANISM IN ORGANO-PHOSPHORUS CHEMISTRY, 1965

7. A. William Johnson. YLID CHEMISTRY, 1966

8. Jan Hamer (Editor). 1,4-CYCLOADDITION REACTIONS, 1967

9. Henri Ulrich. CYCLOADDITION REACTIONS OF HETEROCUMULENES, 1967

10. M. P. Cava and M. J. Mitchell. CYCLOBUTADIENE AND RELATED COMPOUNDS, 1967

11. Reinhard W. Hoffman. DEHYDROBENZENE AND CYCLOALKYNES, 1967

12. Stanley R. Sandler and Wolf Karo. ORGANIC FUNCTIONAL GROUP PREPARATIONS, VOLUME I, 1968; VOLUME II, 1971; VOLUME III, 1972

13. Robert J. Cotter and Markus Matzner. RING-FORMING POLYMERIZATIONS, PART A, 1969; PART B, 1; B, 2, 1972

14. R. H. DeWolfe. CARBOXYLIC ORTHO ACID DERIVATIVES, 1970

15. R. Foster. ORGANIC CHARGE-TRANSFER COMPLEXES, 1969

16. James P. Snyder (Editor). NONBENZENOID AROMATICS, VOLUME I, 1969; VOLUME II, 1971

*Deceased.

17. C. H. Rochester. ACIDITY FUNCTIONS, 1970

18. Richard J. Sundberg. THE CHEMISTRY OF INDOLES, 1970

19. A. R. Katritzky and J. M. Lagowski. CHEMISTRY OF THE HETEROCYCLIC N-OXIDES, 1970

20. Ivar Ugi (Editor). ISONITRILE CHEMISTRY, 1971

21. G. Chiurdoglu (Editor). CONFORMATIONAL ANALYSIS, 1971

22. Gottfried Schill. CATENANES, ROTAXANES, AND KNOTS, 1971

23. M. Liler. REACTION MECHANISMS IN SULPHURIC ACID AND OTHER STRONG ACID SOLUTIONS, 1971

24. J. B. Stothers. CARBON-13 NMR SPECTROSCOPY, 1972

25. Maurice Shamma. THE ISOQUINOLINE ALKALOIDS: CHEMISTRY AND PHARMACOLOGY, 1972

26. Samuel P. McManus (Editor). ORGANIC REACTIVE INTERMEDIATES, 1973

27. H.C. Van der Plas. RING TRANSFORMATIONS OF HETEROCYCLES, VOLUMES 1 AND 2, 1973

28. Paul N. Rylander. ORGANIC SYNTHESES WITH NOBLE METAL CATALYSTS, 1973

29. Stanley R. Sandler and Wolf Karo. POLYMER SYNTHESES, VOLUME I, 1974; VOLUME II, 1977

30. Robert T. Blickenstaff, Anil C. Ghosh, and Gordon C. Wolf. TOTAL SYNTHESIS OF STEROIDS, 1974

31. Barry M. Trost and Lawrence S. Melvin, Jr. SULFUR YLIDES: EMERGING SYNTHETIC INTERMEDIATES, 1975

32. Sidney D. Ross, Manuel Finkelstein, and Eric J. Rudd. ANODIC OXIDATION, 1975

33. Howard Alper (Editor). TRANSITION METAL ORGANOMETALLICS IN ORGANIC SYNTHESIS, VOLUME 1, 1976

34. R. A. Jones and G. P. Bean. THE CHEMISTRY OF PYRROLES, 1976

35. Alan P. Marchand and Roland E. Lehr (Editors). PERICYCLIC REACTIONS, VOLUME I, 1977; VOLUME II, 1977

36. Pierre Crabbé (Editor). PROSTAGLANDIN RESEARCH, 1977

A
B 7
C 8
D 9
E 0
F 1
G 2
H 3
I 4
J 5